Lecture Notes in Statistics

Bo-Cheng Wei

Springer
Singapore
Berlin
Heidelberg
New York
Barcelona
Budapest
Hong Kong
London
Milan
Paris
Santa Clara
Tokyo

Bo-Cheng Wei

Exponential Family Nonlinear Models

Springer

Bo-Cheng Wei
Department of Mathematics
Southeast University
Nanjing 210018
China

Library of Congress Cataloging-in-Publication Data

Wei, Bo-Cheng, 1937-
Exponential family nonlinear models / Bo-Cheng Wei.
p. cm
Includes bibliographical references and indexes.
ISBN 9813083298
1. Exponential families (statistics) 2. Nonlinear theories.
I. Title.
QA276.7.W44 1997
519.5'2—dc21

97-24695
CIP

ISBN 981-3083-29-8

Printed in Singapore

Typesetting: Camera-ready by author
SPIN 10672825 5 4 3 2 1

Preface

This book is devoted to a study of statistical behavior of exponential family nonlinear models, a subject very much in its infancy. These models are the natural extension of generalized linear models and normal nonlinear regression models. The latter two fields have been well developed in the past two decades and several books were published in the 1980s (e.g. Ratkowsky 1983; 1990; Gallant 1987; Bates and Watts 1988; McCullagh and Nelder 1989; Seber and Wild 1989; Wei 1989). However, we have not seen much work on exponential family nonlinear models and there is still no book in this field. In this book, the statistical basis will be introduced, and then some interesting topics for these models will be discussed.

There are seven chapters and two appendices in this book. The first two chapters respectively introduce the basic statistical properties of exponential families and exponential family nonlinear models to provide a sound foundation for the succeeding chapters. Chapter 3 presents a geometric framework for exponential family nonlinear models. We call this the modified BW geometric framework which may be viewed as a modification of that presented by Bates and Watts (1980) for nonlinear regression models. Based on this modified BW geometric framework, Chapters 4 and 5 study some statistical inference from the geometric point of view. Chapter 4 investigates some asymptotic properties of estimation related to the statistical curvatures. Chapter 5 discusses the quadratic approximations of confidence regions of parameters in terms of the curvature measures. In Chapter 6, we are devoted to a study of diagnostics and influence analysis. The basic diagnostic statistics are introduced, and then some specific problems, such as local inference analysis with random perturbations, generalized leverage and its applications, and the test for varying dispersion, are discussed. Chapter 7 explores the possibility of extending the materials of the first six chapters to more general models. The curved exponential families, the multinomial nonlinear models, the embedded models in regular parametric families, the quasi-likelihood models and the covariance structure models are discussed. Appendix A outlines some useful results of matrices and arrays, which are very helpful as a supplement to reading this book. Appendix B gives detailed proofs for some asymptotic results used in Chapter 4 — for readers who are interested in the large sample theory.

This is a theoretical book. It is assumed that readers have a solid background in statistical methods and linear algebra, and in particular, a knowledge of linear regression analysis such as Draper and Smith (1981). Some geometric concepts of Euclidean space are quite helpful when reading Chapters 3 to 5.

I would like to express my sincere thanks to a number of people who have contributed to this book, including Kai-Tai Fang, Wing-Kam Fung, Yue-Qing Hu, Sik-Yum Lee, Guobing Lu, Kai-Fung Lu, Jian-Qing Shi, Fanghuan Wan, Shu-Jia Wang, Xu-Ping Zhong, Hong-Tu Zhu and Zhong-Yi Zhu. I am especially grateful to Professor K.T. Fang and Dr. Jian-Qing Shi for their very valuable comments, suggestions and advice, which greatly improved the presentation of the book. The Departments of Statistics of the Chinese University of Hong Kong and the University of Hong Kong are acknowledged for providing computer facilities. Finally, I thank The National Natural Science Foundation of China and The Natural Science Foundation of Jiangsu Province for their support during the writing of the book.

Bo-Cheng Wei

February 1998

Contents

1 Exponential Family **1**

1.1 Exponential Family 1

1.2 Likelihood 7

1.3 Likelihood Ratio and Deviance 9

2 Exponential Family Nonlinear Models **13**

2.1 Definition 13

2.2 Likelihood 19

2.3 Computation and Examples 24

3 Geometric Framework **29**

3.1 Introduction 29

3.2 Geometry in Expectation Parameter Space 31

3.2.1 Curvature arrays 32

3.2.2 Directional curvatures and maximum curvatures . . . 35

3.2.3 Parameter transformation 40

3.3 Geometry in Natural Parameter Space 44

3.4 Some Specific Models 49

3.4.1 Normal nonlinear regression models 49

3.4.2 Generalized linear models 49

3.4.3 One-parameter model and Efron curvature 51

4 Some Second Order Asymptotics **55**

4.1 Stochastic Expansions 56

4.2 Approximate Bias and Variance 64

4.3 Information Loss 71

4.4 Observed Information and Fisher Information 76

5 Confidence Regions **83**

5.1 Introduction 84

5.2 Likelihood Region in Terms of Curvatures 86

5.2.1 Tangent space projection of likelihood region 86

5.2.2 Influence of curvature measures 90

5.3 Confidence Regions for Parameter Subsets 93

5.3.1 Region based on likelihood ratio statistic 94

5.3.2 Region based on score statistic 96

6 Diagnostics and Influence Analysis **101**

6.1 Introduction 101

6.2 Analysis of Diagnostic Models 105

6.2.1 Diagnostic models 105

6.2.2 Regression coefficients 106

6.2.3 Deviance and dispersion parameter 109

6.2.4 Score statistic of outlier 112

6.3 Influence Diagnostics Based on Case Deletion 116

6.3.1 Diagnostics based on weighted least squares 116

6.3.2 Diagnostics based on deviance 119

6.4 Local Influence Analysis 123

6.4.1 Perturbed models 123

6.4.2 Random perturbation scheme 128

6.5 Generalized Leverage 136

6.5.1 Definition and computation 136

6.5.2 Generalized leverage and local influence 140

6.6 Diagnostics for Varying Dispersion 141

6.6.1 Likelihood ratio and score statistics 142

6.6.2 Adjusted likelihood ratio and score statistics 147

6.6.3 Examples 155

6.6.4 Simulation study 158

7 Extension **165**

7.1 Curved Exponential Families 166

7.1.1 Introduction and geometric framework 166

7.1.2 Statistical analysis 167
7.2 Multinomial Nonlinear Models 170
7.3 Embedded Models 174
7.4 Quasi-likelihood Models 177
7.5 Covariance Structure Models 179

Appendix **185**

A Matrix and Array **185**
A.1 Projection in Weighted Inner Product Space 185
A.2 Array Multiplication 188
A.3 Differentiation of Vector and Matrix 191

B Some Asymptotic Results **195**
B.1 Notation and Preparation 195
B.2 Necessary Condition of Consistency 196
B.3 Consistency and Asymptotic Normality 199
B.4 Asymptotic Numerical Stability of Iterations 208

Bibliography **211**

Author Index **223**

Subject Index **227**

Chapter 1

Exponential Family

Exponential family is the most commonly used family in statistics. In this chapter, the basic statistical properties and some useful lemmas are summarized for this family. Section 1.1 introduces definition, notation and moments generating functions. Some related problems are also discussed. Section 1.2 deals with the likelihood problems. The score function, the observed informations and the Fisher informations are introduced. In Section 1.3, we deal with the likelihood ratio statistic, the deviance and the Kullback-Leibler distance. Two useful lemmas are also given. Readers familiar with these topics may find the pace in this chapter a bit too leisurely and may wish to skim through it. However, it is important to lay a sound foundation carefully at the beginning in order to facilitate the understanding of more subtle and involved arguments which will appear in succeeding chapters. In this chapter, we omit some technical details. For a more comprehensive treatment of the materials discussed, the reader is referred to other literature, such as Barndorff-Nielsen (1978), Morris (1982; 1983), Jorgensen (1987; 1992a,b).

1.1 Exponential Family

A family of distributions $P_{\theta,\phi}$ of an n-vector of random variable $\boldsymbol{Y} = (y_1, \cdots, y_n)^T$, which has the probability density function

$$p(\boldsymbol{y}; \boldsymbol{\theta}, \phi) = \exp[\phi\{\boldsymbol{y}^T\boldsymbol{\theta} - b(\boldsymbol{\theta}) - c(\boldsymbol{y}, \phi)\}] \tag{1.1}$$

with respect to a σ-finite measure ξ, is called an exponential family; where $b(\cdot)$ and $c(\cdot, \cdot)$ are some specific functions, $\boldsymbol{\theta} = (\theta_1, \cdots, \theta_n)^T$ is the natural parameter defined in a natural parameter space $\Theta \subset R^n$, and T denotes the transpose of a vector or matrix. The dispersion parameter ϕ defined in an open subset Φ of real line R is also denoted by $\phi = \sigma^{-2}$ which is usually treated as a nuisance parameter. As introduced by Jorgensen (1987), we use the notation $\boldsymbol{Y} \sim ED(\boldsymbol{\theta}, \phi)$ or $\boldsymbol{Y} \sim ED(\boldsymbol{\mu}, \sigma^2)$ to denote the exponential family (1.1), where $\boldsymbol{\mu} = \mathrm{E}(\boldsymbol{Y})$ is the mean vector. In recent literature (e.g. Smyth 1989; Cordeiro and McCullagh 1991), (1.1) is often denoted by its equivalent form as

$$p(\boldsymbol{y}; \boldsymbol{\theta}, \phi) = \exp[\phi\{\boldsymbol{y}^T\boldsymbol{\theta} - b(\boldsymbol{\theta}) - c(\boldsymbol{y})\} - \frac{1}{2}s(\boldsymbol{y}, \phi)] \tag{1.2}$$

for some specific functions $b(\cdot), c(\cdot)$ and $s(\cdot,\cdot)$. We also prefer to use this form instead of (1.1) in most of the cases. If ϕ is known, then (1.1) or (1.2) defines a linear exponential family (Barndorff-Nielsen 1978; Morris 1982). In this case, Θ is a convex subset of R^n and the cumulant generating function $b(\boldsymbol{\theta})$ is convex and analytical in the interior Θ^0 of Θ. When ϕ is unknown, Jorgensen (1987) called (1.1) or (1.2) the exponential dispersion model to stress the dispersion parameter ϕ. For simplicity, we call (1.1) or (1.2) the exponential family no matter if ϕ is known or unknown, as is often used in the literature. Model (1.2) has two commonly encountered special cases as follows.

(a) Both y and θ are scalars, in particular, y usually represents one of observations, that is $y = y_i, \theta = \theta_i$ are scalars and (1.2) may depend on an independent known variable $\boldsymbol{x}_i$ and has the density function in the following form

$$p(y_i; \theta_i, \phi) = \exp[\phi\{y_i\theta_i - b(\theta_i) - c(y_i)\} - \frac{1}{2}s(y_i, \phi)]. \tag{1.3}$$

(b) The components of $\boldsymbol{Y}$ in (1.2) are independent and each component y_i has the exponential family distribution in the form (1.3). Obviously, the distribution of $\boldsymbol{Y} = (y_1, \cdots, y_n)^T$ still has the form (1.2) and can be denoted by

$$p(\boldsymbol{y}; \boldsymbol{\theta}, \phi) = \exp\left[\sum_{i=1}^{n} [\phi\{y_i\theta_i - b(\theta_i) - c(y_i)\} - \frac{1}{2}s(y_i, \phi)]\right]. \tag{1.4}$$

To compare this with (1.2), it is easily seen that $b(\boldsymbol{\theta}) = \sum_i b(\theta_i)$, $c(\boldsymbol{y}) = \sum_i c(y_i)$ and $s(\boldsymbol{y}, \phi) = \sum_i s(y_i, \phi)$, respectively. In this book, most of our discussions will be based on the form of (1.4).

In summary, the exponential family (1.1) is equivalent to the exponential family (1.2), while (1.3) and (1.4) are special cases of (1.2). Table 1.1 lists six commonly used members of exponential families in practice (McCullagh and Nelder 1989).

Table 1.1 also shows that the function $s(\boldsymbol{y}, \phi)$ may have the form

$$s(\boldsymbol{y}, \phi) = s(\phi) + t(\boldsymbol{y}) \tag{1.5}$$

in some cases. If this equation holds, (1.2) is actually a linear exponential family (Barndorff-Nielsen 1978). In fact, we can set $\tilde{\boldsymbol{\theta}} = (\phi\boldsymbol{\theta}^T, \phi)^T$ and $\tilde{\boldsymbol{y}} = (\boldsymbol{y}^T, -c(\boldsymbol{y}))^T$, then (1.2) can be rewritten as

$$p(\tilde{\boldsymbol{y}}; \boldsymbol{\theta}) = \exp[\tilde{\boldsymbol{y}}^T\tilde{\boldsymbol{\theta}} - \{\phi b(\boldsymbol{\theta}) + \frac{1}{2}s(\phi)\} - \frac{1}{2}t(\boldsymbol{y})];$$

this is obviously a linear exponential family with $\tilde{b}(\tilde{\boldsymbol{\theta}}) = \phi b(\boldsymbol{\theta}) + \frac{1}{2}s(\phi)$, which should be convex in $\Theta \times \Phi$ and analytical in the interior of $\Theta \times \Phi$.

Combining (1.4) and (1.5), we summarize the commonly used exponential families in practice as

$$p(\boldsymbol{y}; \boldsymbol{\theta}, \phi) = \exp\left[\sum_{i=1}^{n} [\phi\{y_i\theta_i - b(\theta_i) - c(y_i)\} - \frac{1}{2}s(\phi) - \frac{1}{2}t(y_i)]\right]. \quad (1.6)$$

In Table 1.1, all six members are of this form.

For exponential family (1.1) or (1.2), it is easy to show that the moment generating function of $\boldsymbol{Y}$ is

$$M(\boldsymbol{\tau}; \boldsymbol{\theta}, \phi) = \mathrm{E}(e^{\boldsymbol{\tau}^T\boldsymbol{Y}}) = \exp[\phi\{b(\boldsymbol{\theta} + \boldsymbol{\tau}/\phi) - b(\boldsymbol{\theta})\}]. \quad (1.7)$$

Taking derivatives for $M(\boldsymbol{\tau}; \boldsymbol{\theta}, \phi)$ in $\boldsymbol{\tau}$ at $\boldsymbol{\tau} = \boldsymbol{0}$, we can get the cumulants of $\boldsymbol{Y}$ of order $(i_1, \cdots, i_n)$ as follows

$$\kappa_{i_1\cdots i_n}(\boldsymbol{\theta}, \phi) = \phi^{1-i_1-\cdots-i_n}\frac{\partial^{i_1+\cdots+i_n}b(\boldsymbol{\theta})}{\partial\theta_1^{i_1}\cdots\partial\theta_n^{i_n}}.$$

In particular, we have

$$\mathrm{E}(\boldsymbol{Y}) = \boldsymbol{\mu} = \dot{\boldsymbol{b}}(\boldsymbol{\theta}), \quad \mathrm{Var}(\boldsymbol{Y}) = \sigma^2\ddot{\boldsymbol{b}}(\boldsymbol{\theta}) = \sigma^2\boldsymbol{V}(\boldsymbol{\mu}), \quad (1.8)$$

where $\boldsymbol{\mu} = (\mu_1, \cdots, \mu_n)^T$ is the expectation parameter defined in a subset $\mathcal{U} \subset R^n$ when $\boldsymbol{\theta} \in \Theta^0$, i.e. $\mathcal{U} = \dot{\boldsymbol{b}}(\Theta^0)$; $\dot{\boldsymbol{b}}(\boldsymbol{\theta}) = \partial b(\boldsymbol{\theta})/\partial\boldsymbol{\theta} = (\partial b(\boldsymbol{\theta})/\partial\theta_i)$ is an n-vector and $\ddot{\boldsymbol{b}}(\boldsymbol{\theta}) = \partial^2 b(\boldsymbol{\theta})/\partial\boldsymbol{\theta}\partial\boldsymbol{\theta}^T = (\partial^2 b(\boldsymbol{\theta})/\partial\theta_i\partial\theta_j)$ is an $n \times n$ matrix. In this book, we use dots over the functions to denote the derivatives. Since $\boldsymbol{\mu} = \dot{\boldsymbol{b}}(\boldsymbol{\theta})$, $\boldsymbol{\theta} = \dot{\boldsymbol{b}}^{-1}(\boldsymbol{\mu})$, where $\dot{\boldsymbol{b}}^{-1}(\cdot)$ is the inverse function of $\dot{\boldsymbol{b}}(\cdot)$, there is a one-to-one mapping between $\boldsymbol{\mu}$ and $\boldsymbol{\theta}$. The variance function $\ddot{\boldsymbol{b}}(\boldsymbol{\theta}) = \ddot{\boldsymbol{b}}(\dot{\boldsymbol{b}}^{-1}(\boldsymbol{\mu}))$ is usually denoted by $\boldsymbol{V}(\boldsymbol{\mu})$ or just $\boldsymbol{V}$ in brief. Note that for models (1.4) and (1.6), $\boldsymbol{V}$s are diagonal matrices.

This is similar to (1.7) where we can find the moment generating function for the "random error" $\boldsymbol{e} = \boldsymbol{Y} - \boldsymbol{\mu} = (e_1, \cdots, e_n)^T$ to get the central moments of $\boldsymbol{Y}$. In fact, we have

$$\mathrm{E}(e^{\boldsymbol{\tau}^T\boldsymbol{e}}) = \exp[\phi\{b(\boldsymbol{\theta} + \boldsymbol{\tau}/\phi) - b(\boldsymbol{\theta})\} - \boldsymbol{\tau}^T\boldsymbol{\mu}].$$

Table 1.1. Characteristics of some exponential families

Model item	Binomial $B(n, p_i)$	Poisson $P(\lambda_i)$	Negative binomial $NB(\nu, p_i)$	Normal $N(\mu_i, \sigma^2)$	Gamma $GA(\mu_i, \nu)$	Inverse Gaussian $GI(\mu_i, \sigma^2)$
ϕ	1	1	1	σ^{-2}	ν	σ^{-2}
θ_i	$\log(p_i/1-p_i)$	$\log\lambda_i$	$\log(1-p_i)$	μ_i	$-\mu_i^{-1}$	$-(2\mu_i^2)^{-1}$
μ_i	$ne^{\theta_i}/(1+e^{\theta_i})$	e^{θ_i}	$\nu e^{\theta_i}/(1-e^{\theta_i})$	θ_i	$-\theta_i^{-1}$	$(-1/2\theta_i)^{1/2}$
$b(\theta_i)$	$n\log(1+e^{\theta_i})$	e^{θ_i}	$\nu\log(1-e^{\theta_i})^{-1}$	$\theta_i^2/2$	$-\log(-\theta_i)$	$-(-2\theta_i)^{1/2}$
$c(y_i)$	$-\log\binom{n}{y_i}$	$\log(y_i!)$	$-\log\binom{\nu+y_i-1}{\nu-1}$	$y_i^2/2$	$-\log(y_i)$	$(2y_i)^{-1}$
$s(y_i, \phi)$	0	0	0	$-\log\phi+\log(2\pi)$	$-2\{\phi\log\phi-\log\Gamma(\phi)\}+2\log y_i$	$-\{\log\phi-\log(2\pi y_i^3)\}$
$\sigma^2 V(\mu_i)$	$n^{-1}\mu_i(n-\mu_i)$	μ_i	$\mu_i+\nu^{-1}\mu_i^2$	σ^2	$\sigma^2\mu_i^2$	$\sigma^2\mu_i^3$
$d_i(y_i, \mu_i)$	$2\Big\{y_i\log(y_i/\mu_i)+$ $(n-y_i)\log\left(\frac{n-y_i}{n-\mu_i}\right)\Big\}$	$2\{y_i\log(y_i/\mu_i)-$ $(y_i-\mu_i)\}$	$2\left[y_i\log\left\{\frac{y_i(\nu+\mu_i)}{\mu_i(\nu+y_i)}\right\}\right.$ $\left.+\nu\log\left(\frac{\nu+y_i}{\nu+\mu_i}\right)\right]$	$(y_i-\mu_i)^2$	$2\left\{\frac{y_i-\mu_i}{\mu_i}-\log\left(\frac{y_i}{\mu_i}\right)\right\}$	$(y_i-\mu_i)^2/\mu_i^2 y_i$
$d_i'(y_i, \mu_i)$	$-2\Big\{y_i\log\frac{\mu_i}{n-\mu_i}+$ $n\log\left(\frac{n-\mu_i}{n}\right)+$ $\log\binom{n}{y_i}\Big\}$	$-2\{y_i\log\mu_i-\mu_i$ $-\log(y_i!)\}$	$-2\Big\{y_i\log\frac{\mu_i}{\nu+\mu_i}+$ $\nu\log\left(\frac{\nu}{\nu+\mu_i}\right)+$ $\log\binom{\nu+y_i-1}{\nu-1}\Big\}$	$(y_i-\mu_i)^2$	$2\left\{\frac{y_i}{\mu_i}-\log\frac{y_i}{\mu_i}\right\}$	$\frac{(y_i-\mu_i)^2}{\mu_i^2 y_i}$
Note	$\mu_i = np_i$ $\sigma^2 = 1$	$\mu_i = \lambda_i$ $\sigma^2 = 1$	$\mu_i = \nu\frac{1-p_i}{p_i}$ $\sigma^2 = 1$	$d_i'(y_i, y_i) = 0$	$d_i'(y_i, y_i) = 2$ $\sigma^2 = \nu^{-1}$	$d'(y_i, y_i) = 0$

By a little calculation, we get

$$\mathrm{E}(e_i) = 0, \qquad \mathrm{E}(e_i e_j) = \sigma^2 V_{ij}, \tag{1.9}$$

$$\mathrm{E}(e_i e_j e_k) = \sigma^4 S_{ijk}, \quad S_{ijk} = \partial^3 b(\boldsymbol{\theta})/\partial\theta_i \partial\theta_j \partial\theta_k, \tag{1.10}$$

and

$$\mathrm{E}(e_i e_j e_k e_l) = \sigma^4 (V_{ij} V_{kl} + V_{ik} V_{jl} + V_{il} V_{jk}) + \sigma^6 \Delta_{ijkl}, \tag{1.11}$$

where $\boldsymbol{V}(\boldsymbol{\mu}) = (V_{ij})$, $\Delta_{ijkl} = \partial^4 b(\boldsymbol{\theta})/\partial\theta_i \partial\theta_j \partial\theta_k \partial\theta_l$; and $i, j, k, l = 1, \cdots, n$.

Now let us verify some formulas about the derivatives of $\boldsymbol{\mu}$ and $\boldsymbol{\theta}$ in (1.1) or (1.2), which will be applied to discuss some problems related to the likelihoods.

Lemma 1.1

For model (1.1) or (1.2), the following derivative formulas of $\boldsymbol{\theta}$ and $\boldsymbol{\mu}$ hold:

$$\frac{\partial \boldsymbol{\mu}}{\partial \boldsymbol{\theta}^T} = \boldsymbol{V}(\boldsymbol{\theta}), \quad \frac{\partial^2 \boldsymbol{\mu}}{\partial \boldsymbol{\theta} \partial \boldsymbol{\theta}^T} = \boldsymbol{b}^{(3)}(\boldsymbol{\theta}) = \boldsymbol{S}, \tag{1.12}$$

$$\frac{\partial \boldsymbol{\theta}}{\partial \boldsymbol{\mu}^T} = \boldsymbol{V}^{-1}(\boldsymbol{\mu}), \qquad \frac{\partial^2 \boldsymbol{\theta}}{\partial \boldsymbol{\mu} \partial \boldsymbol{\mu}^T} = -[\boldsymbol{V}^{-1}][\boldsymbol{V}^{-1} \boldsymbol{S} \boldsymbol{V}^{-1}], \tag{1.13}$$

where $\boldsymbol{b}^{(3)}(\boldsymbol{\theta}) = \boldsymbol{S}$, $\partial^2 \boldsymbol{\mu}/\partial\boldsymbol{\theta}\partial\boldsymbol{\theta}^T$ and $\partial^2\boldsymbol{\theta}/\partial\boldsymbol{\mu}\partial\boldsymbol{\mu}^T$ are $n \times n \times n$ arrays with elements $\partial^3 b(\boldsymbol{\theta})/\partial\theta_i\partial\theta_j\partial\theta_k$, $\partial^2\mu_i/\partial\theta_j\partial\theta_k$ and $\partial^2\theta_i/\partial\mu_j\partial\mu_k$, respectively; $[\cdot][\cdot]$ denotes the array multiplication, see Appendix A or Seber and Wild (1989, p.691) for details.

Proof. It follows from (1.8) that $\boldsymbol{\mu} = \dot{b}(\boldsymbol{\theta})$, $\boldsymbol{V}(\boldsymbol{\theta}) = \ddot{b}(\boldsymbol{\theta})$, hence $\partial\boldsymbol{\mu}/\partial\boldsymbol{\theta}^T = \boldsymbol{V}(\boldsymbol{\theta})$, $\partial\boldsymbol{\theta}/\partial\boldsymbol{\mu}^T = \boldsymbol{V}^{-1}(\boldsymbol{\mu})$ and $\partial^2\boldsymbol{\mu}/\partial\boldsymbol{\theta}\partial\boldsymbol{\theta}^T = \boldsymbol{b}^{(3)}(\boldsymbol{\theta}) = \boldsymbol{S}$ hold. To prove the second formula of (1.13), taking derivative with respect to $\boldsymbol{\mu}$ for the equation $\partial\boldsymbol{\mu}/\partial\boldsymbol{\mu}^T = (\partial\boldsymbol{\mu}/\partial\boldsymbol{\theta}^T)\,(\partial\boldsymbol{\theta}/\partial\boldsymbol{\mu}^T) = \boldsymbol{I}_n$ (where $\boldsymbol{I}_n$ is an identity matrix), and using the formula (A.8) from Appendix A, we get

$$\left(\frac{\partial \boldsymbol{\theta}}{\partial \boldsymbol{\mu}^T}\right)^T \left(\frac{\partial^2 \boldsymbol{\mu}}{\partial \boldsymbol{\theta}\partial\boldsymbol{\theta}^T}\right)\left(\frac{\partial \boldsymbol{\theta}}{\partial \boldsymbol{\mu}^T}\right) + \left[\frac{\partial \boldsymbol{\mu}}{\partial \boldsymbol{\theta}^T}\right]\left[\frac{\partial^2 \boldsymbol{\theta}}{\partial \boldsymbol{\mu}\partial\boldsymbol{\mu}^T}\right] = 0.$$

It follows from (1.12) and $\partial\boldsymbol{\theta}/\partial\boldsymbol{\mu}^T = \boldsymbol{V}^{-1}$ that the above equation is just

$$\boldsymbol{V}^{-1}\boldsymbol{S}\boldsymbol{V}^{-1} + [\boldsymbol{V}]\left[\frac{\partial^2 \boldsymbol{\theta}}{\partial \boldsymbol{\mu}\partial\boldsymbol{\mu}^T}\right] = \boldsymbol{0},$$

which yields the last equation of (1.13) and the lemma is proved. ∎

Here and henceforth, ∎ denotes the end of an important paragraph, such as assumption, definition, lemma, theorem etc.

Note that the vector differentiation and the array multiplication will be used frequently throughout this book. The notation, the formulas and the technical details are given in Appendix A, see also A10 and B4 in the Appendices of Seber and Wild (1989).

Finally, we give a useful lemma for the specific model (1.6), which will be used at the end of this chapter and elsewhere.

Lemma 1.2

If model (1.2) satisfies (1.5) and the derivatives of $c(\cdot)$ and $s(\cdot)$ exist, then the derivatives of $b(\cdot)$ and $c(\cdot)$ satisfy $\dot{\boldsymbol{b}}(\cdot) = \dot{\boldsymbol{c}}^{-1}(\cdot)$, the latter is the inverse of $\dot{\boldsymbol{c}}(\cdot)$ (McCullagh 1983, p.60; Smyth 1989, p.49).

Proof. For model (1.2) with (1.5) and $\boldsymbol{Y} \sim ED(\boldsymbol{\theta}, \phi)$, it follows from $\mathrm{E}\{\partial \log p(\boldsymbol{Y}; \boldsymbol{\theta}, \phi)/\partial\phi\} = 0$ that

$$\boldsymbol{\theta}^T \mathrm{E}(\boldsymbol{Y}) - b(\boldsymbol{\theta}) - \mathrm{E}\{c(\boldsymbol{Y})\} - \frac{1}{2}\dot{s}(\phi) = 0$$

and

$$\mathrm{E}\{c(\boldsymbol{Y})\} = \boldsymbol{\theta}^T \dot{\boldsymbol{b}}(\boldsymbol{\theta}) - b(\boldsymbol{\theta}) - \frac{1}{2}\dot{s}(\phi). \tag{1.14}$$

Now let $\boldsymbol{Y}^1, \cdots \boldsymbol{Y}^N$ be independent and identically distributed observations from (1.2) with (1.5), and $\overline{\boldsymbol{Y}} = N^{-1}\sum_k \boldsymbol{Y}^k$ is the mean vector of these observations. It follows from (1.7) that the moment generating function of $\overline{\boldsymbol{Y}}$ is $\exp[N\phi\{b(\boldsymbol{\theta} + \boldsymbol{\tau}/N\phi) - b(\boldsymbol{\theta})\}]$ and hence the distribution of $\overline{\boldsymbol{Y}}$ is also an exponential family with $\overline{\boldsymbol{Y}} \sim ED(\boldsymbol{\theta}, N\phi)$. Then from (1.14) we have

$$\mathrm{E}[c(\overline{\boldsymbol{Y}})] = \boldsymbol{\theta}^T \dot{\boldsymbol{b}}(\boldsymbol{\theta}) - b(\boldsymbol{\theta}) - \frac{1}{2}\dot{s}(N\phi). \tag{1.15}$$

By the law of large numbers, $\overline{\boldsymbol{Y}} \to \boldsymbol{\mu}$ (*a.e.*), and $c(\overline{\boldsymbol{Y}}) \to c(\boldsymbol{\mu})$ (*a.e.*) as $N \to \infty$, where "*a.e.*" means "convergence almost everywhere". Hence we have $\mathrm{E}[c(\overline{\boldsymbol{Y}})] \to c(\boldsymbol{\mu}) = c(\dot{\boldsymbol{b}}(\boldsymbol{\theta}))$. Comparing this result with (1.15), we have

$$c(\dot{\boldsymbol{b}}(\boldsymbol{\theta})) - \boldsymbol{\theta}^T \dot{\boldsymbol{b}}(\boldsymbol{\theta}) + b(\boldsymbol{\theta}) = -\frac{1}{2}\lim_{N\to\infty} \dot{s}(N\phi).$$

The right-hand side of the above equation is independent of $\boldsymbol{\theta}$, thus by taking derivative with respect to $\boldsymbol{\theta}$ we have

$$\dot{\boldsymbol{c}}^T(\dot{\boldsymbol{b}}(\boldsymbol{\theta}))\ddot{\boldsymbol{b}}(\boldsymbol{\theta}) - \boldsymbol{\theta}^T \ddot{\boldsymbol{b}}(\boldsymbol{\theta}) = \boldsymbol{0}$$

which gives $\dot{\boldsymbol{c}}(\dot{\boldsymbol{b}}(\boldsymbol{\theta})) = \boldsymbol{\theta}$, hence $\dot{\boldsymbol{b}}(\boldsymbol{\theta}) = \dot{\boldsymbol{c}}^{-1}(\boldsymbol{\theta})$ holds. ∎

1.2 Likelihood

Since there is a one-to-one mapping between μ and θ, one may denote the log-likelihood of Y in terms of μ, as often used in the literature (e.g. Barndorff-Nielsen 1978; Morris 1982; 1983; Jorgensen 1987; McCullagh and Nelder 1989). For model (1.2), the log-likelihood of Y for the parameter θ or μ is usually denoted by $l(\mu, y) = \log p(y; \theta, \phi)$ for the fixed nuisance parameter ϕ. If the random character of the log-likelihood is to be stressed, y is replaced by Y. Hence we have

$$l(\mu, y) = \phi\{y^T\theta - b(\theta) - c(y)\} - \frac{1}{2}s(y, \phi), \tag{1.16}$$

where $\theta = \dot{b}^{-1}(\mu)$ and ϕ are implicit in $l(\mu, y)$. To obtain the likelihood equation and the Fisher information matrix, and also for further discussions, we need to find the derivatives of $l(\mu, y)$. It follows from Lemma 1.1 that the score functions of Y for θ and μ are respectively

$$\dot{l}_\theta = \frac{\partial l}{\partial \theta} = \phi(Y - \mu) = \phi e \tag{1.17}$$

and

$$\dot{l}_\mu = \frac{\partial l}{\partial \mu} = \phi V^{-1}(\mu)(Y - \mu) = \phi V^{-1} e, \tag{1.18}$$

where we use e to denote the "random error" $Y - \mu$ in later discussions. The characteristic of e has been given in (1.9) to (1.11). It is easily seen from (1.17) or (1.18) that the maximum likelihood estimators of μ and θ are $\tilde{\mu} = Y$ and $\tilde{\theta} = \dot{b}^{-1}(\tilde{\mu})$, respectively. For simplicity, we often denote $\dot{b}^{-1}(\cdot)$ by $q(\cdot)$ in later discussions, hence $\tilde{\theta} = q(Y)$.

Lemma 1.3

For model (1.2), the observed informations, that is minus the second order derivatives of the log-likelihood function of Y for θ and μ, can be respectively represented as

$$-\ddot{l}_{\theta\theta} = -\frac{\partial^2 l}{\partial\theta\partial\theta^T} = \phi\ddot{b}(\theta) = \phi V(\mu) \tag{1.19}$$

and

$$-\ddot{l}_{\mu\mu} = -\frac{\partial^2 l}{\partial\mu\partial\mu^T} = \phi V^{-1} + \phi[e^T V^{-1}][V^{-1} S V^{-1}]. \tag{1.20}$$

Further, the Fisher information matrices of Y for θ and μ can be expressed as

$$J_\theta(Y) = \phi V(\theta), \quad J_\mu(Y) = \phi V^{-1}(\mu). \tag{1.21}$$

Proof. Formula (1.19) can be obtained directly from (1.17). To prove (1.20), we use (1.13) of Lemma 1.1. Since $\partial l/\partial \boldsymbol{\mu}^T = (\partial l/\partial \boldsymbol{\theta}^T)(\partial \boldsymbol{\theta}/\partial \boldsymbol{\mu}^T)$, it follows from (A.8) of Appendix A that

$$-\ddot{l}_{\mu\mu} = \left(\frac{\partial \boldsymbol{\theta}}{\partial \boldsymbol{\mu}^T}\right)^T \left(-\frac{\partial^2 l}{\partial \boldsymbol{\theta}\partial \boldsymbol{\theta}^T}\right)\left(\frac{\partial \boldsymbol{\theta}}{\partial \boldsymbol{\mu}^T}\right) - \left[\left(\frac{\partial l}{\partial \boldsymbol{\theta}}\right)^T\right]\left[\frac{\partial^2 \boldsymbol{\theta}}{\partial \boldsymbol{\mu}\partial \boldsymbol{\mu}^T}\right].$$

It follows from Lemma 1.1 and (1.19) that

$$\begin{aligned} -\ddot{\boldsymbol{l}}_{\mu\mu} &= \boldsymbol{V}^{-1}(\phi \boldsymbol{V})\boldsymbol{V}^{-1} + [\phi(\boldsymbol{y}-\boldsymbol{\mu})^T][[\boldsymbol{V}^{-1}][\boldsymbol{V}^{-1}\boldsymbol{S}\boldsymbol{V}^{-1}]] \\ &= \phi \boldsymbol{V}^{-1} + \phi[\mathbf{e}^T \boldsymbol{V}^{-1}][\boldsymbol{V}^{-1}\boldsymbol{S}\boldsymbol{V}^{-1}] \end{aligned}$$

and (1.20) is proved. Further, by taking expectation for equations (1.19) and (1.20), we obtain (1.21). ∎

Finally, we consider the dispersion parameter $\phi = \sigma^{-2}$. It is easily seen from (1.16) that

$$\begin{aligned} \dot{l}_\phi &= \frac{\partial l}{\partial \phi} = \{\boldsymbol{y}^T\boldsymbol{\theta} - b(\boldsymbol{\theta}) - c(\boldsymbol{y})\} - \frac{1}{2}\dot{s}(\boldsymbol{y},\phi), \\ -\ddot{l}_{\phi\phi} &= -\frac{\partial^2 l}{\partial \phi^2} = \frac{1}{2}\ddot{s}(\boldsymbol{y},\phi), \end{aligned}$$

and

$$-\ddot{\boldsymbol{l}}_{\phi\theta} = \boldsymbol{y} - \boldsymbol{\mu}, \quad -\ddot{\boldsymbol{l}}_{\phi\mu} = \boldsymbol{V}^{-1}(\boldsymbol{y}-\boldsymbol{\mu}),$$

where $\dot{s}(\boldsymbol{y},\phi)$ and $\ddot{s}(\boldsymbol{y},\phi)$ are the first two derivatives of $s(\boldsymbol{y},\phi)$ in ϕ. We assume that $c(\boldsymbol{Y})$ and $s(\boldsymbol{Y},\phi)$ satisfy $E(\dot{l}_\phi) = 0$. It is easily seen that the dispersion parameter $\phi = \sigma^{-2}$ is orthogonal to parameters $\boldsymbol{\theta}$ and $\boldsymbol{\mu}$ in the sense of Cox and Reid (1987), that is $E(-\ddot{\boldsymbol{l}}_{\phi\theta}) = \mathbf{0}$, $E(-\ddot{\boldsymbol{l}}_{\phi\mu}) = \mathbf{0}$. The maximum likelihood estimator $\tilde{\phi}$ of ϕ satisfies

$$\dot{s}(\boldsymbol{Y},\tilde{\phi}) = 2\{\boldsymbol{Y}^T\tilde{\boldsymbol{\theta}} - b(\tilde{\boldsymbol{\theta}}) - c(\boldsymbol{Y})\}.$$

Now we consider a commonly encountered situation in which (1.5) or (1.6) is satisfied. In this case, $\dot{s}(\boldsymbol{y},\phi)$ and $\ddot{s}(\boldsymbol{y},\phi)$ will be replaced by $\dot{s}(\phi)$ and $\ddot{s}(\phi)$, respectively. Then $\tilde{\phi}$ can be denoted by

$$\tilde{\phi} = \dot{s}^{-1}[2\{\boldsymbol{Y}^T\tilde{\boldsymbol{\theta}} - b(\tilde{\boldsymbol{\theta}}) - c(\boldsymbol{Y})\}].$$

Further, the Fisher information of $\boldsymbol{Y}$ for ϕ is of a quite simple form

$$J_\phi(\boldsymbol{Y}) = -\ddot{l}_{\phi\phi} = \frac{1}{2}\ddot{s}(\phi). \tag{1.22}$$

Here we assume that $\ddot{s}(\phi) > 0$ in Φ.

1.3 Likelihood Ratio and Deviance

The deviance is an important and well-known statistic in discussing the problems related to exponential families. We introduce the deviance starting with the likelihood ratio statistic.

For model (1.2) with the fixed dispersion parameter ϕ, we consider the testing of hypothesis

$$H_0 : \boldsymbol{\mu} = \boldsymbol{\mu}_0; \qquad H_1 : \boldsymbol{\mu} \neq \boldsymbol{\mu}_0.$$

Since $\tilde{\boldsymbol{\mu}} = \boldsymbol{y}$, the likelihood ratio statistic of this test is

$$\begin{aligned} LR(\boldsymbol{\mu}_0) &= 2\{l(\tilde{\boldsymbol{\mu}}, \boldsymbol{y}) - l(\boldsymbol{\mu}_0, \boldsymbol{y})\} \\ &= 2\phi\{\boldsymbol{y}^T\boldsymbol{\theta} - b(\boldsymbol{\theta})\}_{\mu=y} - 2\phi\{\boldsymbol{y}^T\boldsymbol{\theta} - b(\boldsymbol{\theta})\}_{\mu=\mu_0} \\ &= 2\phi\{\boldsymbol{y}^T q(\boldsymbol{y}) - b(q(\boldsymbol{y}))\} - 2\phi\{\boldsymbol{y}^T q(\boldsymbol{\mu}_0) - b(q(\boldsymbol{\mu}_0))\}, \end{aligned}$$

where $\boldsymbol{\theta} = \dot{\boldsymbol{b}}^{-1}(\boldsymbol{\mu}) = \boldsymbol{q}(\boldsymbol{\mu})$ is used. It is well known that $LR(\boldsymbol{\mu}_0)$ asymptotically has $\chi^2(n)$ distribution (Cox and Hinkley 1974, p.314). The above equation holds for any $\boldsymbol{\mu}$, so we may write

$$LR(\boldsymbol{\mu}) = D^*(\boldsymbol{y}, \boldsymbol{\mu}) = \phi D(\boldsymbol{y}, \boldsymbol{\mu}) \sim \chi^2(n) \ (asy.)$$

where

$$\begin{aligned} D(\boldsymbol{y}, \boldsymbol{\mu}) &= 2\{\boldsymbol{y}^T q(\boldsymbol{y}) - b(q(\boldsymbol{y}))\} - 2\{\boldsymbol{y}^T q(\boldsymbol{\mu}) - b(q(\boldsymbol{\mu}))\} \\ &= 2\{\boldsymbol{y}^T\boldsymbol{\theta} - b(\boldsymbol{\theta}) - c(\boldsymbol{y})\}_{\mu=y} - 2\{\boldsymbol{y}^T\boldsymbol{\theta} - b(\boldsymbol{\theta}) - c(\boldsymbol{y})\}. \quad (1.23) \end{aligned}$$

$D(\boldsymbol{y}, \boldsymbol{\mu})$ is called the deviance, which is an important statistic in generalized linear models and in exponential family nonlinear models (see Chapter 2). In fact, the deviance $D(\boldsymbol{y}, \boldsymbol{\mu})$ is the kernel part of the log-likelihood function $l(\boldsymbol{\mu}, \boldsymbol{y})$ in (1.16):

$$l(\boldsymbol{\mu}, \boldsymbol{y}) = -\frac{1}{2}\phi D(\boldsymbol{y}, \boldsymbol{\mu}) - \frac{1}{2}s(\boldsymbol{y}, \phi) + \phi\{\boldsymbol{y}^T\boldsymbol{\theta} - b(\boldsymbol{\theta}) - c(\boldsymbol{y})\}_{\mu=y}.$$

Furthermore, we have the following Hoeffding representation:

Lemma 1.4

The exponential family (1.1) or (1.2) and its log-likelihood function can be represented as

$$p(\boldsymbol{y}; \boldsymbol{\mu}, \phi) = p(\boldsymbol{y}; \boldsymbol{y}, \phi) \exp\{-\frac{1}{2}\phi D(\boldsymbol{y}, \boldsymbol{\mu})\} \qquad (1.24)$$

and

$$l(\boldsymbol{\mu}, \phi; \boldsymbol{y}) = l(\boldsymbol{y}, \phi; \boldsymbol{y}) - \frac{1}{2}\phi D(\boldsymbol{y}, \boldsymbol{\mu}). \tag{1.25}$$

Proof. It follows from (1.1), (1.2) and (1.23) that

$$\begin{aligned}\frac{p(\boldsymbol{y}; \boldsymbol{\mu}, \phi)}{p(\boldsymbol{y}; \boldsymbol{y}, \phi)} &= \exp[\phi\{\boldsymbol{y}^T\boldsymbol{\theta} - b(\boldsymbol{\theta})\} - \phi\{\boldsymbol{y}^T\boldsymbol{\theta} - b(\boldsymbol{\theta})\}_{\mu=y}] \\ &= \exp\{-\frac{1}{2}\phi D(\boldsymbol{y}, \boldsymbol{\mu})\},\end{aligned}$$

which yields (1.24) and (1.25). ∎

This lemma shows that the log-likelihood of $\boldsymbol{Y}$ for $\boldsymbol{\mu}$ or $\boldsymbol{\theta}$ is actually proportional to the deviance $D(\boldsymbol{y}, \boldsymbol{\mu})$. Therefore we may find the maximum likelihood estimator of $\boldsymbol{\mu}$ or $\boldsymbol{\theta}$ by minimizing $D(\boldsymbol{y}, \boldsymbol{\mu})$ instead of maximizing $l(\boldsymbol{\mu}, \phi; \boldsymbol{y})$ while the maximum likelihood estimator of ϕ can be easily obtained from (1.25). It is easily seen from (1.23) that

$$\min_{\mu} D(\boldsymbol{y}, \boldsymbol{\mu}) = D(\boldsymbol{y}, \boldsymbol{y}) = 0.$$

Further, from (1.23) we can get the derivatives of $D(\boldsymbol{y}, \ \boldsymbol{\mu})$ as follows:

$$\begin{aligned}\dot{\boldsymbol{D}}_\theta &= \frac{\partial\{D(\boldsymbol{y}, \boldsymbol{\mu})\}}{\partial\theta} = -2(\boldsymbol{y} - \boldsymbol{\mu}) = -2\phi^{-1}\dot{\boldsymbol{l}}_\theta, \\ \dot{\boldsymbol{D}}_\mu &= \frac{\partial\{D(\boldsymbol{y}, \boldsymbol{\mu})\}}{\partial\boldsymbol{\mu}} = -2\boldsymbol{V}^{-1}(\boldsymbol{y} - \boldsymbol{\mu}) = -2\phi^{-1}\dot{\boldsymbol{l}}_\mu.\end{aligned}$$

In exponential family, there is a close relationship between the deviance and the Kullback-Leibler distance, which is defined as

$$K(\boldsymbol{\mu}_1, \boldsymbol{\mu}_2) = \mathrm{E}_{\mu_1}\left\{\log\frac{p(\boldsymbol{y}; \boldsymbol{\mu}_1, \phi)}{p(\boldsymbol{y}; \boldsymbol{\mu}_2, \phi)}\right\},$$

where $\boldsymbol{\mu}_1$ and $\boldsymbol{\mu}_2$ are two values of $\boldsymbol{\mu}$ and their counterparts of $\boldsymbol{\theta}$ are denoted by $\boldsymbol{\theta}_1$ and $\boldsymbol{\theta}_2$, respectively. From (1.1) or (1.2) we have

$$K(\boldsymbol{\mu}_1, \boldsymbol{\mu}_2) = \phi\{\boldsymbol{\mu}_1^T\boldsymbol{\theta}_1 - b(\boldsymbol{\theta}_1)\} - \phi\{\boldsymbol{\mu}_1^T\boldsymbol{\theta}_2 - b(\boldsymbol{\theta}_2)\}.$$

Taking $\boldsymbol{\mu}_1 = \boldsymbol{y}, \ \ \boldsymbol{\mu}_2 = \boldsymbol{\mu}$ and comparing the result with (1.23), we have

$$K(\boldsymbol{y}, \boldsymbol{\mu}) = \frac{1}{2}\phi D(\boldsymbol{y}, \boldsymbol{\mu}) = \frac{1}{2}LR(\boldsymbol{\mu}).$$

By this equation, (1.24) can also be written as

$$p(\boldsymbol{y}; \boldsymbol{\mu}, \phi) = p(\boldsymbol{y}; \boldsymbol{y}, \phi)\exp\{-K(\boldsymbol{y}, \boldsymbol{\mu})\}.$$

Now we discuss some specific properties for models (1.4) and (1.6). For these models, the deviance can be represented as

$$\begin{aligned} D(y, \mu) &= \sum_{i=1}^{n} [2\{y_i\theta_i - b(\theta_i) - c(y_i)\}_{\mu_i = y_i} - 2\{y_i\theta_i - b(\theta_i) - c(y_i)\}] \\ &= \sum_{i=1}^{n} d_i(y_i, \mu_i), \end{aligned}$$

where

$$\begin{aligned} d_i(y_i, \mu_i) &= 2\{y_i\theta_i - b(\theta_i) - c(y_i)\}_{\mu_i = y_i} - 2\{y_i\theta_i - b(\theta_i) - c(y_i)\} \\ &= 2\{y_i q(y_i) - b(q(y_i))\} - 2\{y_i\theta_i - b(\theta_i)\} \end{aligned}$$

and $q(y_i) = \dot{b}^{-1}(y_i)$. Table 1.1 gives the deviances for some commonly encountered exponential families.

Equations (1.4) and (1.23) show that the main part of exponential family distribution is the quantities $2\{y_i\theta_i - b(\theta_i) - c(y_i)\}$, $i = 1, \cdots, n$. Smyth (1989) proposed to use the notation

$$d_i'(y_i, \mu_i) = -2\{y_i\theta_i - b(\theta_i) - c(y_i)\},$$
$$D'(y, \mu) = \sum_{i=1}^{n} d_i'(y_i; \mu_i).$$

Then we have

$$d_i(y_i, \mu_i) = d_i'(y_i, \mu_i) - d_i'(y_i, y_i), \tag{1.26}$$
$$D(y, \mu) = D'(y, \mu) - D'(y, y).$$

From (1.24), the exponential family (1.4) can be represented as

$$p(y; \mu, \phi) = \exp\{-\frac{1}{2}\phi D'(y, \mu) - \frac{1}{2}\Sigma_{i=1}^{n} s(y_i, \phi)\}. \tag{1.27}$$

Furthermore, for model (1.6) we can reduce (1.27) according to the following lemma.

Lemma 1.5

For model (1.6), $d_i'(y_i, y_i)$ is a constant, say k, which is independent of y_i and satisfies

$$d_i(y_i, \mu_i) = d_i'(y_i, \mu_i) - k, \tag{1.28}$$
$$D(y, \mu) = D'(y, \mu) - nk,$$

and

$$p(\mathbf{y};\boldsymbol{\mu},\phi)=\exp\{-\frac{1}{2}\phi D(\mathbf{y},\boldsymbol{\mu})\}\cdot\exp\{-\frac{n}{2}(s(\phi)+k\phi)-\frac{1}{2}\Sigma_{i=1}^{n}t(y_i)\}. \tag{1.29}$$

Proof. It follows from $\theta_i=\dot{b}^{-1}(\mu_i)=q(\mu_i)$ that

$$d_i'(y_i,y_i)=-2\{y_iq(y_i)-b(q(y_i))-c(y_i)\}.$$

Taking derivative for $d_i'(y_i,y_i)$ with respect to y_i yields

$$\begin{aligned}\frac{d}{dy_i}\{d_i'(y_i,y_i)\} &= -2\{q(y_i)+y_i\dot{q}(y_i)-\dot{b}(q(y_i))\dot{q}(y_i)-\dot{c}(y_i)\}\\ &= -2\{q(y_i)-\dot{c}(y_i)\},\end{aligned}$$

here we use the result $\dot{b}(q(y_i))=\dot{b}(\dot{b}^{-1}(y_i))=y_i$. Further, it follows from Lemma 1.2 that $\dot{c}(y_i)=\dot{b}^{-1}(y_i)=q(y_i)$ and we get $(d/dy_i)\{d_i'(y_i,y_i)\}=0$ which states that $d_i'(y_i,y_i)$ is a constant, say k. Thus (1.28) follows from (1.26), and (1.29) can be obtained from (1.27) and (1.28). ∎

Table 1.1 shows that $d_i'(y_i,y_i)=k=0$ for normal and inverse Gaussian families. In this case we have

$$\begin{aligned}d_i(y_i,\mu_i)=d_i'(y_i,\mu_i)&=-2\{\,y_i\theta_i-b(\theta_i)-c(y_i)\},\\ D(\mathbf{y},\boldsymbol{\mu})&=D'(\mathbf{y},\boldsymbol{\mu}).\end{aligned}$$

For gamma family, $d_i'(y_i,y_i)=2$, hence

$$\begin{aligned}d_i(y_i,\mu_i)&=-2\{y_i\theta_i-b(\theta_i)-c(y_i)\}-2,\\ D(\mathbf{y},\boldsymbol{\mu})&=D'(\mathbf{y},\boldsymbol{\mu})-2n.\end{aligned}$$

Finally, we briefly introduce the weighted exponential family. In some practical situations, the dispersion parameter $\phi=\sigma^{-2}$ in (1.4) may be weighted by some known weights w_i (McCullagh and Nelder 1989, p.29; Smyth 1989), that is $y_i\sim ED(\mu_i,\sigma^2w_i^{-1})$, $i=1,\cdots,n$. In this case, (1.4) becomes

$$p(\mathbf{y};\boldsymbol{\theta},\phi)=\exp\left[\sum_{i=1}^{n}\left[\phi w_i\{y_i\theta_i-b(\theta_i)-c(y_i)\}-\frac{1}{2}s(y_i,\phi w_i)\right]\right]. \tag{1.30}$$

Obviously, the statistical behavior of this family is very similar to family (1.4) except that each log-likelihood of y_i in (1.4) is weighted by w_i, and so are equations (1.16) to (1.29). For instance, the deviance $d_i'(y_i,\mu_i)$ defined near (1.26) should be

$$d_i'(y_i,\mu_i)=-2w_i\{y_i\theta_i-b(\theta_i)-c(y_i)\}.$$

The other equations are treated similarly.

Chapter 2

Exponential Family Nonlinear Models

Exponential family nonlinear models are natural extensions of generalized linear models and normal nonlinear regression models. In recent years, a number of authors have been concerned about the inference of these models. Cordeiro and Paula (1989) studied the improved likelihood ratio statistics; Cook and Tsai (1990) discussed the cubic approximations of confidence regions; Pazman (1991) presented a saddle point approximation of distribution of maximum likelihood estimator from the geometric point of view; Wei and Shi (1994) studied some diagnostics problems; while Jorgensen (1983) and McCullagh (1983) studied some general models which included exponential family nonlinear models as their special cases. However, the research of exponential family nonlinear models is still very much in its infancy. A large number of inference problems remain to be explored.

In this chapter, we introduce the basis of exponential family nonlinear models as the preparation for succeeding chapters. Section 2.1 reviews the definition of these models and points out the relationship between exponential family nonlinear models and some other models such as generalized linear models, normal nonlinear regression models and curved exponential families. Section 2.2 deals with the likelihood problems. The score functions, the observed information and the Fisher information are introduced. In Section 2.3, we discuss the computation methods of maximum likelihood estimation.

2.1 Definition

Suppose that the components of $\boldsymbol{Y} = (y_i, \cdots, y_n)^T$ are independent random variables, in which each y_i has exponential family distribution (1.3) and may depend on an independent known variable $\boldsymbol{x}_i$ $(i = 1, \cdots, n)$. Then the distribution of $\boldsymbol{Y}$ has the form (1.4). Now we introduce the exponential family nonlinear models based on (1.4) as follows.

Definition 2.1

Suppose that the parameters of interest in (1.4) are $\boldsymbol{\beta} = (\beta_1, \cdots, \beta_p)^T$ defined in a subset $\mathcal{B}$ of R^p $(p < n)$ and the distributions of y_i depending on $\boldsymbol{x}_i$ satisfy the following constraint conditions:

$$\eta_i \triangleq g(\mu_i) = f(\boldsymbol{x}_i; \boldsymbol{\beta}) \triangleq f_i(\boldsymbol{\beta}), \quad i = 1, \cdots, n, \tag{2.1}$$

where $g(\cdot)$ is a known monotonic link function; $f(\cdot;\cdot)$ is a known function with an unknown vector parameter $\boldsymbol{\beta}$ and a known explanatory variable of q-vector $\boldsymbol{x}_i$, respectively (in what follows, the notation $\triangleq$ denotes a convention); then the models (2.1) with (1.4) are called the exponential family nonlinear models or generalized nonlinear models. ∎

Cordeiro and Paula (1989) first called these models the exponential family nonlinear models, so did Cook and Tsai (1990) and Pazman (1991). We also prefer to use this name to stress the exponential family. For these models, the density function (1.4) may be denoted by $p(\boldsymbol{y}; \boldsymbol{\theta}(\boldsymbol{\beta}), \phi)$ or $p(\boldsymbol{y}; \boldsymbol{\mu}(\boldsymbol{\beta}), \phi)$; and also by $p(\boldsymbol{y}; \boldsymbol{\beta}, \phi)$ for simplicity.

It is easily seen from this definition that the model (2.1) with (1.4) includes many commonly encountered regression models as their special cases. The following two are of special interest:

(a) If $f(\boldsymbol{x}_i; \boldsymbol{\beta}) = \boldsymbol{x}_i^T \boldsymbol{\beta}$, then (2.1) represent the well-known generalized linear models (GLM), which have been fully studied in the past two decades (e.g. McCullagh and Nelder 1989; Firth 1991).

(b) If $g(\mu_i) = \mu_i$, then (2.1) represent a general class of nonlinear regression models. In particular, if $g(\mu_i) = \mu_i$ with $b(\theta_i) = \theta_i^2/2$ (see Table 1.1), then (2.1) becomes a normal nonlinear regression model, which has been used often in recent years (e.g. Bates and Watts 1988; Seber and Wild 1989; Ratkowsky 1990).

In this book, we shall present some unified theories for exponential family nonlinear models, which must be valid to quite general classes of linear and nonlinear regression models, including generalized linear models and normal nonlinear regression models. In particular, We shall pay more attention to the nonlinear models with gamma distributions and inverse Gaussian distributions.

For further discussions, we assume these models satisfy the following assumption.

Assumption A

(a) When $\beta_1 \neq \beta_2$, $\phi_1 \neq \phi_2$, $\xi\{p(Y;\beta_1,\phi_1) \neq p(Y;\beta_2,\phi_2)\} > 0$.

(b) Let $l(\beta,\phi;y) = \log p(y;\theta(\beta),\phi)$. For fixed β and ϕ, $\partial l/\partial\beta_a$ $(a = 1,\cdots,p)$ and $\partial l/\partial\phi$ are linearly independent.

(c) The moments of random variables $\partial l/\partial\beta_a$ $(a = 1,\cdots,p)$ and $\partial l/\partial\phi$ exist at least up to the third order.

(d) The partial derivatives $\partial/\partial\beta_a$ $(a = 1,\cdots,p), \partial/\partial\phi$ and the integration with respect to the measure $\xi(dy)$ can always be interchanged for any integrable function $h(y;\theta,\phi)$. In particular, $\xi(dy) = dy$, the Lebesgue measure in R^n is assumed in most of the situations.

(e) The function $g(\cdot)$ is differentiable up to the third order. $f(x_i;\beta)$ in (2.1) as a function of x_i is defined in a compact subset $\mathcal{X}$ in R^q; as a function of β is defined in an open subset $\mathcal{B}$ in R^p and is differentiable up to the third order. All the above derivatives are continuous in $\mathcal{X} \times \mathcal{B}$.

(f) $\inf_\Theta \ddot{b}(\theta) > 0$, $\sup_\Theta |b^{(3)}(\theta)| < +\infty$. ∎

In this book, the above regularity conditions, (a) to (f), are required for models (2.1) and (1.4).

Definition (2.1) shows that the parameter β may be connected with the natural parameter θ_i and the expectation parameter μ_i as follows.

(a) From $\mu_i = \dot{b}(\theta_i)$, $g(\mu_i) = g(\dot{b}(\theta_i)) = f(x_i;\beta)$, we have

$$\theta_i = \theta_i(\beta) = \dot{b}^{-1} \circ g^{-1} \circ f(x_i;\beta), \quad \theta_i(\cdot) = \dot{b}^{-1} \circ g^{-1} \circ f(\cdot\,;\cdot),$$

where $\circ$ denotes the product of two functions (as mappings). If the link function $g(\cdot)$ meets certain conditions such that $\theta_i = \eta_i = f(x_i;\beta)$, which means that $\dot{b}^{-1} \circ g^{-1}$ is an identity function, that is $g(\cdot) = \dot{b}^{-1}(\cdot)$; then both function $g(\cdot)$ and the model (2.1) are called the canonical link, which is much easier to deal with than the non-canonical link (see Fahrmeir and Kaufmann 1985; McCullagh and Nelder 1989 for details).

(b) Equation (2.1) can also be denoted by the expectation parameter μ_i as

$$\mu_i = \mu_i(\beta) = g^{-1} \circ f(x_i;\beta), \quad \mu_i(\cdot) = g^{-1} \circ f(\cdot\,;\cdot).$$

As used by many authors in the literature, one usually studies the statistical behavior of β in terms of the expectation parameter $\mu = \mu(\beta)$ (e.g.

McCullagh 1983; Efron 1986; Jorgensen 1987; Cordeiro and Paula 1989; McCullagh and Nelder 1989). In this book, we also prefer to use $\boldsymbol{\mu} = \boldsymbol{\mu}(\boldsymbol{\beta})$ in most of the situations. Under the parameter $\boldsymbol{\mu}$, the model (2.1) with (1.4) may be represented as

$$p(\boldsymbol{y}; \boldsymbol{\mu}(\boldsymbol{\beta}), \phi) = \exp\left[\sum_{i=1}^{n} [\phi\{y_i\theta_i - b(\theta_i) - c(y_i)\} - \frac{1}{2}s(y_i, \phi)]\right],$$

$$\mu_i = \mu_i(\boldsymbol{\beta}) \quad \text{or} \quad \theta_i = \theta_i(\boldsymbol{\beta}); \quad i = 1, \cdots, n. \tag{2.2}$$

In later discussions, the vector notation is often used, then (2.1) can be written as

$$\boldsymbol{\eta} = \boldsymbol{g}(\boldsymbol{\mu}) = \boldsymbol{f}(\boldsymbol{\beta}); \quad \boldsymbol{\mu} = \boldsymbol{\mu}(\boldsymbol{\beta}) \tag{2.3}$$

where

$$\boldsymbol{\eta} = (\eta_1, \cdots, \eta_n)^T, \ \boldsymbol{g}(\boldsymbol{\mu}) = (g(\mu_1), \cdots, g(\mu_n))^T,$$

$$\boldsymbol{f}(\boldsymbol{\beta}) = (f_1(\boldsymbol{\beta}), \cdots, f_n(\boldsymbol{\beta}))^T, \ \ f_i(\boldsymbol{\beta}) = f(\boldsymbol{x}_i; \boldsymbol{\beta}),$$

and

$$\boldsymbol{\mu}(\boldsymbol{\beta}) = \boldsymbol{g}^{-1} \circ \boldsymbol{f}(\boldsymbol{\beta}) = (\mu_1(\boldsymbol{\beta}), \cdots, \mu_n(\boldsymbol{\beta}))^T.$$

Efron (1975; 1978), Amari (1982a,b; 1985) and other authors have studied curved exponential families which seem similar to exponential family nonlinear models. However, it should be emphasized that there is a substantial difference between these two models. The curved exponential family introduced by Efron (1975) and Amari (1982a) is based on (1.1) with $\phi = 1$, $\boldsymbol{\theta} = \boldsymbol{\theta}(\boldsymbol{\beta})$ and independently and identically distributed (iid) observations, say $\boldsymbol{Y}^m = (y_1^m, \cdots, y_n^m)^T, m = 1, \cdots, M$, where each $\boldsymbol{Y}^m$ has an exponential family distribution of the form (1.1) with fixed n. They studied the statistical behavior of $\boldsymbol{\beta}$ when M is sufficiently large. But exponential family nonlinear models are actually univariate regression models, in which the observations are just $y_1, \cdots, y_n$, which are independent but not necessarily identically distributed (non-iid) and each y_i is a scalar with an exponential family distribution (1.3). One may study the statistical behavior of $\boldsymbol{\beta}$ when n is sufficiently large. Many authors are concerned about this regression case (e.g. McCullagh 1983; Cordeiro and Paula 1989; Cook and Tsai 1990); we also deal with this situation.

Finally, we remind the reader that for exponential family nonlinear models (2.1), the properties of exponential families shown in Chapter 1 still hold at any fixed value $\boldsymbol{\mu} = \boldsymbol{\mu}(\boldsymbol{\beta})$. In particular, Lemma 1.3 still holds for (2.1) at $\boldsymbol{\mu} = \boldsymbol{\mu}(\boldsymbol{\beta})$. In this case, we have

$$p(\boldsymbol{y}; \boldsymbol{\mu}(\boldsymbol{\beta}), \phi) = p(\boldsymbol{y}; \boldsymbol{y}, \phi) \exp\{-\frac{1}{2}\phi D(\boldsymbol{y}, \boldsymbol{\mu}(\boldsymbol{\beta}))\}. \tag{2.4}$$

This equation shows that we can find the maximum likelihood estimator $\hat{\beta}$ of β by minimizing $D(\boldsymbol{y}, \boldsymbol{\mu}(\boldsymbol{\beta}))$ instead of maximizing $l(\boldsymbol{\mu}(\boldsymbol{\beta}), \phi; \boldsymbol{y})$; and the maximum likelihood estimator $\hat{\phi}$ of ϕ can be obtained from (2.4) by using $\hat{\boldsymbol{\mu}} = \boldsymbol{\mu}(\hat{\boldsymbol{\beta}})$.

Similarly, equations (1.26) to (1.29) still hold at $\boldsymbol{\mu} = \boldsymbol{\mu}(\boldsymbol{\beta})$. Furthermore, we can find the mean and the variance of deviance based on these equations for exponential family nonlinear models.

Lemma 2.1

For model (2.1) with (2.2), the mean and the variance of $D'(\boldsymbol{Y}, \boldsymbol{\mu}(\boldsymbol{\beta}))$ at any $\boldsymbol{\beta}$ can be represented as

$$\mathrm{E}\{D'(\boldsymbol{Y}, \boldsymbol{\mu}(\boldsymbol{\beta}))\} = -\sum_{i=1}^{n} \mathrm{E}\{\dot{s}(y_i, \phi)\},$$

$$\mathrm{Var}\{D'(\boldsymbol{Y}, \boldsymbol{\mu}(\boldsymbol{\beta}))\} = 2\sum_{i=1}^{n} \mathrm{E}\{\ddot{s}(y_i, \phi)\},$$

where $D'(\boldsymbol{y}, \boldsymbol{\mu})$ is given near (1.26); see also (1.23).

Proof. Taking the first two derivatives for $\int p(\boldsymbol{y}; \boldsymbol{\mu}(\boldsymbol{\beta}), \phi) d\boldsymbol{y} = 1$ with respect to the parameter ϕ and using the formula (1.27), we get

$$\int p(\boldsymbol{y}; \boldsymbol{\mu}(\boldsymbol{\beta}), \phi)\{-\frac{1}{2}D'(\boldsymbol{y}, \boldsymbol{\mu}(\boldsymbol{\beta})) - \frac{1}{2}\Sigma_{i=1}^{n} \dot{s}(y_i, \phi)\} d\boldsymbol{y} = 0,$$

$$\int p(\boldsymbol{y}; \boldsymbol{\mu}(\boldsymbol{\beta}), \phi)\,[\{-\frac{1}{2}D'(\boldsymbol{y}, \boldsymbol{\mu}(\boldsymbol{\beta})) - \frac{1}{2}\Sigma_{i=1}^{n} \dot{s}(y_i, \phi)\}^2 - \frac{1}{2}\Sigma_{i=1}^{n} \ddot{s}(y_i, \phi)]\, d\boldsymbol{y} = 0.$$

The lemma follows immediately from these. ∎

Applying this lemma to the case of $n = 1$, we get the following corollary.

Corollary

For any $\boldsymbol{\beta}$ and

$$d_i'(y_i, \mu_i(\boldsymbol{\beta})) = -2\{y_i\theta_i - b(\theta_i) - c(y_i)\},$$

we have

$$\mathrm{E}\{d_i'(y_i, \mu_i(\boldsymbol{\beta}))\} = -\mathrm{E}\{\dot{s}(y_i, \phi)\},$$

and

$$\mathrm{Var}\{d_i'(y_i, \mu_i(\boldsymbol{\beta}))\} = 2\mathrm{E}\{\ddot{s}(y_i, \phi)\}.$$ ∎

In particular, for the specific model (1.6), we can get more preferable results as follows.

Lemma 2.2

For exponential family nonlinear model (2.1) with the distribution (1.6), the means and the variances of $D(\boldsymbol{Y}, \boldsymbol{\mu}(\boldsymbol{\beta}))$ and $d(y_i, \mu_i(\boldsymbol{\beta}))$ at any $\boldsymbol{\beta}$ can be respectively represented as

$$\begin{aligned} \mathrm{E}\{D(\boldsymbol{Y}, \boldsymbol{\mu}(\boldsymbol{\beta}))\} &= -n\{\dot{s}(\phi) + k\}, \\ \mathrm{Var}\{D(\boldsymbol{Y}, \boldsymbol{\mu}(\boldsymbol{\beta})\} &= 2n\ddot{s}(\phi); \\ \mathrm{E}\{d_i(y_i, \mu_i(\boldsymbol{\beta}))\} &= -\{\dot{s}(\phi) + k\}, \\ \mathrm{Var}\{d_i(y_i, \mu_i(\boldsymbol{\beta})\} &= 2\ddot{s}(\phi); \end{aligned}$$

where $d_i'(y_i, y_i) = k$ is a constant.

Proof. It is easy to get these results by using Lemmas 1.5 and 2.1. ∎

It is worthy to note the fact that for model (2.1) with (1.6), the mean and the variance of the deviance depend only on the dispersion parameter but do not depend on the parameter $\boldsymbol{\beta}$.

Example 2.1 (Normal, inverse Gaussian and gamma nonlinear models).

Table 1.1 shows that for normal model and inverse Gaussian model, $k = 0, s(\phi) = -\log\phi$, then from Lemma 2.2 we have

$$\mathrm{E}\{D(\boldsymbol{Y}, \boldsymbol{\mu}(\boldsymbol{\beta}))\} = n\sigma^2, \quad \mathrm{Var}\{D(\boldsymbol{Y}, \boldsymbol{\mu}(\boldsymbol{\beta}))\} = 2n\sigma^4,$$

where $\sigma^2 = \phi^{-1}$, $D(\boldsymbol{Y}, \boldsymbol{\mu}) = \sum_i (y_i - \mu_i)^2$ for normal model and

$$D(\boldsymbol{Y}, \boldsymbol{\mu}) = \sum_{i=1}^{n} \{(y_i - \mu_i)^2 / \mu_i^2 y_i\}$$

for inverse Gaussian model.

For gamma model, $k = 2$, $s(\phi) = -2\{\phi\log\phi - \log\Gamma(\phi)\}$, where $\Gamma(\cdot)$ is the gamma function. Then Lemma 2.2 gives

$$\begin{aligned} \mathrm{E}\{D(\boldsymbol{Y}, \boldsymbol{\mu}(\boldsymbol{\beta}))\} &= 2n\{\log\phi - \psi(\phi)\}, \\ \mathrm{Var}\{D(\boldsymbol{Y}, \boldsymbol{\mu}(\boldsymbol{\beta}))\} &= 4n\{\dot{\psi}(\phi) - \phi^{-1}\} \end{aligned}$$

where

$$D(\boldsymbol{Y}, \boldsymbol{\mu}) = 2\sum_{i=1}^{n} \{\mu_i^{-1}(y_i - \mu_i) - \log(y_i/\mu_i)\},$$

$\psi(\phi) = \partial\{\log\Gamma(\phi)\}/\partial\phi$ is the diagamma function (or psi function):

$$\psi(x) = \sum_{k=0}^{\infty} \left(\frac{1}{k+1} - \frac{1}{k+x} \right) - C,$$

and $C = 0.57721566\cdots$ is the Eular constant. ∎

2.2 Likelihood

Now let $\hat{\beta}$ and $\hat{\phi}$ be the maximum likelihood estimators of β and ϕ for exponential family nonlinear models (2.1) and the corresponding quantities of θ and μ be denoted by $\hat{\theta} = \theta(\hat{\beta})$ and $\hat{\mu} = \mu(\hat{\beta})$, respectively. Equation (2.4) shows that we can find $\hat{\beta}$, $\hat{\phi} = \phi(\hat{\beta})$ separately, and usually compute $\hat{\beta}$ first. To provide an iterative scheme for obtaining $\hat{\beta}$, and also for some theoretical purposes, we start with deriving the first two derivatives of log-likelihood of $\boldsymbol{Y}$ for the parameter β. For simplicity, we denote $l(\mu(\beta), y)$ or $l(\theta(\beta), y)$ by $l(\beta)$ for fixed ϕ, that is (1.16) and (1.27) are respectively denoted by

$$l(\beta) = l(\mu(\beta), y) = \phi\{y^T\theta - b(\theta) - c(y)\} - \frac{1}{2}s(y, \phi), \quad \mu = \mu(\beta), \tag{2.5}$$

and

$$\begin{aligned} l(\beta) &= -\frac{1}{2}\phi D'(y, \mu(\beta)) - \frac{1}{2}s(y, \phi) \\ &= -\frac{1}{2}\phi D(y, \mu(\beta)) - \frac{1}{2}s(y, \phi) - \frac{1}{2}\phi D'(y, y). \end{aligned}$$

The first two derivatives of $l(\beta)$ in β are denoted by $\dot{l}(\beta)$ and $\ddot{l}(\beta)$, respectively. Note that in model (2.1) with (2.2), the variance function $\ddot{b}(\theta) = V$ is a diagonal matrix. We usually write

$$V = \mathrm{diag}(V_i), \quad V_i = V_{ii}$$

for short ($i = 1, \cdots, n$). Further, since $\theta = \dot{b}^{-1}(\mu)$, $\mu = \mu(\beta)$, the variance function is also written as $V(\mu)$ or $V(\beta)$. Similar convention will be used for other quantities. Moreover, we set

$$D(\beta) = \frac{\partial \mu(\beta)}{\partial \beta^T}, \quad W(\beta) = \frac{\partial^2 \mu(\beta)}{\partial \beta \partial \beta^T},$$

$$D_\theta(\beta) = \frac{\partial \theta(\beta)}{\partial \beta^T}, \quad W_\theta(\beta) = \frac{\partial^2 \theta(\beta)}{\partial \beta \partial \beta^T},$$

where D and D_θ are $n \times p$ matrices, and W and W_θ are $n \times p \times p$ arrays, respectively. By Assumption A, all of these derivatives exist. There is a close relationship among these derivatives:

$$D_\theta = V^{-1}D, \quad W_\theta = [V^{-1}][W - \Gamma], \tag{2.6}$$

where

$$\Gamma = D^T V^{-1} S V^{-1} D. \tag{2.7}$$

Proof. It follows from Lemma 1.1, and (A.7) and (A.8) of Appendix A that

$$\begin{aligned}
\boldsymbol{D}_\theta &= \frac{\partial\boldsymbol{\theta}}{\partial\boldsymbol{\beta}^T} = \frac{\partial\boldsymbol{\theta}}{\partial\boldsymbol{\mu}^T}\frac{\partial\boldsymbol{\mu}}{\partial\boldsymbol{\beta}^T} = \boldsymbol{V}^{-1}\boldsymbol{D},\\
\boldsymbol{W}_\theta &= \frac{\partial^2\boldsymbol{\theta}}{\partial\boldsymbol{\beta}\partial\boldsymbol{\beta}^T} = \left(\frac{\partial\boldsymbol{\mu}}{\partial\boldsymbol{\beta}^T}\right)^T\left(\frac{\partial^2\boldsymbol{\theta}}{\partial\boldsymbol{\mu}\partial\boldsymbol{\mu}^T}\right)\left(\frac{\partial\boldsymbol{\mu}}{\partial\boldsymbol{\beta}^T}\right) + \left[\frac{\partial\boldsymbol{\theta}}{\partial\boldsymbol{\mu}^T}\right]\left[\frac{\partial^2\boldsymbol{\mu}}{\partial\boldsymbol{\beta}\partial\boldsymbol{\beta}^T}\right]\\
&= -\boldsymbol{D}^T[\boldsymbol{V}^{-1}][\boldsymbol{V}^{-1}\boldsymbol{S}\boldsymbol{V}^{-1}]\boldsymbol{D} + [\boldsymbol{V}^{-1}][\boldsymbol{W}]\\
&= [\boldsymbol{V}^{-1}][\boldsymbol{W} - \boldsymbol{\Gamma}].
\end{aligned}$$

∎

From the above equations, we get the following lemma.

Lemma 2.3

For exponential family nonlinear models (2.1), the score function, the observed information related to the log-likelihood function $l(\boldsymbol{\beta}) = l(\boldsymbol{\mu}(\boldsymbol{\beta}), \boldsymbol{y})$ can be represented as

$$\dot{l}(\boldsymbol{\beta}) = \phi\boldsymbol{D}^T(\boldsymbol{\beta})\boldsymbol{V}^{-1}(\boldsymbol{\beta})(\boldsymbol{y} - \boldsymbol{\mu}(\boldsymbol{\beta})) = \phi\boldsymbol{D}_\theta^T(\boldsymbol{\beta})(\boldsymbol{y} - \boldsymbol{\mu}(\boldsymbol{\beta})), \tag{2.8}$$

$$\begin{aligned}
-\ddot{l}(\boldsymbol{\beta}) &= \phi\boldsymbol{D}^T\boldsymbol{V}^{-1}\boldsymbol{D} - \phi[\boldsymbol{e}^T\boldsymbol{V}^{-1}][\boldsymbol{W} - \boldsymbol{\Gamma}]\\
&= \phi\boldsymbol{D}_\theta^T\boldsymbol{V}\boldsymbol{D}_\theta - \phi[\boldsymbol{e}^T][\boldsymbol{W}_\theta].
\end{aligned} \tag{2.9}$$

Proof. It follows from equations (1.17) to (1.20) and (A.10) of Appendix A that

$$\begin{aligned}
\dot{l}(\boldsymbol{\beta}) &= \frac{\partial l}{\partial\boldsymbol{\beta}} = \left(\frac{\partial\boldsymbol{\theta}}{\partial\boldsymbol{\beta}^T}\right)^T\left(\frac{\partial l}{\partial\boldsymbol{\theta}}\right) = \phi\boldsymbol{D}_\theta^T(\boldsymbol{\beta})(\boldsymbol{y} - \boldsymbol{\mu}(\boldsymbol{\beta})),\\
\ddot{l}(\boldsymbol{\beta}) &= \frac{\partial^2 l}{\partial\boldsymbol{\beta}\partial\boldsymbol{\beta}^T} = \left(\frac{\partial\boldsymbol{\theta}}{\partial\boldsymbol{\beta}^T}\right)^T\left(\frac{\partial^2 l}{\partial\boldsymbol{\theta}\partial\boldsymbol{\theta}^T}\right)\left(\frac{\partial\boldsymbol{\theta}}{\partial\boldsymbol{\beta}^T}\right) + \left[\left(\frac{\partial l}{\partial\boldsymbol{\theta}}\right)^T\right]\left[\frac{\partial^2\boldsymbol{\theta}}{\partial\boldsymbol{\beta}\partial\boldsymbol{\beta}^T}\right]\\
&= -\phi\boldsymbol{D}_\theta^T\boldsymbol{V}\boldsymbol{D}_\theta + \phi[\boldsymbol{e}^T][\boldsymbol{W}_\theta].
\end{aligned}$$

The first equations of (2.8) and (2.9) can be obtained from (2.6) and the above equations. ∎

For further discussions, we set the following assumption.

Assumption B

The true parameter of $\boldsymbol{\beta}$ to be estimated is an interior point of $\mathcal{B}$. The maximum likelihood estimators $\hat{\boldsymbol{\beta}}$ and $\hat{\phi}$ of $\boldsymbol{\beta}$ and ϕ exist and are unique in $\mathcal{B}$ and Φ, respectively. Further, we assume that $-\ddot{l}(\boldsymbol{\beta}) > 0$ and $-\partial^2 l/\partial\phi^2 > 0$

hold in some neighborhoods of $\hat{\boldsymbol{\beta}}$ and $\hat{\phi}$, respectively. ∎

Under Assumptions A and B, we have the following conclusions:

(a) The maximum likelihood estimator $\hat{\boldsymbol{\beta}}$ in (2.1) satisfies

$$\boldsymbol{D}_\theta^T(\hat{\boldsymbol{\beta}})(\boldsymbol{Y} - \mu(\hat{\boldsymbol{\beta}})) = \boldsymbol{D}^T(\hat{\boldsymbol{\beta}})V^{-1}(\hat{\boldsymbol{\beta}})(\boldsymbol{Y} - \boldsymbol{\mu}(\hat{\boldsymbol{\beta}})) = \mathbf{0}. \tag{2.10}$$

Notice that there is no $\hat{\phi}$ involving this equation, thus $\hat{\boldsymbol{\beta}}$ does not depend on $\hat{\phi}$.

(b) The Fisher information matrix of $\boldsymbol{Y}$ for $\boldsymbol{\beta}$ in exponential family nonlinear models is

$$\boldsymbol{J}_\beta(\boldsymbol{Y}) \triangleq \boldsymbol{J}(\boldsymbol{\beta}) = \phi \boldsymbol{D}^T \boldsymbol{V}^{-1} \boldsymbol{D} = \phi \boldsymbol{D}_\theta^T \boldsymbol{V} \boldsymbol{D}_\theta. \tag{2.11}$$

(c) The dispersion parameter ϕ is orthogonal to $\boldsymbol{\beta}$ in the sense of Cox and Reid (1987), that is $E(-\partial^2 l/\partial\boldsymbol{\beta}\partial\phi) = \mathbf{0}$.

Now let us consider the maximum likelihood estimator $\hat{\phi}$ of the dispersion parameter ϕ. There is a close relationship between $\hat{\phi}$ and the deviance. In fact, it follows from (1.27) that

$$\dot{l}_\phi = \frac{\partial l}{\partial \phi} = -\frac{1}{2}D'(\boldsymbol{y}, \boldsymbol{\mu}) - \frac{1}{2}\sum_{i=1}^n \dot{s}(y_i, \phi).$$

Thus $\hat{\phi}$ satisfies

$$\sum_{i=1}^n \dot{s}(y_i, \hat{\phi}) = -D'(\boldsymbol{Y}, \hat{\boldsymbol{\mu}}) \tag{2.12}$$

and we may find $\hat{\phi}$ after obtaining $\hat{\boldsymbol{\beta}}$. In particular, for model (2.1) with (1.6), $\hat{\phi}$ satisfies

$$\begin{aligned} -\dot{s}(\hat{\phi}) &= n^{-1}D'(\boldsymbol{Y}, \hat{\boldsymbol{\mu}}) = n^{-1}D(\boldsymbol{Y}, \hat{\boldsymbol{\mu}}) + k \\ &= \frac{1}{n}\sum_{i=1}^n \{d_i(y_i, \hat{\mu}_i) + k\}. \end{aligned} \tag{2.13}$$

Further, the observed information and the Fisher information of $\boldsymbol{Y}$ for ϕ in model (2.1) with (1.6) can be represented as

$$-\ddot{l}_{\phi\phi} = J_\phi(\boldsymbol{Y}) = \frac{1}{2}n\ddot{s}(\phi). \tag{2.14}$$

After obtaining $\hat{\boldsymbol{\beta}}$, we can get $\hat{\boldsymbol{\mu}} = \mu(\hat{\boldsymbol{\beta}})$, $\hat{\boldsymbol{\theta}} = \boldsymbol{\theta}(\hat{\boldsymbol{\beta}})$ and the deviance $D(\boldsymbol{Y}, \hat{\boldsymbol{\mu}})$. Then $\hat{\phi}$ can be computed from (2.12) or (2.13). Now let us see some examples.

Example 2.2 (Normal, inverse Gaussian and gamma nonlinear models).

By Table 1.1, for normal model and inverse Gaussian model, $s(\phi) = -\log\phi$, $\dot{s}(\phi) = -\phi^{-1}$ and $k = 0$, hence from (2.13) we have

$$\hat{\phi}^{-1} = \hat{\sigma}^2 = n^{-1}D(\boldsymbol{Y}, \hat{\boldsymbol{\mu}}) = \frac{1}{n}\sum_{i=1}^{n} d_i(y_i, \hat{\mu}_i).$$

In particular, for inverse Gaussian model:

$$\hat{\sigma}^2 = \frac{1}{n}\sum_{i=1}^{n}\{(y_i - \hat{\mu}_i)^2/\hat{\mu}_i^2 y_i\}.$$

In the case of gamma model, $s(\phi) = -2\{\phi\log\phi - \log\Gamma(\phi)\}$, $\dot{s}(\phi) = -2\{1 + \log\phi - \psi(\phi)\}$ and $k = 2$, hence from (2.13) we have

$$\log\hat{\phi} - \psi(\hat{\phi}) = \frac{1}{2n}D(\boldsymbol{Y}, \hat{\boldsymbol{\mu}}),$$

where $\psi(\phi)$ is the diagamma function (see Example 2.1). An approximation of $\hat{\phi}$ based on this equation was given by Cordeiro and McCullagh (1991)

$$\hat{\phi} = \frac{n[1 + \{1 + 2D(\boldsymbol{Y}, \hat{\boldsymbol{\mu}})/3n\}^{1/2}]}{2D(\boldsymbol{Y}, \hat{\boldsymbol{\mu}})}.$$ ∎

There are some other methods to estimate the dispersion parameter $\sigma^2 = \phi^{-1}$. The following are summarized by Jorgensen (1987).

(a) $\hat{\sigma}_1^2 = (n-p)^{-1}D(\boldsymbol{Y}, \hat{\boldsymbol{\mu}})$ which is an asymptotic unbiased estimator of σ^2.

(b) $\hat{\sigma}_2^2 = (n-p)^{-1}\boldsymbol{X}^2$, where $\boldsymbol{X} = (\boldsymbol{Y} - \hat{\boldsymbol{\mu}})^T\boldsymbol{V}(\hat{\boldsymbol{\mu}})^{-1}(\boldsymbol{Y} - \hat{\boldsymbol{\mu}})$ is the generalized Pearson statistic. This is actually a moment estimator.

(c) $\hat{\sigma}_3^2$ maximizes the following modified profile likelihood for parameter σ^2 (Barndorff-Nielsen, 1983): $L^0(\sigma^2) = \sigma^p p(\boldsymbol{y};, \hat{\boldsymbol{\mu}}, \sigma^2)$, where $p(\boldsymbol{y}; \boldsymbol{\mu}, \sigma^2)$ is the density of the distribution $ED(\boldsymbol{\mu}, \sigma^2)$. ∎

Finally, we consider some problems of testing of a hypothesis, which will be discussed further in Chapter 5.

(a) Under the test $H_0 : \boldsymbol{g}(\boldsymbol{\mu}) = \boldsymbol{f}(\boldsymbol{\beta})$ for a fixed ϕ, the likelihood ratio statistic is

$$LR(\hat{\boldsymbol{\mu}}) = 2\{l(\tilde{\boldsymbol{\mu}}, \boldsymbol{Y}) - l(\hat{\boldsymbol{\mu}}, \boldsymbol{Y})\}.$$

Since $\tilde{\mu} = \boldsymbol{Y}$, $LR(\hat{\mu})$ is namely (see (1.23))

$$LR(\hat{\mu}) = D^*(\boldsymbol{Y}, \hat{\mu}) = \phi D(\boldsymbol{Y}, \hat{\mu}).$$

Hence $D(\boldsymbol{Y}, \hat{\mu})$ asymptotically has $\phi^{-1}\chi^2(n-p)$ distribution.

(b) Under the test $H_0 : \boldsymbol{\beta} = \boldsymbol{\beta}_0$ for a fixed ϕ, the likelihood ratio statistic is

$$\begin{aligned} LR(\boldsymbol{\beta}_0) &= 2\{l(\mu(\hat{\boldsymbol{\beta}}), \boldsymbol{Y}) - l(\mu(\boldsymbol{\beta}_0), \boldsymbol{Y})\} \\ &\triangleq 2\{l(\hat{\boldsymbol{\beta}}) - l(\boldsymbol{\beta}_0)\}, \end{aligned} \tag{2.15}$$

which asymptotically has $\chi^2(p)$ distribution for a fixed ϕ. It is easily seen that

$$\begin{aligned} LR(\boldsymbol{\beta}_0) &= 2\{l(\tilde{\mu}, \boldsymbol{Y}) - l(\mu(\boldsymbol{\beta}_0), \boldsymbol{Y})\} - 2\{l(\tilde{\mu}, \boldsymbol{Y}) - l(\mu(\hat{\boldsymbol{\beta}}), \boldsymbol{Y})\} \\ &= \phi D(\boldsymbol{Y}, \mu(\boldsymbol{\beta}_0)) - \phi D(\boldsymbol{Y}, \mu(\hat{\boldsymbol{\beta}})). \end{aligned}$$

(c) Under the test $H_0 : \boldsymbol{\beta}_2 = \boldsymbol{\beta}_{20}$ for a fixed ϕ, where $\boldsymbol{\beta}_2$ is the last p_2 parameters of $\boldsymbol{\beta}$, then the likelihood ratio statistic is (Cox and Hinkley 1974, p.322)

$$LR_s(\boldsymbol{\beta}_{20}) = 2\{l(\hat{\boldsymbol{\beta}}) - l(\tilde{\boldsymbol{\beta}}_0)\},$$

where $\boldsymbol{\beta} = (\boldsymbol{\beta}_1^T, \boldsymbol{\beta}_2^T)^T$, $\tilde{\boldsymbol{\beta}}_0 = (\tilde{\boldsymbol{\beta}}_1^T(\boldsymbol{\beta}_{20}), \boldsymbol{\beta}_{20}^T)^T$ and $\tilde{\boldsymbol{\beta}}_1(\boldsymbol{\beta}_{20})$ maximizes $l(\boldsymbol{\beta}_1, \boldsymbol{\beta}_{20})$ for the fixed $\boldsymbol{\beta}_{20}$. Then $LR_s(\boldsymbol{\beta}_{20})$ asymptotically has $\chi^2(p_1)$ distribution for a fixed ϕ, where $p_1 = p - p_2$ is the dimension of $\boldsymbol{\beta}_1$ (Cox and Hinkley 1974, p.323).

The above equation holds for any fixed $\boldsymbol{\beta}_{20}$. In general, we can introduce the profile likelihood of $\boldsymbol{Y}$ for a parameter subset $\boldsymbol{\beta}_2$. The profile likelihood is frequently used in the literature and is defined as

$$l_p(\boldsymbol{\beta}_2) = l(\tilde{\boldsymbol{\beta}}_1(\boldsymbol{\beta}_2), \boldsymbol{\beta}_2),$$

where $\tilde{\boldsymbol{\beta}}_1(\boldsymbol{\beta}_2)$ maximizes $l(\boldsymbol{\beta}_1, \boldsymbol{\beta}_2)$ for each value of $\boldsymbol{\beta}_2$. Now let $\hat{\boldsymbol{\beta}} = (\hat{\boldsymbol{\beta}}_1^T, \hat{\boldsymbol{\beta}}_2^T)^T$, then it is easily seen that

$$\tilde{\boldsymbol{\beta}}_1(\hat{\boldsymbol{\beta}}_2) = \hat{\boldsymbol{\beta}}_1 \quad \text{and} \quad l_p(\hat{\boldsymbol{\beta}}_2) = l(\hat{\boldsymbol{\beta}}_1, \hat{\boldsymbol{\beta}}_2) = l(\hat{\boldsymbol{\beta}}).$$

So $LR_s(\boldsymbol{\beta}_{20})$ given above can be expressed as

$$LR_s(\boldsymbol{\beta}_{20}) = 2\{l_p(\hat{\boldsymbol{\beta}}_2) - l_p(\boldsymbol{\beta}_{20})\}.$$

This equation holds for any $\boldsymbol{\beta}_2 = \boldsymbol{\beta}_{20}$, so we can write

$$LR_s(\boldsymbol{\beta}_2) = 2\{l_p(\hat{\boldsymbol{\beta}}_2) - l_p(\boldsymbol{\beta}_2)\}.$$

This expression can be applied to construct the confidence regions for $\boldsymbol{\beta}_2$ (see Chapter 5 for details).

2.3 Computation and Examples

Since parameters $\boldsymbol{\beta}$ are orthogonal to the dispersion parameter ϕ, we may compute $\hat{\boldsymbol{\beta}}$ and $\hat{\phi}$ separately. In practice, we can compute $\hat{\boldsymbol{\beta}}$ using the Gauss-Newton iterative method based on $\dot{\boldsymbol{l}}(\boldsymbol{\beta})$ and $\ddot{\boldsymbol{l}}(\boldsymbol{\beta})$ or $\boldsymbol{J}_{\beta}(\boldsymbol{Y})$ given in (2.8), (2.9) and (2.11). After obtaining $\hat{\boldsymbol{\beta}}, \hat{\phi}$ can be computed by solving equation (2.12) or (2.13). Now we first focus on introducing the computation method to get $\hat{\boldsymbol{\beta}}$.

Taking a first order Taylor series approximation to the likelihood equation $\dot{\boldsymbol{l}}(\hat{\boldsymbol{\beta}}) = \mathbf{0}$, we obtain

$$\dot{\boldsymbol{l}}(\hat{\boldsymbol{\beta}}) \approx \dot{\boldsymbol{l}}(\boldsymbol{\beta}_0) + \ddot{\boldsymbol{l}}(\boldsymbol{\beta}_0)(\hat{\boldsymbol{\beta}} - \boldsymbol{\beta}_0) \approx \mathbf{0}$$

$$\hat{\boldsymbol{\beta}} \approx \boldsymbol{\beta}_0 + \{-\ddot{\boldsymbol{l}}(\boldsymbol{\beta}_0)\}^{-1}\dot{\boldsymbol{l}}(\boldsymbol{\beta}_0).$$

Therefore the Gauss-Newton iteration procedure can be expressed as

$$\boldsymbol{\beta}^{i+1} = \boldsymbol{\beta}^i + \{-\ddot{\boldsymbol{l}}(\boldsymbol{\beta}^i)\}^{-1}\dot{\boldsymbol{l}}(\boldsymbol{\beta}^i), \quad i = 0, 1, 2, \cdots$$

where $\boldsymbol{\beta}^i$ is the i-th iteration. In this iteration equation, the observed information $-\ddot{\boldsymbol{l}}(\boldsymbol{\beta})$ is usually replaced by the expected information $\boldsymbol{J}_{\beta}(\boldsymbol{Y}) = \mathrm{E}\,(-\ddot{\boldsymbol{l}}(\boldsymbol{\beta})) = \phi \boldsymbol{D}^T\boldsymbol{V}^{-1}\boldsymbol{D}$. This replacement is convenient and acceptable (e.g. Jennrich 1969; Seber and Wild 1989; see also Theorem 2.1). Then the Gauss-Newton iteration procedure of $\hat{\boldsymbol{\beta}}$ is given by

$$\boldsymbol{\beta}^{i+1} = \boldsymbol{\beta}^i + \{(\boldsymbol{D}^T\boldsymbol{V}^{-1}\boldsymbol{D})^{-1}\boldsymbol{D}^T\boldsymbol{V}^{-1}\boldsymbol{e}\}^i, \tag{2.16}$$

where $\{\cdot\}^i$ denotes the i-th iteration, that is $\boldsymbol{D}, \boldsymbol{V}$ and $\boldsymbol{e} = \boldsymbol{Y} - \boldsymbol{\mu}(\boldsymbol{\beta})$ are all evaluated at $\boldsymbol{\beta} = \boldsymbol{\beta}^i$. This iteration equation can be rewritten as

$$\boldsymbol{\beta}^{i+1} = \{(\boldsymbol{D}^T\boldsymbol{V}^{-1}\boldsymbol{D})^{-1}\boldsymbol{D}^T\boldsymbol{V}^{-1}\boldsymbol{Z}\}^i, \tag{2.17}$$

$$\boldsymbol{Z} = \boldsymbol{D}\boldsymbol{\beta} + \boldsymbol{e}. \tag{2.18}$$

Equation (2.17) shows that the iteration procedure can be regarded in form as a formula of the generalized least squares estimator for the following general nonlinear regression

$$\boldsymbol{Z} = \boldsymbol{\mu}(\boldsymbol{\beta}) + \boldsymbol{\delta}, \quad \boldsymbol{\delta} \sim (\mathbf{0}\ ,\ \boldsymbol{V}),$$

where $\boldsymbol{\delta} \sim (\mathbf{0}, \boldsymbol{V})$ means $\mathrm{E}(\boldsymbol{\delta}) = \mathbf{0}$, $\mathrm{Var}(\boldsymbol{\delta}) = \boldsymbol{V}$. For further discussions, the reader is referred to the literature, such as Gallant (1987), del Pino (1989), Seber and Wild (1989). In general, the convergence of iteration (2.16) is quite quick, but it strongly depends on the choice of the initial value $\boldsymbol{\beta}^0$. Jennrich (1969) gave a set of regularity conditions (see also Chapter 4) under which the Gauss-Newton iteration procedure of nonlinear least squares estimator has asymptotically numerical stability, i.e. $\boldsymbol{\beta}^i \to \hat{\boldsymbol{\beta}}$ $(a.e.)$ as $i \to \infty$, when the initial value β^0 is in a certain neighborhood of β_0. Jennrich's result can be extended to our iteration procedure which is based on the generalized nonlinear least squares estimator (2.17). Now we give a theorem that will be proved in Appendix B.4.

Theorem 2.1

For model (2.2), if Assumptions A to C hold (Assumption C will be given in Section 4.1), then there exists a neighborhood $N(\delta_0)$ of β_0 with radius δ_0 and an integer $n_0(Y)$ for almost every Y, such that $\beta^i \to \hat{\beta}_n(Y)$ as $i \to +\infty$, for any $n \geq n_0(Y)$ and $\beta^0 \in N(\delta_0)$, where β_0 is the true value of β to be estimated, $\hat{\beta}_n(Y)$ is the maximum likelihood estimator of β_0 based on $Y = (y_1, \cdots, y_n)^T$ and β^0 is the initial value of the iteration procedure according to (2.16). ∎

We now assume that β^i in (2.16) converges to $\hat{\beta}$. Then it follows from (2.17) and (2.18) that

$$\hat{\beta} = \{(D^T V^{-1} D)^{-1} D^T V^{-1} Z\}_{\hat{\beta}}, \tag{2.19}$$

$$Z = D\hat{\beta} + \hat{e}, \tag{2.20}$$

where D, V and Z are all evaluated at $\hat{\beta}$, and $\hat{e} = e(\hat{\beta}) = Y - \mu(\hat{\beta})$. These equations are quite useful in applications. Once equation (2.19) has been established, many important results are immediate consequences (Andersen 1992). In (2.19), the elements of D are regarded as "explanatory variables" and Z can be viewed as the "response", then $\hat{\beta}$ may be viewed as the "generalized least squares estimator" using the elements of V^{-1} as weights. In other words, $\hat{\beta}$ can be regarded in form as the generalized least squares estimator of the following linear model:

$$Z = D\beta + e, \quad e \sim (0, V); \tag{2.21}$$

or the ordinary least squares estimator of following linear model

$$V^{-\frac{1}{2}} Z = V^{-\frac{1}{2}} D\beta + \varepsilon, \quad \varepsilon \sim (0, I_n), \tag{2.22}$$

where I_n is an identity matrix. But it should be stressed that D, V and Z in (2.21) and (2.22) are actually evaluated at $\hat{\beta}$.

As an example of this argument, we can naturally define the Pearson residuals as (Andersen 1992)

$$rp = V^{-\frac{1}{2}} Z - V^{-\frac{1}{2}} D\hat{\beta} = V^{-\frac{1}{2}} \hat{e}.$$

Hence we have

$$rp \triangleq (rp_i) = V^{-\frac{1}{2}}(\hat{\beta})(Y - \mu(\hat{\beta})),$$

$$rp_i = V_i^{-\frac{1}{2}}(\hat{\beta})(y_i - \mu_i(\hat{\beta})). \tag{2.23}$$

Similar method and several statistics for generalized linear models have been introduced by Pregibon (1981). Based on (2.21) and (2.22), we can study some diagnostics problems for exponential family nonlinear models (see Chapter 6 for details).

In practice, the modified Gauss-Newton iteration method is also often used (Hartley 1961; Gallent 1987). By this method, the likelihood $l(\beta^i)$ will increase towards its maximum $l(\hat{\beta})$. To introduce this method, we need the following lemma.

Lemma 2.4

For any β satisfying $\dot{l}(\beta) \neq \mathbf{0}$, there exists a value $\lambda^* > 0$, such that $l(\beta + \lambda G(\beta)) > l(\beta)$ for $0 < \lambda < \lambda^*$, where $G(\beta)$ is the second term of (2.16), that is

$$G(\beta) = \{D^T(\beta)V^{-1}(\beta)D(\beta)\}^{-1}D^T(\beta)V^{-1}(\beta)(Y - \mu(\beta)).$$

Proof. Let $q(\lambda) = l(\beta + \lambda G(\beta))$, $\lambda > 0$ then we have

$$\begin{aligned} l(\beta + \lambda G(\beta)) - l(\beta) &= q(\lambda) - q(0) \\ &= \dot{q}(0)\lambda + \lambda\alpha \\ &= \{\dot{q}(0) + \alpha\}\lambda \end{aligned}$$

where $\alpha \to 0$ as $\lambda \to 0$. It follows from (2.8) that

$$\dot{q}(0) = \{\dot{l}(\beta)\}^T G(\beta) = \phi e^T V^{-1} D(D^T V^{-1} D)^{-1} D^T V^{-1} e.$$

Obviously, $\dot{q}(0) > 0$ always holds. So there exists a value $\lambda^* > 0$, such that $\dot{q}(0) + \alpha > 0$ when $0 < \lambda < \lambda^*$. This results in $l(\beta + \lambda G(\beta)) > l(\beta)$. ∎

By this lemma, we can summarize the modified Gauss-Newton iteration method as follows.

(a) Choose an initial value β_0 and compute $G_0 = G(\beta_0)$. Find a $0 < \lambda_0 < 1$, such that

$$l(\beta_0 + \lambda_0 G_0) > l(\beta_0).$$

(b) Let $\beta_1 = \beta_0 + \lambda_0 G_0$. Compute $G_1 = G(\beta_1)$ and find a $0 < \lambda_1 < 1$, such that

$$l(\beta_1 + \lambda_1 G_1) > l(\beta_1).$$

(c) Let $\beta_2 = \beta_1 + \lambda_1 G_1, \cdots$

It is easily seen that by this procedure, we have $l(\beta_{i+1}) > l(\beta_i)$ for $i = 1, 2, \cdots$, and may have $l(\beta_i) \to l(\hat{\beta})$ under some regularity conditions (Gallant 1987). There are several ways for choosing values λ_i at each iteration. Gallant (1987) suggested quite simple values of λs, which are $\lambda = 2^{-1}, 2^{-2}, 2^{-3}, \cdots$.

Now let us look at some examples for computing parameter estimators.

Example 2.3 (Product sales data).

These data were given by Whitmore (1986). We call them the "product sales data" for short. In this data set, x_i represents the projected sales amounts of the i-th product reported by a market survey organization and y_i is the corresponding actual sales amounts of a company ($i = 1, \cdots, 20$). Whitmore (1986) suggested an inverse Gaussian fit by using

$$y_i \sim IG(\beta x_i^\gamma,\ \kappa^{-1} x_i^{-\rho}), \qquad i = 1, \cdots, 20, \tag{2.24}$$

that is $\mathrm{E}(y_i) = \mu_i = \beta x_i^\gamma$, $\mathrm{Var}(y_i) = \sigma_i^2 V(\mu_i)$ with $\sigma_i^{-2} = \kappa x_i^\rho$ and $V(\mu_i) = \mu_i^3$. Here we set $\rho = 0$ to simplify the computation. In this case $\sigma_i^2 = \kappa^{-1}$ for all i. Then (2.24) becomes $y_i \sim IG(\beta x_i^\gamma, \kappa^{-1})$ which is an inverse Gaussian nonlinear model with $\mu_i = \beta x_i^\gamma$ and $\mathrm{Var}(y_i) = \sigma^2 \mu_i^3$ ($\sigma^2 = \kappa^{-1}$). We can compute $\hat{\beta}, \hat{\gamma}$ based on (2.19) and get the deviance $D(\boldsymbol{Y}, \hat{\boldsymbol{\mu}})$ from Table 1.1. Then $\hat{\kappa}$ can be obtained from Example 2.2. The results are summarized in Table 2.1, where "s.e." denotes the standard error of the estimator.

Table 2.1. Estimates of product sales data

Parameter	Estimate	s.e.
β	1.1890	0.0708
γ	0.9998	0.0109
$D(\boldsymbol{Y}, \hat{\boldsymbol{\mu}})$	0.003549	$\hat{\kappa}$=5635.47

Example 2.4 (European rabbit data No. 1).

These data were originally given by Dudzinski and Mykytowycz (1961) and studied by Ratkowsky (1983) based on the ordinary nonlinear regression model. The data set consists of 71 observations in which $y = \log Z$ and Z denotes the dry weight of eye lens (in milligrams) for the European rabbit in Australia and x is the corresponding age (in days) of the rabbit. Here we fit the data once again using the following inverse Gaussian nonlinear model: $y_i = \log Z_i$, $y_i \sim IG(\mu_i, \sigma^2)$ and

$$\mu_i = \alpha - \frac{\beta}{x_i + \gamma}, \qquad i = 1, \cdots, 71. \tag{2.25}$$

The estimates of parameters based on (2.19) are given in Table 2.2. These results are quite similar to those of Ratkowsky (1983, p.109). See Chapter 6 for further discussions.

Table 2.2. Estimates of European rabbit data

Parameter	Estimate	s.e.
α	5.634	0.0250
β	128.536	6.0945
γ	36.787	2.2091
σ^2	$10^{-4} \times 0.49$	

Example 2.5 (Leukemia data No. 1).

The leukemia data were studied by Cook and Weisberg (1982, p.193) and Lee (1987; 1988). The data consisted of a sample of 33 patients who died of acute leukemia. There are two covariate variables in these data. First, the count of white blood cells (WBC) is a main measure of the patient's initial condition, more severe conditions being reflected by higher counts. Second, each patient is classified as either AG (= 1) positive or AG (= 0) negative, where AG indicates the presence or absence of a certain morphologic characteristic in the WBC. The response y is the number (1 or 0) of patients surviving at least 52 weeks for each combination of WBC and AG category. The sample size is $n = 30$ (there are three patients in case 15 and two patients in case 30). Cook and Weisberg (1982) fitted these data using linear logistic regression model. As an alternative, Lee (1988) considered a transformation of the covariate WBC. Following Lee (1988), we consider the following nonlinear logistic regression model:

$$\text{logit}(p_i) = \beta_0 + \beta_1 AG_i + \beta_2 WBC_i^{\lambda}, \quad n = 1, \cdots, 30, \tag{2.26}$$

where $p_i = P\{y_i = 1\}$ and $\text{logit}(p_i) = \log(p_i/1 - p_i)$. The estimates of parameters are summarized in Table 2.3. There is no indication from this summary that the model is inadequate, though the asymptotic standard errors of the estimates appear relatively large (see Chapter 6 for further discussions).

Table 2.3. Estimates of leukemia data

Parameter	Estimate	s.e.
β_0	-8.69	15.28
β_1	33.46	55.07
β_2	2.58	1.13
λ	-0.18	0.43
$D(\mathbf{Y}, \hat{\boldsymbol{\mu}})$	22.84	

Chapter 3

Geometric Framework

In this chapter, we present a differential geometric framework for exponential family nonlinear models. Section 3.1 briefly reviews the developments of differential geometric method in statistical inference in recent years. In Section 3.2, which is the main part of this chapter, we modify the geometric framework of Bates and Watts (1980) in nonlinear regression models so that it can be applied to exponential family nonlinear models in expectation parameter space. The curvature arrays, the directional curvatures and the maximum curvatures are introduced. The effects of parameterizations on the curvatures are also discussed. Section 3.3 introduces the dual geometry in natural parameter space. An important relationship between two kinds of curvature arrays in dual spaces is derived. In Section 3.4, we give some examples.

In this chapter, and also in other chapters, we need a few geometric concepts such as surface, tangent space, curvature and so on. The reader is referred to a standard textbook of differential geometry, such as, Millman and Parker (1977) or certain statistical books, such as, Seber and Wild (1989, Appendix B). In particular, the second order derivatives of a vector and its projection onto a linear space, which detailed in Appendix A, are used throughout this book.

3.1 Introduction

The interface between differential geometry and statistics is a subject in which statisticians and mathematicians have been interested since Efron's (1975) pioneer work appeared. The use of the differential geometric method to investigate the properties of statistical models has been well developed in the past two decades (e.g. review papers by Barndorff-Nielsen, Cox and Reid 1986; Kass 1989; books written by Amari 1985; Seber and Wild 1989;

Murray and Rice 1993). The differential geometric method has some specific advantages in statistical analysis. There are at least three aspects by our understanding: (a) provide a preferable measure for assessing the nonlinearity of a statistical model; (b) provide a heuristic and intuitive second order approximation rather than a linear approximation (e.g. Efron 1975; Amari 1985; Seber and Wild 1989; see also Chapters 4 and 5 in this book); and (c) provide some new tools and insights into statistical analysis. There are two kinds of differential geometric frameworks which are widely accepted, and proved to be efficient and successful in statistical analysis. One framework was presented by Efron (1975) and Amari (1982a) for curved exponential families and some other parametric families by introducing a Riemannian manifold corresponding to those families. Many contributions to the statistical inference have been made along this line (e.g. Efron 1978; Efron and Hinkley 1978; Amari 1982b; 1985; 1989; Amari and Kumon 1983; 1988; Eguchi 1983; Kumon and Amari 1983; 1984; Kass 1984; 1989; Barndorff-Nielsen 1986; 1988; Wei 1986; 1988; Amari *et al.* 1987; Moolgavkar and Venzon 1987; Ross 1987a,b; Wei and Zhao 1987; Mitchell 1988; Vos 1989; 1991; Ravishanker, Melnick and Tsai 1990; Efron and Johnstone 1991; Okamoto, Amari and Takeuchi 1991; Critchley, Marriott and Salmon 1993; 1994; Kass and Elizabeth 1994; Zhu and Wei 1996). The other geometric framework was presented by Bates and Watts (1980) for nonlinear regression models by introducing two kinds of curvature measures in Euclidean space. Because the Euclidean geometry is relatively simple and the regression model is quite useful, many authors have concentrated on this approach. A lot of contributions also have been made along this line (e.g. Bates and Watts 1981; 1988; Hamilton, Watts and Bates 1982; Hougaard 1982; 1985; 1986; Bates, Hamilton and Watts 1983; Ratkowsky 1983; 1990; Tsai 1983; Cook and Tsai 1985; 1990; Hamilton and Watts 1985; Cook and Goldberg 1986; Cook, Tsai and Wei 1986; Hamilton 1986; Clarke 1987a,b; Cook 1987; Pazman 1987; 1990; Seber and Wild 1989; Wei 1989; 1991; 1994; 1995; Cook and Weisberg 1990; St. Laurent and Cook 1992; 1993). Furthermore, Cook (1986) applied the ideas of geometric method of Bates and Watts (1980) for assessing the local influence of small perturbations of a statistical model. Many authors have been interested in the use of this approach (e.g. Thomas 1990; Weissfeld 1990; Escobar and Meeker 1992; Schall and Dunne 1992; Wu and Luo 1993a,b; Wei and Hickernell 1996).

Compare these two geometric frameworks, each of which has its own advantages and characteristics. In the framework of Bates and Watts (BW), the geometry is relatively simple and intuitive. Therefore it might be easy to understand and deal with. But there has been little work done outside of the class of normal nonlinear regression models (Cook and Tsai 1990). In the framework of Efron and Amari (EA), it can be applied to quite a general class of models which include the curved exponential families. But the geometry based on the Riemannian manifold and the tensor notation might be a little complicated to most statisticians. Besides, as we have seen, the statistical inference based on EA geometric framework usually requires iid observations (Efron 1975; Amari 1985). We have not seen the work on regression by using EA framework. As pointed out by Amari (1985), the BW geometric structures can be regarded as a special case of EA geometry. Indeed, it is true from a mathematical viewpoint. But we do not think the BW geometry can be replaced by the EA geometry because BW geometry still has some preferable characteristics, such as (a) it uses Euclidean space rather than Riemannian manifold; (b) it uses orthonormal basis rather than general basis; (c) it uses matrix notation rather than tensor notation; and (d) it is easy to deal with the regression problems.

In this book, we present a modified BW (MBW) geometric framework and try to combine the advantages of both BW and EA geometries, at least, for certain statistical models and some inference problems. Briefly, the model we study is analogous to that of the EA framework and the geometry we use is analogous to that of the BW framework. The key point of modification to the BW geometric framework is that we introduce a weighted inner product to Euclidean space based on the Fisher information matrix, so that our MBW framework can be applied to quite a general class of models, such as exponential family nonlinear models, quasi-likelihood nonlinear models, covariance structure models, the embedded models in regular parametric families and so on. In this chapter, we only discuss exponential family nonlinear models. Chapter 7 will deal with more general models.

For exponential family nonlinear models, to investigate the statistical behavior of parameter $\boldsymbol{\beta}$, one can use either the expectation parameter $\boldsymbol{\mu} = \boldsymbol{\mu}(\boldsymbol{\beta})$ or the natural parameter $\boldsymbol{\theta} = \boldsymbol{\theta}(\boldsymbol{\beta})$. Similarly, we can construct the differential geometric framework both in the expectation parameter space and in the natural parameter space. But as is often used in the literature (e.g. Morris 1982: 1983; Jorgensen 1983; 1987; McCullagh 1987; McCullagh and Nelder 1989). the expectation parameter might be more preferable. One of the reasons is that the observed vector $\boldsymbol{Y}$ can be naturally viewed as a vector in the expectation parameter space. We will pay more attention to the geometric framework in expectation parameter space.

3.2 Geometry in Expectation Parameter Space

We start with the geometric interpretation of the maximum likelihood estimator of $\boldsymbol{\beta}$ in exponential family nonlinear models (2.2). As shown in (2.10), the maximum likelihood estimator $\hat{\boldsymbol{\beta}}$ must satisfy

$$\boldsymbol{D}^T(\hat{\boldsymbol{\beta}})\boldsymbol{V}^{-1}(\hat{\boldsymbol{\beta}})\boldsymbol{e}(\hat{\boldsymbol{\beta}}) = \boldsymbol{0}, \quad \boldsymbol{e}(\boldsymbol{\beta}) = \boldsymbol{Y} - \boldsymbol{\mu}(\boldsymbol{\beta}). \tag{3.1}$$

To see the geometric interpretation of this equation more clearly, we first review a special case for which (2.3) reduces to

$$g(\boldsymbol{\mu}) = \boldsymbol{\mu} = \boldsymbol{\mu}(\boldsymbol{\beta}), \quad \boldsymbol{Y} \sim N(\boldsymbol{\mu}(\boldsymbol{\beta}),\ \sigma^2 \boldsymbol{I}_n).$$

This is just the normal nonlinear regression model. In this case, $\boldsymbol{V}(\boldsymbol{\mu}) = \boldsymbol{I}_n$ is an identity matrix and (3.1) reduces to

$$\boldsymbol{D}^T(\hat{\boldsymbol{\beta}})(\boldsymbol{Y} - \boldsymbol{\mu}(\hat{\boldsymbol{\beta}})) = 0. \tag{3.2}$$

As pointed out implicitly by Bates and Watts (1988; see also Ratkowsky 1983), this equation shows that under the ordinary inner product in Euclidean space R^n, the residual vector $\hat{\boldsymbol{e}} = \boldsymbol{Y} - \boldsymbol{\mu}(\hat{\boldsymbol{\beta}})$ is orthogonal to the space spanned by the column vectors of $\boldsymbol{D}(\hat{\boldsymbol{\beta}})$. That is the tangent space at $\hat{\boldsymbol{\beta}}$ of the "solution locus" $\boldsymbol{\mu} = \boldsymbol{\mu}(\boldsymbol{\beta})$ defined by Bates and Watts (1980), which is a p-dimensional surface in R^n. Similarly, we can extend this heuristic geometric interpretation to the general exponential family nonlinear models. In fact, if we define an inner product in R^n with weight matrix $\boldsymbol{V}^{-1}(\hat{\boldsymbol{\beta}})$, i.e. $\langle \boldsymbol{a}, \boldsymbol{b} \rangle = \boldsymbol{a}^T \boldsymbol{V}^{-1} \boldsymbol{b}$ for $\boldsymbol{a}, \boldsymbol{b} \in R^n$, then (3.1) also shows that under this inner product the residual vector $\hat{\boldsymbol{e}} = \boldsymbol{Y} - \boldsymbol{\mu}(\hat{\boldsymbol{\beta}})$ is orthogonal to the column vectors of $\boldsymbol{D}(\hat{\boldsymbol{\beta}})$ which spans the tangent space of the surface $\boldsymbol{\mu} = \boldsymbol{\mu}(\boldsymbol{\beta})$ at $\hat{\boldsymbol{\beta}}$. Bates and Watts (1980) presented a geometric framework for nonlinear regression models based on the geometric interpretation of (3.2). Similarly, we can also introduce a geometric framework for our exponential family nonlinear models based on the geometric interpretation of (3.1). Now let us give the details.

3.2.1 Curvature arrays

We first describe the geometric structure, then introduce definitions of two kinds of curvature arrays in expectation parameter space.

For exponential family nonlinear models (2.2), take $\boldsymbol{\mu}$ as a set of coordinates in expectation parameter space $\mathcal{U} \subset R^n$, the solution locus π is defined as $\boldsymbol{\mu} = \boldsymbol{\mu}(\boldsymbol{\beta})$ (see (2.3)), which represents a p-dimensional surface in R^n. To introduce a tangent space associated with the solution locus π, we let

$$c_a: \quad \boldsymbol{c}_a(t) = \boldsymbol{\mu}(\beta_1, \cdots \beta_{a-1}, \beta_a + t, \beta_{a+1}, \cdots \beta_p), \quad a = 1, \cdots, p,$$

where t is a real number. c_as are called the coordinate curves of π at $\boldsymbol{\beta}$. Obviously, the tangent vector of c_a at $\boldsymbol{\beta}$ (i.e. $t = 0$) is the derivative of $\boldsymbol{c}_a(t)$ at $t = 0$, denoted by $\dot{\boldsymbol{c}}_a(0)$. It is easily seen that $\dot{\boldsymbol{c}}_a(0) = \partial \boldsymbol{\mu} / \partial \beta_a$ at $\boldsymbol{\beta}$ $(a = 1, \cdots, p)$, which are just the column vectors of $\partial \boldsymbol{\mu}(\boldsymbol{\beta}) / \partial \boldsymbol{\beta}^T = \boldsymbol{D}(\boldsymbol{\beta})$. The space spanned by $\dot{\boldsymbol{c}}_a(0), a = 1, \cdots, p$, which are tangent vectors of the coordinate curves at $\boldsymbol{\beta}$, is called the tangent space of π at $\boldsymbol{\beta}$ and denoted by $T_{\boldsymbol{\beta}}$. Note that by Assumption C given in Section 4.1, $\boldsymbol{D}(\boldsymbol{\beta})$ is of full rank in column, hence tangent vectors $\dot{c}_a(0)$ $(a = 1, \cdots, p)$ are linearly independent. So the tangent space T_β is spanned by the columns of $\boldsymbol{D}(\boldsymbol{\beta})$ and is a p-dimensional linear subspace of R^n, which can be viewed as a linear approximation of the solution locus π at $\boldsymbol{\beta}$ in R^n. To get the second order approximation of the solution locus $\boldsymbol{\mu} = \boldsymbol{\mu}(\boldsymbol{\beta})$, the second order derivatives of $\boldsymbol{\mu} = \boldsymbol{\mu}(\boldsymbol{\beta})$ are needed, which will naturally result in the curvatures. To do this, we first introduce a weighted inner product in Euclidean space R^n.

Following Efron (1975) and Amari (1982a), we define the Fisher information inner product for any two vectors $\boldsymbol{a}$ and $\boldsymbol{b}$ in R^n as

$$\langle \boldsymbol{a}, \boldsymbol{b} \rangle = \boldsymbol{a}^T \boldsymbol{V}^{-1} \boldsymbol{b}. \tag{3.3}$$

Appendix A gives a summary for this inner product. Notice that the Fisher information of $\boldsymbol{Y}$ for $\boldsymbol{\mu}$ is $\boldsymbol{J}_\mu(\boldsymbol{Y}) = \phi \boldsymbol{V}^{-1}(\boldsymbol{\mu})$ (see (1.21)), but here we skip the nuisance parameter ϕ which is assumed to be known or estimated separately.

Under the defined inner product, the associated normal space of the tangent space T_β is a linear subspace T'_β of R^n for which $\boldsymbol{a} \in T'_\beta$ if and only if $\boldsymbol{a}^T \boldsymbol{V}^{-1} \boldsymbol{b} = 0$ holds for any $\boldsymbol{b} \in T_\beta$ and T'_β is an $(n-p)$-dimensional linear subspace of R^n. Now we can get the orthonormal basis for the tangent space T_β and the normal space T'_β. Suppose that the QR decomposition of $\boldsymbol{D}(\boldsymbol{\beta})$ under the inner product (3.3) is given by (see Appendix A)

$$\boldsymbol{D}(\boldsymbol{\beta}) = (\boldsymbol{Q}, \boldsymbol{N}) \begin{pmatrix} \boldsymbol{R} \\ \boldsymbol{0} \end{pmatrix} = \boldsymbol{Q}\boldsymbol{R}, \tag{3.4}$$

where $\boldsymbol{R}$ is a $p \times p$ nonsingular upper triangular matrix with positive diagonal elements and the column vectors of $\boldsymbol{Q}$ and $\boldsymbol{N}$ are respectively orthonormal basis for the tangent space and the normal space of the solution locus $\boldsymbol{\mu} = \boldsymbol{\mu}(\boldsymbol{\beta})$ at $\boldsymbol{\beta}$ in R^n. The matrices $\boldsymbol{Q}$ and $\boldsymbol{N}$ satisfy

$$\boldsymbol{Q}^T \boldsymbol{V}^{-1} \boldsymbol{Q} = \boldsymbol{I}_p, \quad \boldsymbol{Q}^T \boldsymbol{V}^{-1} \boldsymbol{N} = \boldsymbol{0}, \quad \boldsymbol{N}^T \boldsymbol{V}^{-1} \boldsymbol{N} = \boldsymbol{I}_{n-p},$$

where $\boldsymbol{I}_p$ and $\boldsymbol{I}_{n-p}$ are identity matrices of order p and $n-p$, respectively.

Based on the above geometric structure, we introduce the following definition.

Definition 3.1

For exponential family nonlinear models (2.2), the intrinsic curvature array $\boldsymbol{A}^I$ and the parameter-effects curvature array $\boldsymbol{A}^P$ in expectation parameter space are respectively defined as

$$\boldsymbol{A}^I = [\boldsymbol{N}^T \boldsymbol{V}^{-1}][\boldsymbol{U}], \quad \boldsymbol{A}^P = [\boldsymbol{Q}^T \boldsymbol{V}^{-1}][\boldsymbol{U}], \tag{3.5}$$

$$\boldsymbol{U} = \boldsymbol{L}^T \boldsymbol{W} \boldsymbol{L}, \quad \boldsymbol{L} = \boldsymbol{R}^{-1}. \qquad \blacksquare$$

By the array multiplication, it is easily seen from this definition that $\boldsymbol{A}^I$ is an $(n-p)\times p\times p$ array and $\boldsymbol{A}^P$ is a $p\times p\times p$ array, respectively. We can see the geometric meaning of this definition by making a reparameterization $\boldsymbol{\gamma} = \boldsymbol{R}(\boldsymbol{\beta}-\boldsymbol{\beta}_0)$ as described by Bates and Watts (1980) for nonlinear regression models. It is easy to show that the first two derivatives of $\boldsymbol{\mu} = \boldsymbol{\mu}(\boldsymbol{\beta}(\boldsymbol{\gamma}))$ in $\boldsymbol{\gamma}$ with $\boldsymbol{\beta} = \boldsymbol{\beta}(\boldsymbol{\gamma}) = \boldsymbol{L}\boldsymbol{\gamma}+\boldsymbol{\beta}_0$ are respectively

$$\frac{\partial \boldsymbol{\mu}}{\partial \boldsymbol{\gamma}^T} = \frac{\partial \boldsymbol{\mu}}{\partial \boldsymbol{\beta}^T}\frac{\partial \boldsymbol{\beta}}{\partial \boldsymbol{\gamma}^T} = \boldsymbol{D}\boldsymbol{L} = \boldsymbol{Q},$$

and

$$\frac{\partial^2 \boldsymbol{\mu}}{\partial \boldsymbol{\gamma}\partial \boldsymbol{\gamma}^T} = \left(\frac{\partial \boldsymbol{\beta}}{\partial \boldsymbol{\gamma}^T}\right)^T\left(\frac{\partial^2 \boldsymbol{\mu}}{\partial \boldsymbol{\beta}\partial \boldsymbol{\beta}^T}\right)\left(\frac{\partial \boldsymbol{\beta}}{\partial \boldsymbol{\gamma}^T}\right) = \boldsymbol{L}^T\boldsymbol{W}\boldsymbol{L} = \boldsymbol{U}.$$

These results state that the array $\boldsymbol{U}$ in (3.5) is just the second order derivative of $\boldsymbol{\mu}(\boldsymbol{\beta}(\boldsymbol{\gamma}))$ with respect to the parameter $\boldsymbol{\gamma}$, while the first order derivative of $\boldsymbol{\mu}(\boldsymbol{\beta}(\boldsymbol{\gamma}))$ is just $\boldsymbol{Q}$ that forms an orthonormal basis for the tangent space of the solution locus π at $\boldsymbol{\beta}(\boldsymbol{\gamma})$. Therefore (3.5) shows that $\boldsymbol{A}^I$ and $\boldsymbol{A}^P$ are just the projections of the second order derivative $\boldsymbol{U}$ onto the orthonormal basis of the normal space T'_β and the tangent space T_β, respectively. To see this more clearly, let $\boldsymbol{A}^I = (A^I_{\kappa ij})$, $\boldsymbol{A}^P = (A^P_{aij})$, $\boldsymbol{U} = (\boldsymbol{U}_{ij})$, $\boldsymbol{N} = (\boldsymbol{N}_\kappa)$, and $\boldsymbol{Q} = (\boldsymbol{Q}_a)$, where $\kappa = 1,\cdots,n-p$; $a,i,j = 1,\cdots,p$. Then $\boldsymbol{U}_{ij}$ denotes an $n\times 1$ vector of $\boldsymbol{U}$ at (i,j), and $\boldsymbol{N}_\kappa$ and $\boldsymbol{Q}_a$ are κ-th and a-th orthonormal basis of T'_β and T_β, respectively. Therefore $A^I_{\kappa ij} = \boldsymbol{N}^T_\kappa \boldsymbol{V}^{-1}\boldsymbol{U}_{ij}$ and $A^P_{aij} = \boldsymbol{Q}^T_a\boldsymbol{V}^{-1}\boldsymbol{U}_{ij}$ are the projections of $\boldsymbol{U}_{ij}$ onto $\boldsymbol{N}_\kappa$ and $\boldsymbol{Q}_a$, respectively. This geometric interpretation is consistent with the usual curvature definition in differential geometry (e.g. Millman and Parker 1977). Therefore the curvature arrays $\boldsymbol{A}^I$ and $\boldsymbol{A}^P$ indicate how the solution locus is far from the tangent space, a linear approximation of the solution locus at the neighborhood of $\boldsymbol{\beta}$. These may provide suitable measures of the severity of nonlinearity of the model.

It is easily seen that the curvature arrays defined by Bates and Watts (1980) can be regarded as a special case of definition (3.5). In fact, if model (2.2) represents the normal nonlinear regression, then we have

$$g(\mu_i) = \mu_i = \theta_i = f(x_i;\beta), \quad b(\theta_i) = \frac{1}{2}\theta_i^2; \tag{3.6}$$

$i = 1,\cdots,n$. Equation (1.8) shows that $\boldsymbol{V} = \boldsymbol{I}_n$ is an identity matrix. Hence the inner product (3.3) is just the ordinary inner product $\langle \boldsymbol{a},\boldsymbol{b}\rangle = \boldsymbol{a}^T\boldsymbol{b}$ and the definition (3.5) exactly coincides with the equation (2.18) of Bates and Watts (1980), as expected.

As pointed out by Bates and Watts (1980), $\boldsymbol{A}^I$ indicates the intrinsic nonlinearity of the model and $\boldsymbol{A}^P$ indicates the apparent nonlinearity caused by the parameterization of the model.

It should be emphasized that the form of definition (3.5) seems similar to the BW geometric framework, but the inner product based on the Fisher information matrix is crucial. This inner product may cause a substantial extension from normal nonlinear regression models to a much wider class of models (see also Chapter 7). The Fisher information inner product (which is called the Riemannian metric in the terminology of EA geometric framework) was also used by Efron (1975; 1978) and Amari (1982a) and discussed by Amari (1985) in detail. At this point, geometric structure is dissimilar to the BW framework but is similar to the EA framework. Besides, our curvature arrays $\boldsymbol{A}^I$ and $\boldsymbol{A}^P$ are also respectively consistent with the embedded normal curvature and the embedded connection of Amari (1985) with $\alpha = -1$ connection (for details, see Amari 1985). However, it is contrary to the EA framework that we use the orthonormal basis for the tangent space and the normal space. Further, we have not introduced connections (see Section 3.3 and Table 3.2 for more discussions).

Finally, we introduce a useful formula as following, which will be used in later discussions. The curvature arrays $\boldsymbol{A}^I$ and $\boldsymbol{A}^P$ satisfy

$$[\boldsymbol{N}][\boldsymbol{A}^I] + [\boldsymbol{Q}][\boldsymbol{A}^P] = \boldsymbol{U}. \tag{3.7}$$

Proof. The projection matrices of the tangent space and the normal space are respectively $\boldsymbol{P}_T = \boldsymbol{Q}\boldsymbol{Q}^T\boldsymbol{V}^{-1}$ and $\boldsymbol{P}_N = \boldsymbol{N}\boldsymbol{N}^T\boldsymbol{V}^{-1}$ satisfying $\boldsymbol{P}_T + \boldsymbol{P}_N = \boldsymbol{I}_n$ (see Appendix A.1). Hence it follows from the rules of array multiplication (see Appendix A.2) that

$$\begin{aligned} \boldsymbol{U} &= [\boldsymbol{Q}\boldsymbol{Q}^T\boldsymbol{V}^{-1} + \boldsymbol{N}\boldsymbol{N}^T\boldsymbol{V}^{-1}][\boldsymbol{U}] \\ &= [\boldsymbol{Q}][[\boldsymbol{Q}^T\boldsymbol{V}^{-1}][\boldsymbol{U}]] + [\boldsymbol{N}][[\boldsymbol{N}^T\boldsymbol{V}^{-1}][\boldsymbol{U}]] \\ &= [\boldsymbol{Q}][\boldsymbol{A}^P] + [\boldsymbol{N}][\boldsymbol{A}^I]. \end{aligned}$$

Then (3.7) is proved. ∎

3.2.2 Directional curvatures and maximum curvatures

Bates and Watts (1980) defined the directional curvatures and the maximum curvatures for nonlinear regression models. In this subsection, we extend their definition to exponential family nonlinear models. Now let $\boldsymbol{\beta}(b) = \boldsymbol{\beta}_0 + b\boldsymbol{h}$ be a straight line in the parameter space $\mathcal{B}$ through a point $\boldsymbol{\beta}_0$, where $\boldsymbol{h} = (h_1, \cdots, h_p)^T$ is a nonzero direction vector and b is a real value. This line generates a curve in solution locus π:

$$c_h: \quad \boldsymbol{\mu}_h(b) = \boldsymbol{\mu}(\boldsymbol{\beta}_0 + b\boldsymbol{h})$$

that is called the lifted line. The tangent vector and the acceleration vector of this curve at $\boldsymbol{\beta}_0$ (i.e. $b = 0$) are respectively denoted by

$$\dot{\boldsymbol{\mu}}_h = \left.\frac{d\{\boldsymbol{\mu}_h(b)\}}{db}\right|_{b=0}, \qquad \ddot{\boldsymbol{\mu}}_h = \left.\frac{d^2\{\boldsymbol{\mu}_h(b)\}}{db^2}\right|_{b=0}.$$

It follows from (A.8) that

$$\frac{d\{\mu_h(b)\}}{db} = \frac{\partial \mu}{\partial \beta^T}\frac{d\beta(b)}{db} = Dh,$$

$$\frac{d^2\{\mu_h(b)\}}{db^2} = \left\{\frac{d\beta(b)}{db}\right\}^T \frac{\partial^2 \mu}{\partial\beta\partial\beta^T}\left\{\frac{\partial\beta(b)}{db}\right\} = h^T W h.$$

Then we have

$$\dot{\mu}_h = Dh, \qquad \ddot{\mu}_h = h^T W h, \tag{3.8}$$

where D and W are evaluated at β_0 (i.e. $b = 0$).

Now let the normal component and the tangential component of $\ddot{\mu}_h$ be denoted by $\ddot{\mu}_h^N$ and $\ddot{\mu}_h^T$, respectively. From this we introduce the following definition.

Definition 3.2

The intrinsic curvature and the parameter-effects curvature along the direction h at β_0 for model (2.2) are defined as

$$K_h^N = \frac{\|\ddot{\mu}_h^N\|}{\|\dot{\mu}_h\|^2}, \qquad K_h^P = \frac{\|\ddot{\mu}_h^T\|}{\|\dot{\mu}_h\|^2}, \tag{3.9}$$

and the corresponding maximum curvatures along all the directions are defined as

$$K^N = \max_{\|h\|=1} K_h^N, \quad K^P = \max_{\|h\|=1} K_h^T, \tag{3.10}$$

where $\|\cdot\|$ denotes the norm of a vector, that is $\|\alpha\| = (\alpha^T V^{-1}\alpha)^{\frac{1}{2}}$ for any given vector α. ∎

It is easily seen that the curvatures of this definition are just the "normed acceleration" on the lifted line c_h. The maximum curvatures K^N and K^P provide a measure for assessing the severity of nonlinearity of a given exponential family nonlinear model. From (3.8), we have

$$K_h^N = \frac{\|P_N(h^T W h)\|}{h^T D^T V^{-1} D h}, \qquad K_h^P = \frac{\|P_T(h^T W h)\|}{h^T D^T V^{-1} D h}. \tag{3.11}$$

These equations show that curvatures K_h^N and K_h^T depend only on the direction h and is independent of its length.

Pazman (1991) has discussed the maximum curvatures for exponential family nonlinear models similar to Definition 3.2.

There is close connection between the curvature arrays defined in (3.5) and the directional curvatures defined in (3.9). From (3.11), we have the following theorem.

Theorem 3.1

Under the notation and definitions stated earlier, we have

$$(K_h^N)^2 = \|N(\tilde{a}^T A^I \tilde{a})\|^2 = (\tilde{a}^T A^I \tilde{a})^T(\tilde{a}^T A^I \tilde{a}), \quad (3.12)$$

$$(K_h^P)^2 = \|Q(\tilde{a}^T A^P \tilde{a})\|^2 = (\tilde{a}^T A^P \tilde{a})^T(\tilde{a}^T A^P \tilde{a}), \quad (3.13)$$

where $\tilde{a} = V^{-\frac{1}{2}} a$ satisfying $\tilde{a}^T \tilde{a} = 1$ and a is a unit vector given by $a = V^{\frac{1}{2}} Rh / \|V^{\frac{1}{2}} Rh\|$.

Proof. We just prove (3.13). It follows from (3.11) and (3.4) that

$$K_h^N = \frac{\|P_N(h^T W h)\|}{h^T R^T R h}.$$

By the rules of array multiplication (Appendix A.2), we have

$$\begin{aligned} P_N(h^T W h) &= [NN^T V^{-1}][h^T W h] \\ &= h^T [NN^T V^{-1}][W] h \\ &= h^T R^T [N][[N^T V^{-1}][L^T W L]] R h \\ &= h^T R^T [N][A^I] R h. \end{aligned}$$

Then K_h^N can be written as

$$\begin{aligned} K_h^N &= \frac{\|h^T R^T V^{\frac{1}{2}} [N][V^{-\frac{1}{2}} A^I V^{-\frac{1}{2}}] V^{\frac{1}{2}} R h\|}{h^T R^T V^{\frac{1}{2}} V^{-1} V^{\frac{1}{2}} R h} \\ &= \|a^T [N][V^{-\frac{1}{2}} A^I V^{-\frac{1}{2}}] a\| \\ &= \|N(\tilde{a}^T A^I \tilde{a})\|. \end{aligned}$$

Equation (3.13) follows from this. Similarly, we can get (3.13). ∎

It is very interesting to note that we may use the ordinary inner product to compute directional curvatures by Theorem 3.1. In fact, if we denote the ordinary inner product by $\langle a, b\rangle' = a^T b$, then it is easily seen that (3.13) and (3.13) can be rewritten as

$$K_h^N = \|\tilde{a}^T A^I \tilde{a}\|', \quad K_h^P = \|\tilde{a}^T A^P \tilde{a}\|' \quad (3.14)$$

where $\|\tilde{a}\|' = 1$ and $\|\cdot\|'$ denotes the norm under the ordinary inner product, i.e. $\|\alpha\|' = (\alpha^T \alpha)^{\frac{1}{2}}$ for a vector α. But note that A^I and A^P given in (3.5) are defined by the weighted inner product. If model (2.2) represents the nonlinear regression (3.6), then result (3.14) exactly coincides with equations (2.19) and (2.20) of Bates and Watts (1980), as expected.

Now we introduce an algorithm for computing the maximum curvatures K^N and K^P which are main measures of the severity of nonlinearity of the model. It is hard to get analytical formulas to compute these curvatures, except for some special cases, but we can provide a simple iterative algorithm for obtaining these values as given by Bates and Watts (1980). Since the algorithms for computing K^N and K^P are exactly the same, in the following, we just introduce the method to compute K^N.

Lemma 3.1

The direction $\tilde{\boldsymbol{a}}$ under which K_h^N achieves its maximum must satisfy

$$\boldsymbol{g} = [(\tilde{\boldsymbol{a}}^T \boldsymbol{A}^I \tilde{\boldsymbol{a}})^T][\boldsymbol{A}^I \tilde{\boldsymbol{a}}] = k\tilde{\boldsymbol{a}}, \qquad k > 0, \tag{3.15}$$

where $\tilde{\boldsymbol{a}}$ is given in Theorem 3.1 and k is a proportional constant.

Proof. We start with (3.10) and (3.14) for proving this lemma. By the Lagrange's method of multiplers, we set

$$F(\tilde{\boldsymbol{a}}, \lambda) = (\tilde{\boldsymbol{a}}^T \boldsymbol{A}^I \tilde{\boldsymbol{a}})^T (\tilde{\boldsymbol{a}}^T \boldsymbol{A}^I \tilde{\boldsymbol{a}}) - \lambda(\tilde{\boldsymbol{a}}^T \tilde{\boldsymbol{a}} - 1).$$

Taking derivative of $F(\tilde{\boldsymbol{a}}, \lambda)$ in $\tilde{\boldsymbol{a}}$ and using array multiplication, we get

$$\frac{\partial F(\tilde{\boldsymbol{a}}, \lambda)}{\partial \tilde{\boldsymbol{a}}} = 4[(\tilde{\boldsymbol{a}}^T \boldsymbol{A}^I \tilde{\boldsymbol{a}})^T][\boldsymbol{A}^I \tilde{\boldsymbol{a}}] - 2\lambda\tilde{\boldsymbol{a}} = \boldsymbol{0},$$

which gives (3.15) with $k = \lambda/2$. Multiplying the above equation by $\tilde{\boldsymbol{a}}^T$, we get $k = (\tilde{\boldsymbol{a}}^T A^I \tilde{\boldsymbol{a}})^T (\tilde{\boldsymbol{a}}^T\ A^I \tilde{\boldsymbol{a}}) > 0$. ∎

This lemma shows that if we choose a direction $\tilde{\boldsymbol{a}}$ in (3.14) such that the vector $\boldsymbol{g} = [(\tilde{\boldsymbol{a}}^T \boldsymbol{A}^I \tilde{\boldsymbol{a}})^T][\boldsymbol{A}^I \tilde{\boldsymbol{a}}]$ is proportional to the vector $\tilde{\boldsymbol{a}}$ and has the same direction as $\tilde{\boldsymbol{a}}$, then we can get the maximum curvatures K^N and K^T from (3.14). One may design an iteration procedure for obtaining the maximum curvatures (Bates and Watts 1980), which is summarized in Figure 3.1.

In this procedure, the values $\tilde{\boldsymbol{a}}_i$ were proposed by Bates and Watts (1980). The convergence criterion δ determines how close $\tilde{\boldsymbol{a}}_i$ and $\tilde{\boldsymbol{g}}_i$ are where both are in the same direction. As they are unit vectors, their ordinary inner product will be 1 if they are exactly in the same direction. Bates and Watts (1980) proposed $\delta = 0.0001$ for the convergence criterion.

It is occasionally possible that the iteration procedure may result in a local maximum rather than the global maximum. To avoid this, it may be useful to give appropriate lower bounds and upper bounds for maximum curvatures K^N and K^P as discussed by Bates and Watts (1980) for nonlinear regression models. We get parallel results for exponential family

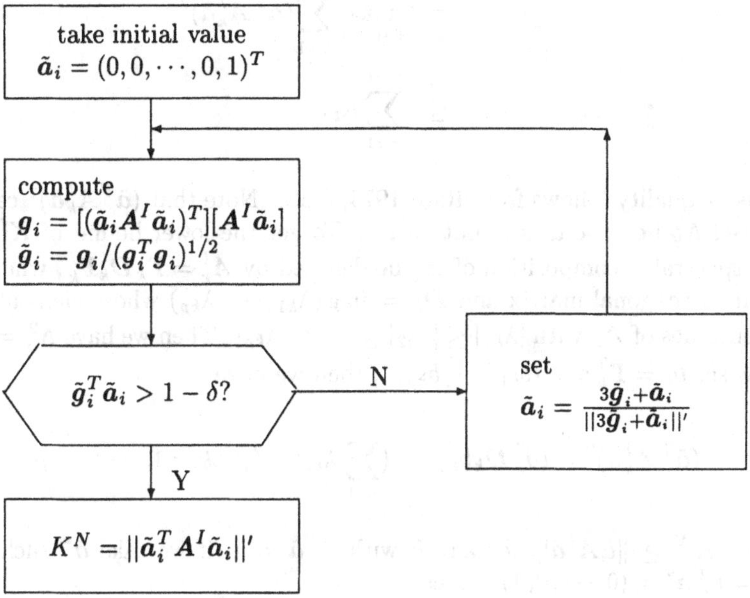

Figure 3.1 ALGORITHM OF MAXIMUM CURVATURE

nonlinear models. Since the formulas are very similar for K^N and K^P, we just give a formula about K^N.

Lemma 3.2

Let the k-th face of A^I be A_k^I whose largest and smallest absolute eigenvalues are respectively Λ_k and $\lambda_k (k = 1, \cdots, n-p)$. Then we have

$$\Lambda_\alpha^2 + \sum_{k \neq \alpha} \lambda_k^2 \leq (K^N)^2 \leq \sum_{k=1}^{n-p} \Lambda_k^2,$$

where α is the index which yields the maximum of $\Lambda_k, k = 1, \cdots, n-p$, i.e.

$$\Lambda_\alpha^2 = \max_{1 \leq k \leq n-p} \Lambda_k^2.$$

Proof. From (3.14) we have

$$
\begin{aligned}
(K^N)^2 &= \max_{\|\tilde{a}\|'=1} (\|\tilde{a}^T A^I \tilde{a}\|')^2 \\
&= \max_{\|\tilde{a}\|'=1} \sum_{k=1}^{n-p} (\tilde{a}^T A_k^I \tilde{a})^2 \\
&\le \sum_{k=1}^{n-p} \Lambda_k^2 .
\end{aligned}
$$

This inequality follows from Rao (1973, p.62). Note that $(\tilde{a}^T A_k^I \tilde{a})^2$ cannot exceed Λ_k^2 because $\tilde{a}$ is a unit vector. To get the lower bound of K^N, let the spectral decomposition of A_k^I be denoted by $A_k^I = \Gamma_k D_k \Gamma_k^T$, where Γ_k is the orthogonal matrix and $D_k = \mathrm{diag}(\lambda_{k1}, \cdots, \lambda_{kp})$ whose elements are eigenvalues of A_k^I with $|\lambda_{k1}| \le |\lambda_{k2}| \le \cdots \le |\lambda_{kp}|$. Then we have $\Lambda_\alpha^2 = \lambda_{\alpha p}^2$. Now set $b_k = \Gamma_k^T \tilde{a} = (b_{k1}, \cdots, b_{kp})^T$, then we have

$$
(\tilde{a}^T A_k^I \tilde{a})^2 = (b_k^T D_k b_k)^2 = (\sum_{i=1}^{p} \lambda_{ki} b_{ki}^2)^2, \quad k = 1, \cdots, n-p.
$$

Since $K^N \ge \|\tilde{a} A^I \tilde{a}\|'$ for any $\tilde{a}$ with $\tilde{a}^T \tilde{a} = 1$, if we take $\tilde{a}^*$ such that $b^* = \Gamma_\alpha^T \tilde{a}^* = (0, \cdots, 0, 1)^T$, then

$$
\begin{aligned}
(K^N)^2 &\ge \|(\tilde{a}^*)^T A^I \tilde{a}^*\|' \\
&= \left\{(b^*)^T D_\alpha b^*\right\}^2 + \sum_{k \ne \alpha} \left\{(\tilde{a}^*)^T A_k^I (\tilde{a}^*)\right\}^2 \\
&\ge \Lambda_\alpha^2 + \sum_{k \ne \alpha} \lambda_k^2 .
\end{aligned}
$$

∎

3.2.3 Parameter transformation

In this subsection, we study the influence of parameterizations on the curvatures defined earlier. We shall show that the intrinsic curvatures K_h^N and A^I are invariant under parameterizations. For the parameter-effects curvature array A^P, a formula corresponding to a transformed set of parameters will be given.

Now let $\beta = B(\gamma)$ be a one-to-one transformation of β in a region $\mathcal{G}$ of γ such that $B(\gamma) \in \mathcal{B}$ as $\gamma \in \mathcal{G}$. We assume that the derivative $\dot{B}_\gamma = \partial\beta/\partial\gamma^T$ is nonsingular in $\mathcal{G}$. The inverse mapping of $B(\gamma)$ and its derivative are denoted by $\gamma = \gamma(\beta)$ and $\dot{\gamma}_\beta = \partial\gamma/\partial\beta^T = \dot{B}_\gamma^{-1}$, respectively. Under this transformation, $\mu = \mu(\beta)$ in (2.3) is also a function of γ

according to $\boldsymbol{\beta} = \boldsymbol{B}(\boldsymbol{\gamma})$. Now let $\boldsymbol{D}_{\boldsymbol{\gamma}} = \partial\boldsymbol{\mu}/\partial\boldsymbol{\gamma}^T$, $\boldsymbol{W}_{\boldsymbol{\gamma}} = \partial^2\boldsymbol{\mu}/\partial\boldsymbol{\gamma}\partial\boldsymbol{\gamma}^T$.

Then it follows from (A.8) of Appendix A that

$$\boldsymbol{D}_\gamma = \frac{\partial\boldsymbol{\mu}}{\partial\boldsymbol{\beta}^T}\frac{\partial\boldsymbol{\beta}}{\partial\boldsymbol{\gamma}^T} = \boldsymbol{D}\dot{\boldsymbol{B}}_\gamma, \tag{3.16}$$

and

$$\begin{aligned} \boldsymbol{W}_\gamma &= \left(\frac{\partial\boldsymbol{\beta}}{\partial\boldsymbol{\gamma}^T}\right)\left(\frac{\partial^2\boldsymbol{\mu}}{\partial\boldsymbol{\beta}\partial\boldsymbol{\beta}^T}\right)\left(\frac{\partial\boldsymbol{\beta}}{\partial\boldsymbol{\gamma}^T}\right) + \left[\frac{\partial\boldsymbol{\mu}}{\partial\boldsymbol{\beta}^T}\right]\left[\frac{\partial^2\boldsymbol{\beta}}{\partial\boldsymbol{\gamma}\partial\boldsymbol{\gamma}^T}\right] \\ &= \dot{\boldsymbol{B}}_\gamma^T\boldsymbol{W}\dot{\boldsymbol{B}}_\gamma + [\boldsymbol{D}][\ddot{\boldsymbol{B}}_\gamma], \end{aligned} \tag{3.17}$$

where $\ddot{\boldsymbol{B}}_\gamma = \partial^2\boldsymbol{\beta}/\partial\boldsymbol{\gamma}\partial\boldsymbol{\gamma}^T$ is a $p \times p \times p$ array.

We first consider the directional curvatures under parameter $\boldsymbol{\gamma}$. Suppose that a straight line in $\mathcal{G}$ through a point $\boldsymbol{\gamma}_0$ is $\boldsymbol{\gamma} = \boldsymbol{\gamma}_0 + b\boldsymbol{g}$, where $\boldsymbol{g}$ is a direction vector and b is a real number. This line generates a lifted line in solution locus:

$$c_g: \quad \boldsymbol{\mu}_g(b) = \boldsymbol{\mu}(\boldsymbol{B}(\boldsymbol{\gamma}_0 + b\boldsymbol{g})).$$

By Definition 3.2 and (3.11), the intrinsic curvature and the parameter-effects curvature at $\boldsymbol{\gamma}_0$ along the direction $\boldsymbol{g}$ are respectively

$$K_g^N = \frac{\|\boldsymbol{P}_N(\boldsymbol{g}^T\boldsymbol{W}_\gamma\boldsymbol{g})\|}{\boldsymbol{g}^T\boldsymbol{D}_\gamma^T\boldsymbol{V}^{-1}\boldsymbol{D}_\gamma\boldsymbol{g}}, \qquad K_g^P = \frac{\|\boldsymbol{P}_T(\boldsymbol{g}^T\boldsymbol{W}_\gamma\boldsymbol{g})\|}{\boldsymbol{g}^T\boldsymbol{D}_\gamma^T\boldsymbol{V}^{-1}\boldsymbol{D}_\gamma\boldsymbol{g}}. \tag{3.18}$$

We notice that $\boldsymbol{\beta} = \boldsymbol{B}(\boldsymbol{\gamma}_0 + b\boldsymbol{g})$ represents a curve in $\mathcal{B}$ through $\boldsymbol{\beta}_0 = \boldsymbol{B}(\boldsymbol{\gamma}_0)$. The tangent vector of this curve at $\boldsymbol{\gamma}_0$ is $\boldsymbol{h} = d\boldsymbol{\beta}/db = \dot{\boldsymbol{B}}_\gamma\boldsymbol{g}$ evaluated at $\boldsymbol{\gamma}_0$ (i.e. $b = 0$). We can show that K_g^N is just the directional curvature along $\boldsymbol{h}$ under parameter $\boldsymbol{\beta}$. We have the following theorem.

Theorem 3.2

Under notation stated earlier, the directional intrinsic curvature is invariant under parameterizations, that is

$$K_g^N = K_h^N, \qquad \boldsymbol{h} = \dot{\boldsymbol{B}}_\gamma\boldsymbol{g}.$$

Proof. We use (3.16) to (3.18) to prove the theorem. It is easily seen from (3.16) that $\boldsymbol{g}^T\boldsymbol{D}_\gamma^T\boldsymbol{V}^{-1}\boldsymbol{D}_\gamma\boldsymbol{g} = \boldsymbol{h}^T\boldsymbol{D}^T\boldsymbol{V}^{-1}\boldsymbol{D}\boldsymbol{h}$. By the rules of array multiplication (Appendix A.2), we have

$$\begin{aligned} \boldsymbol{P}_N(\boldsymbol{g}^T\boldsymbol{W}_\gamma\boldsymbol{g}) &= \boldsymbol{g}^T[\boldsymbol{P}_N][\boldsymbol{W}_\gamma]\boldsymbol{g} \\ &= \boldsymbol{g}^T\{[\boldsymbol{P}_N][\dot{\boldsymbol{B}}_\gamma^T\boldsymbol{W}\dot{\boldsymbol{B}}_\gamma] + [\boldsymbol{P}_N][[\boldsymbol{D}][\ddot{\boldsymbol{B}}_\gamma]]\}\boldsymbol{g}. \end{aligned}$$

Using $\boldsymbol{P}_N\boldsymbol{D} = \boldsymbol{0}$, $\dot{\boldsymbol{B}}_\gamma \boldsymbol{g} = \boldsymbol{h}$, we have

$$\boldsymbol{P}_N(\boldsymbol{g}^T \boldsymbol{W}_\gamma \boldsymbol{g}) = [\boldsymbol{P}_N][\boldsymbol{h}^T \boldsymbol{W} \boldsymbol{h}] = \boldsymbol{P}_N(\boldsymbol{h}^T \boldsymbol{W} \boldsymbol{h}).$$

Substituting the above results into (3.18), we get

$$K_g^N = \frac{\|\boldsymbol{P}_N(\boldsymbol{h}^T \boldsymbol{W} \boldsymbol{h})\|}{\boldsymbol{h}^T \boldsymbol{D}^T \boldsymbol{V}^{-1} \boldsymbol{D} \boldsymbol{h}} = K_h^N. \qquad \blacksquare$$

For the parameter-effects curvature, it is not necessary to have $K_g^P = K_h^P$. In fact, it follows from (3.17) and (3.18) that

$$\boldsymbol{P}_T(\boldsymbol{g}^T \boldsymbol{W}_\gamma \boldsymbol{g}) = \boldsymbol{P}_T(\boldsymbol{h}^T \boldsymbol{W} \boldsymbol{h}) + \boldsymbol{D}(\boldsymbol{g}^T \ddot{\boldsymbol{B}}_\gamma \boldsymbol{g}).$$

The second term in this equation does not need to vanish.

For the curvature arrays, we have the following theorem.

Theorem 3.3

Suppose that $\boldsymbol{\beta} = \boldsymbol{B}(\boldsymbol{\gamma})$ is a one-to-one transformation of $\boldsymbol{\beta}$ in a region $\mathcal{G}$ of $\boldsymbol{\gamma}$ such that $\boldsymbol{B}(\boldsymbol{\gamma}) \in \mathcal{B}$ as $\boldsymbol{\gamma} \in \mathcal{G}$. The curvatures under parameters $\boldsymbol{\beta}$ and $\boldsymbol{\gamma}$ are denoted by $\boldsymbol{A}^I$, $\boldsymbol{A}^P$ and $\boldsymbol{A}_\gamma^I$, $\boldsymbol{A}_\gamma^P$, respectively. Then the intrinsic curvature array is invariant, that is $\boldsymbol{A}_\gamma^I = \boldsymbol{A}^I$, and the parameter-effects curvature arrays $\boldsymbol{A}_\gamma^P$ and $\boldsymbol{A}^P$ satisfy

$$\begin{aligned} \boldsymbol{A}_\gamma^P &= \boldsymbol{A}^P + \boldsymbol{A}^G, \\ \boldsymbol{A}^G &= [\boldsymbol{R}][\boldsymbol{L}^T \dot{\boldsymbol{B}}_\gamma^{-T} \ddot{\boldsymbol{B}}_\gamma \dot{\boldsymbol{B}}_\gamma^{-1} \boldsymbol{L}] \\ &= -[\boldsymbol{R}][\boldsymbol{L}^T[\dot{\boldsymbol{\gamma}}_\beta^{-1}][\ddot{\boldsymbol{\gamma}}_\beta]\boldsymbol{L}], \end{aligned} \qquad (3.19)$$

where $\ddot{\boldsymbol{\gamma}}_\beta = \partial^2 \boldsymbol{\gamma}/\partial \boldsymbol{\beta} \partial \boldsymbol{\beta}^T$ and $\dot{\boldsymbol{B}}_\gamma^{-T} = (\dot{\boldsymbol{B}}_\gamma^{-1})^T$.

Proof. Under the parameter $\boldsymbol{\gamma}$, it is easily seen that the solution locus π_γ is $\boldsymbol{\mu} = \boldsymbol{\mu}(\boldsymbol{B}(\boldsymbol{\gamma}))$ whose tangent space T_γ at $\boldsymbol{\gamma}$ is spanned by columns of $\partial \boldsymbol{\mu}/\partial \boldsymbol{\gamma}^T = \boldsymbol{D}_\gamma$ at $\boldsymbol{\gamma}$. It follows from (3.4) and (3.16) that $\boldsymbol{D}_\gamma = \boldsymbol{Q}\boldsymbol{R}\dot{\boldsymbol{B}}_\gamma$. This result shows that we can still choose the columns of $\boldsymbol{Q}$ as the orthonormal basis for T_γ, and the QR decomposition of $\boldsymbol{D}_\gamma$ may be given by $\boldsymbol{D}_\gamma = \boldsymbol{Q}_\gamma \boldsymbol{R}_\gamma$ with $\boldsymbol{Q}_\gamma = \boldsymbol{Q}$, $\boldsymbol{R}_\gamma = \boldsymbol{R}\dot{\boldsymbol{B}}_\gamma$ and $\boldsymbol{L}_\gamma = \boldsymbol{R}_\gamma^{-1} = \dot{\boldsymbol{B}}_\gamma^{-1}\boldsymbol{L}$ (Bates and Watts 1981). Obviously, the columns of $\boldsymbol{N}$ in (3.4) still form an orthonormal basis for the normal space T_γ' of π_γ and hence $\boldsymbol{N}_\gamma = \boldsymbol{N}$. Now we can find curvature arrays $\boldsymbol{A}_\gamma^I$ and $\boldsymbol{A}_\gamma^P$ by these statements. In fact, from equations (3.5), (3.17) and $\boldsymbol{N}^T \boldsymbol{V}^{-1} \boldsymbol{D} = \boldsymbol{0}$, we obtain the following formulas

$$\begin{aligned}
\boldsymbol{A}_\gamma^I &= [\boldsymbol{N}_\gamma^T \boldsymbol{V}^{-1}][\boldsymbol{L}_\gamma^T \boldsymbol{W}_\gamma \boldsymbol{L}_\gamma] \\
&= [\boldsymbol{N}^T \boldsymbol{V}^{-1}][\boldsymbol{L}^T \dot{\boldsymbol{B}}_\gamma^{-T} \boldsymbol{W}_\gamma \dot{\boldsymbol{B}}^{-1} \boldsymbol{L}] \\
&= [\boldsymbol{N}^T \boldsymbol{V}^{-1}][\boldsymbol{L}^T \boldsymbol{W} \boldsymbol{L}] = \boldsymbol{A}^I. \\
\boldsymbol{A}_\gamma^P &= [\boldsymbol{Q}_\gamma^T \boldsymbol{V}_\gamma^{-1}][\boldsymbol{L}_\gamma^T \boldsymbol{W}_\gamma \boldsymbol{L}_\gamma] \\
&= [\boldsymbol{Q}^T \boldsymbol{V}^{-1}][\boldsymbol{L}^T \boldsymbol{W} \boldsymbol{L}] + [\boldsymbol{Q}^T \boldsymbol{V}^{-1} \boldsymbol{D}][\boldsymbol{L}^T \dot{\boldsymbol{B}}_\gamma^{-T} \ddot{\boldsymbol{B}}_\gamma \dot{\boldsymbol{B}}_\gamma^{-1} \boldsymbol{L}] \\
&= \boldsymbol{A}^P + [\boldsymbol{R}][\boldsymbol{L}^T \dot{\boldsymbol{B}}_\gamma^{-T} \ddot{\boldsymbol{B}}_\gamma \dot{\boldsymbol{B}}_\gamma^{-1} \boldsymbol{L}] = \boldsymbol{A}^P + \boldsymbol{A}^G.
\end{aligned}$$

Further, since $\boldsymbol{I}_p = \partial\gamma/\partial\gamma^T = (\partial\gamma/\partial\boldsymbol{\beta}^T)(\partial\boldsymbol{\beta}/\partial\gamma^T)$, it follows from (A.8) that

$$\left(\frac{\partial\boldsymbol{\beta}}{\partial\gamma^T}\right)^T \left(\frac{\partial^2\gamma}{\partial\boldsymbol{\beta}\partial\boldsymbol{\beta}^T}\right)\left(\frac{\partial\boldsymbol{\beta}}{\partial\gamma^T}\right) + \left[\frac{\partial\gamma}{\partial\boldsymbol{\beta}^T}\right]\left[\frac{\partial^2\boldsymbol{\beta}}{\partial\gamma\partial\gamma^T}\right] = \boldsymbol{0},$$

which gives

$$\dot{\boldsymbol{B}}_\gamma^T \ddot{\gamma}_\beta \dot{\boldsymbol{B}}_\gamma + [\dot{\gamma}_\beta][\ddot{\boldsymbol{B}}_\gamma] = \boldsymbol{0},$$

$$\dot{\boldsymbol{B}}_\gamma^{-T} \ddot{\boldsymbol{B}}_\gamma \dot{\boldsymbol{B}}^{-1} = -[\dot{\gamma}_\gamma^{-1}][\ddot{\gamma}_\beta],$$

$$\boldsymbol{A}_\gamma^P = \boldsymbol{A}^P - [\boldsymbol{R}][\boldsymbol{L}^T[\dot{\gamma}_\beta^{-1}][\ddot{\gamma}_\beta]\boldsymbol{L}]. \qquad \blacksquare$$

Equation (3.19) may help us to find certain reparameterization under which $\boldsymbol{A}_\gamma^P = \boldsymbol{0}$ or as small as possible. In fact, it follows from (3.19) that if $\boldsymbol{A}_\gamma^P = \boldsymbol{0}$, then $\gamma = \gamma(\boldsymbol{\beta})$ must satisfy

$$\boldsymbol{A}^P = [\boldsymbol{Q}^T \boldsymbol{V}^{-1}][\boldsymbol{L}^T \boldsymbol{W} \boldsymbol{L}] = [\boldsymbol{R}][\boldsymbol{L}^T[\dot{\gamma}_\beta^{-1}][\ddot{\gamma}_\beta]\boldsymbol{L}],$$

which results in

$$[\dot{\gamma}_\beta^{-1}][\ddot{\gamma}_\beta] = [\boldsymbol{L}\boldsymbol{Q}^T \boldsymbol{V}^{-1}][\boldsymbol{W}] = \boldsymbol{T},$$

or

$$\ddot{\gamma}_\beta = [\dot{\gamma}_\beta][\boldsymbol{T}],$$

$$\boldsymbol{T} = [(\boldsymbol{D}^T \boldsymbol{V}^{-1} \boldsymbol{D})^{-1} \boldsymbol{D}^T \boldsymbol{V}^{-1}][\boldsymbol{W}]. \tag{3.20}$$

The above equations give a system of differential equations for obtaining transformations $\gamma = \gamma(\boldsymbol{\beta})$ which produce zero parameter-effects curvature. Obviously, it is not easy to solve these equations; see Bates and Watts (1981) for more discussions. Note that $\boldsymbol{T}$ in (3.20) is easily obtained because it just consists of the weighted least squares coefficients of the acceleration vectors (i.e. $\boldsymbol{W}$) regressed on the first order derivative of $\boldsymbol{\mu}(\boldsymbol{\beta})$ (i.e. $\boldsymbol{D}$) with the weight matrix $\boldsymbol{V}^{-1}$.

3.3 Geometry in Natural Parameter Space

The geometric framework introduced in Section 3.2 is based on the expectation parameter space. We can also introduce an associated dual geometry based on the natural parameter $\boldsymbol{\theta}$ and the corresponding Fisher information matrix $\phi \boldsymbol{V}(\boldsymbol{\theta})$. The duality in exponential family was first studied by Barndorff-Nielsen (1978). The dual geometry of exponential family was studied by Efron (1978), Amari (1982a) and discussed by Amari (1985) in detail. In this section, we introduce the dual geometry based on the modified BW geometric framework for exponential family nonlinear models. Since the original BW geometric framework has not considered the dual geometry (it is not necessary), our basic idea actually comes from Efron (1978) and Amari (1985).

For exponential family nonlinear models (2.2), by similar consideration as shown in Section 3.2, we start with the geometric interpretation of the maximum likelihood estimator of $\boldsymbol{\beta}$ in terms of the natural parameter $\boldsymbol{\theta}$. It follows from (2.10) that $\hat{\boldsymbol{\beta}}$ must satisfy

$$\boldsymbol{D}_\theta^T(\hat{\boldsymbol{\beta}})(\boldsymbol{Y} - \boldsymbol{\mu}(\hat{\boldsymbol{\beta}})) = \boldsymbol{0}.$$

Since the Fisher information matrix of $\boldsymbol{Y}$ for $\boldsymbol{\theta}$ is $\phi \boldsymbol{V}(\boldsymbol{\theta})$ (see (1.21)), this equation may be rewritten as

$$\boldsymbol{D}_\theta^T(\hat{\boldsymbol{\beta}})\boldsymbol{V}(\hat{\boldsymbol{\beta}})\boldsymbol{e}^*(\hat{\boldsymbol{\beta}}) = \boldsymbol{0}, \qquad \boldsymbol{e}^*(\boldsymbol{\beta}) = \boldsymbol{Y}^* - \boldsymbol{\theta}(\boldsymbol{\beta}). \tag{3.21}$$

Here we introduce a pseudo observed vector $\boldsymbol{Y}^*$ and a pseudo residual vector $\boldsymbol{e}^*(\boldsymbol{\beta})$ given by (Jorgensen 1983)

$$\begin{aligned} \boldsymbol{e}^*(\boldsymbol{\beta}) &= \boldsymbol{V}^{-1}(\boldsymbol{\beta})(\boldsymbol{Y} - \boldsymbol{\mu}(\boldsymbol{\beta})) = \boldsymbol{Y}^* - \boldsymbol{\theta}(\boldsymbol{\beta}), \\ \boldsymbol{Y}^* &= \boldsymbol{\theta}(\boldsymbol{\beta}) + \boldsymbol{V}^{-1}(\boldsymbol{\beta})(\boldsymbol{Y} - \boldsymbol{\mu}(\boldsymbol{\beta})). \end{aligned}$$

It is easily seen that $\mathrm{E}(\boldsymbol{Y}^*) = \boldsymbol{\theta}(\boldsymbol{\beta})$ and $\mathrm{Var}(\boldsymbol{Y}^*) = \sigma^2 \boldsymbol{V}^{-1}(\boldsymbol{\beta})$. After obtaining (3.21), which is quite similar to (3.1), the geometric interpretation of $\hat{\boldsymbol{\beta}}$ in terms of the natural parameter becomes quite similar to that in terms of the expectation parameter introduced in Section 3.2. Equation (3.21) shows that under the weighted inner product (with weight matrix $\boldsymbol{V}(\hat{\boldsymbol{\beta}})$, see (3.22)), the "residual vector" $\hat{e}^* = \boldsymbol{e}^*(\hat{\boldsymbol{\beta}})$ is orthogonal to all column vectors of $\boldsymbol{D}_\theta(\hat{\boldsymbol{\beta}}) = (\partial \boldsymbol{\theta}/\partial \boldsymbol{\beta}^T)_{\hat{\boldsymbol{\beta}}}$. This means that $\hat{e}^*$ is orthogonal to the tangent space of the surface $\boldsymbol{\theta} = \boldsymbol{\theta}(\boldsymbol{\beta})$ in $\Theta \subset R^n$ at $\hat{\boldsymbol{\beta}}$ (see Section 3.2 for details).

Similar to Section 3.2, we can introduce a geometric framework in natural parameter space Θ based on the above geometric interpretation. Now take $\boldsymbol{\theta}$ as a coordinate in natural parameter space $\Theta \subset R^n$. The solution locus π_θ is defined as $\boldsymbol{\theta} = \boldsymbol{\theta}(\boldsymbol{\beta})$ in $\Theta \subset R^n$, which is a p-dimensional surface in R^n. The tangent space $T(\boldsymbol{\beta})$ of π_θ at $\boldsymbol{\beta}$ is a linear subspace which is spanned by the tangent vectors of the coordinate curves of the solution locus π_θ (see Subsection 3.2.1). It is easy to show that $T(\boldsymbol{\beta})$ is just spanned by the columns of $\partial\boldsymbol{\theta}(\boldsymbol{\beta})/\partial\boldsymbol{\beta}^T = \boldsymbol{D}_\theta(\boldsymbol{\beta})$ at $\boldsymbol{\beta}$. For any two vectors $\boldsymbol{a}$ and $\boldsymbol{b}$ in R^n, we define the Fisher information inner product as

$$\langle \boldsymbol{a}, \boldsymbol{b} \rangle_\theta = \boldsymbol{a}^T \boldsymbol{V} \boldsymbol{b}. \tag{3.22}$$

Under this inner product, the associated normal space of the tangent space is denoted by $T'(\boldsymbol{\beta})$ which is an $(n-p)$-dimensional linear subspace of R^n orthogonal to $T(\boldsymbol{\beta})$. Now we can get an orthonormal basis for the tangent space $T(\boldsymbol{\beta})$ and the normal space $T'(\boldsymbol{\beta})$. Suppose that the QR decomposition of $\boldsymbol{D}_\theta(\boldsymbol{\beta})$ under the inner product (3.22) is (see Appendix A)

$$\boldsymbol{D}_\theta = (\boldsymbol{Q}_\theta, \boldsymbol{N}_\theta) \begin{pmatrix} \boldsymbol{R}_\theta \\ \boldsymbol{0} \end{pmatrix} = \boldsymbol{Q}_\theta \boldsymbol{R}_\theta,$$

where $\boldsymbol{R}_\theta$ is a $p \times p$ nonsingular upper triangular matrix with positive diagonal elements and the column vectors of $\boldsymbol{Q}_\theta$ and $\boldsymbol{N}_\theta$ are respectively orthonormal basis for the tangent space and the normal space of the solution locus $\boldsymbol{\theta} = \boldsymbol{\theta}(\boldsymbol{\beta})$ at $\boldsymbol{\beta}$ in R^n. The matrices $\boldsymbol{Q}_\theta$ and $\boldsymbol{N}_\theta$ satisfy

$$\boldsymbol{Q}_\theta^T \boldsymbol{V} \boldsymbol{Q}_\theta = \boldsymbol{I}_p, \qquad \boldsymbol{Q}_\theta^T \boldsymbol{V} \boldsymbol{N}_\theta = \boldsymbol{0}, \qquad \boldsymbol{N}_\theta^T \boldsymbol{V} \boldsymbol{N}_\theta = \boldsymbol{I}_{n-p}.$$

Then we can define the following curvature arrays similar to Definition 3.1.

Definition 3.3

For exponential family nonlinear models (2.2), the intrinsic curvature array $\boldsymbol{A}_\theta^I$ and the parameter-effects curvature array $\boldsymbol{A}_\theta^P$ in natural parameter space are respectively defined as

$$\boldsymbol{A}_\theta^I = [\boldsymbol{N}_\theta^T \boldsymbol{V}][\boldsymbol{U}_\theta], \qquad \boldsymbol{A}_\theta^P = [\boldsymbol{Q}_\theta^T \boldsymbol{V}][\boldsymbol{U}_\theta]; \tag{3.23}$$

$$\boldsymbol{U}_\theta = \boldsymbol{L}_\theta^T \boldsymbol{W}_\theta \boldsymbol{L}_\theta, \quad \boldsymbol{L}_\theta = \boldsymbol{R}_\theta^{-1}. \qquad \blacksquare$$

It is easily seen that this definition is quite similar to Definition 3.1 except that here we use the solution locus $\boldsymbol{\theta} = \boldsymbol{\theta}(\boldsymbol{\beta})$ to replace its duality $\boldsymbol{\mu} = \boldsymbol{\mu}(\boldsymbol{\beta})$ and we use the inner product (3.22) to replace its duality (3.3). In the following, we give an important relationship between these two kinds of curvatures.

Theorem 3.4

The curvature arrays defined in (3.5) and (3.23) satisfy

$$A_\theta^I = A^I - \Gamma^I, \qquad \Gamma^I = [N^T V^{-1}][L^T \Gamma L]; \tag{3.24}$$
$$A_\theta^P = A^P - \Gamma^P, \qquad \Gamma^P = [Q^T V^{-1}][L^T \Gamma L]; \tag{3.25}$$

where $\Gamma = D^T V^{-1} S V^{-1} D$ and $S = b^{(3)}(\theta)$ are given in (2.7) and (1.12), respectively.

Proof. It follows from (2.6) and (3.4) that

$$D_\theta = V^{-1} D = V^{-1} Q R = Q_\theta R_\theta;$$

here $Q_\theta = V^{-1} Q$ satisfying $Q_\theta^T V Q_\theta = Q^T V^{-1} Q = I_p$ and $R_\theta = R$. Thus this equation is just the QR decomposition of D_θ. Further, it is also easy to show that $N_\theta = V^{-1} N$ satisfying $N_\theta^T V N_\theta = N^T V^{-1} N = I_{n-p}$, $N_\theta^T V Q_\theta = N^T V^{-1} Q = 0$. So we have

$$Q_\theta = V^{-1} Q, \qquad N_\theta = V^{-1} N.$$

Hence it follows from (2.6) and (2.7) that

$$A_\theta^I = [(V^{-1}N)^T V][L_\theta^T [V^{-1}][W - \Gamma] L_\theta].$$

It follows from Appendix A.2 and $L_\theta = L$ that

$$\begin{aligned} A_\theta^I &= [N^T V^{-1}][L^T (W - \Gamma) L] \\ &= A^I - \Gamma^I. \end{aligned}$$

Then (3.24) follows. Similarly, we can get (3.25). ∎

By the same manner as described in Subsection 3.2.2, we can also define the directional curvatures and the maximum curvatures in natural parameter space. All the definitions and the formulas shown there still hold in natural parameter space. The necessary change is that all the quantities should be replaced by the corresponding dualities. As an example, we list the directional normal curvature in natural parameter space as follows. The dualities corresponding to (3.9) and (3.11) are

$$(K_h^N)_\theta = \frac{\|\ddot{\theta}_h^N\|_\theta}{\|\dot{\theta}_h\|_\theta^2} = \frac{\|(P_N)_\theta (h^T W_\theta h)\|_\theta}{h^T D_\theta^T V D_\theta h}. \tag{3.26}$$

Here the norm of a vector α is defined as $\|\alpha\|_\theta^2 = \alpha^T V \alpha$, $(P_N)_\theta = N_\theta N_\theta^T V$, and D_θ, W_θ are the first two derivatives of $\theta(\beta)$.

The duality between the geometry of natural parameter space and the geometry of expectation parameter space is summarized in Table 3.1.

We have introduced a geometric framework for exponential family nonlinear models by introducing the Fisher information inner products in expectation parameter space and its duality, in natural parameter space. The

Table 3.1. Duality between expectation parameter space and natural parameter space

	Expectation parameters space	Natural parameters space
Parameter	$\mu = \dot{b}(\theta), \dfrac{\partial \mu}{\partial \theta^T} = V$	$\theta = \dot{b}^{-1}(\mu), \dfrac{\partial \theta}{\partial \mu^T} = V^{-1}$
Observed vector	$\mathrm{E}(Y) = \mu(\beta)$ $\mathrm{Var}(Y) = \sigma^2 V(\beta)$	$\mathrm{E}(Y^*) = \theta(\beta)$ $\mathrm{Var}(Y^*) = \sigma^2 V^{-1}(\beta)$
Fisher information	$\sigma^{-2} V^{-1}(\mu)$	$\sigma^{-2} V(\theta)$
Likelihood equation	$D^T(\hat{\beta}) V^{-1}(\hat{\beta}) e(\hat{\beta}) = 0$	$D_\theta^T(\hat{\beta}) V(\hat{\beta}) e^*(\hat{\beta}) = 0$
Inner product	$\langle a, b \rangle = a^T V^{-1} b$	$\langle a, b \rangle_\theta = a^T V b$
Solution locus	$\mu = \mu(\beta)$	$\theta = \theta(\beta)$
Derivative	$\dfrac{\partial \mu}{\partial \beta^T} = D = QR$	$\dfrac{\partial \theta}{\partial \beta^T} = D_\theta = Q_\theta R_\theta$
Curvature array	$A^I = [N^T V^{-1}][U]$	$A_\theta^I = [N_\theta^T V][U_\theta]$
Directional curvature	$K_h^N = \dfrac{\Vert\ddot{\mu}_h^N\Vert}{\Vert\dot{\mu}_h\Vert^2}$	$(K_h^N)_\theta = \dfrac{\Vert\ddot{\theta}_h^N\Vert_\theta}{\Vert\dot{\theta}_h\Vert_\theta^2}$

main difference between our modified BW geometric framework and the BW geometric framework is the introduction of the Fisher information inner product. As pointed out by Efron (1978), this inner product plays a fundamental role for extending the Euclidean geometry from normal family to exponential families. The dual geometry that was not considered in the BW framework is an inevitable result of the Fisher information inner product. Our framework based on the Fisher information matrix is applicable to models much more general than nonlinear regression models (see Chapter 7 for more details).

From a mathematical viewpoint, both BW and MBW geometric framework may be regarded as a special case of the EA geometric framework (Amari 1985). In fact, the exponential family (1.1) is "m-flat" in μ-coordinate (with known ϕ) under $\alpha = -1$ connection by Amari's (1985) terminology and the curvature arrays A^I and A^P defined in (3.5) for exponential family nonlinear models (2.2) are just the embedded normal curvature and the embedded connection respectively under the orthonormal basis in EA framework with $\alpha = -1$. Similarly, family (1.1) is "e-flat" in θ-coordinate (with known ϕ) under $\alpha = 1$ connection by Amari's (1985) terminology and the curvature arrays defined in (3.23) for model (2.2) are just the embedded normal curvature and the embedded connection respectively under the orthonormal basis in the EA framework with $\alpha = 1$. This means that the connections of $\alpha = \pm 1$ have been actually used (implicitly) in the MBW framework. The main difference between the MBW framework and the EA framework is that we have not introduced the general α connections, which are a family of affine connections. Of course, this must restrict the use of our MBW framework. However, this also makes the discussions simplified. In particular, we can use the matrix notation in Euclidean space rather than use the tensor notation in Riemannian manifold. The other reason we have not introduced the general connections is that if we restrict our interests in some specific problems, such as the statistical inference related to the maximum likelihood estimators or quasi-likelihood estimators and the like, the connections of $\alpha = \pm 1$ might be enough for the general use. In fact, in the literature relating to EA framework, the frequent usages of connections are also $\alpha = \pm 1$ connections and the others are rarely used.

The comparisons of MBW, BW and EA geometries are summarized in Table 3.2. This table shows that the MBW framework is the same as the BW framework in the first five items of the table. The embedded curvature array of the MBW framework is also quite similar to that of the BW framework. But for the last three items of the table, the MBW framework has something similar to the EA framework and is quite dissimilar to the BW framework.

Table 3.2. Comparison of MBW, BW and EA geometries

	BW	MBW	EA
Enveloping space	Euclidean	Euclidean	Riemannian
Notation	matrix, array	matrix, array	tensor
Basis of tangent space	orthonormal	orthonormal	general
Directional curvatures	defined	defined	undefined
Maximum curvatures	defined	defined	undefined
Embedded curvatures	$\boldsymbol{A}^I$	$\boldsymbol{A}^I$ and its duality $\boldsymbol{A}_\theta^I$	a family of α-curvatures
Inner product	ordinary	Fisher information inner product	Fisher information inner product
Connections	undefined	$\alpha = \pm 1$ connection	a family of α connections
Duality	undefined	$\alpha = -1$ connection versus $\alpha = 1$ connection	α connection versus $-\alpha$ connection

3.4 Some Specific Models

3.4.1 Normal nonlinear regression models

Now we consider the following nonlinear model:

$$y_i = f(\boldsymbol{x}_i;\boldsymbol{\beta}) + \varepsilon_i, \qquad \varepsilon_i \sim N(0,\sigma^2), \tag{3.27}$$

$i = 1,\cdots,n$. The probability density function of y_i can be represented as

$$p(y_i;\boldsymbol{\beta},\sigma) = \exp\{\frac{1}{\sigma^2}(y_i\mu_i - \frac{1}{2}\mu_i^2 - \frac{1}{2}y_i^2) - \frac{1}{2}\log(2\pi\sigma^2)\},$$

where $\mu_i = \mathrm{E}(y_i) = f(\boldsymbol{x}_i;\boldsymbol{\beta})$ and $g(\mu_i) = \mu_i$. It is easily seen that the natural parameter θ_i is equal to μ_i and $b(\theta_i) = b(\mu_i) = (1/2)\mu_i^2$. Therefore for normal nonlinear regression models, the natural parameter space coincides with the expectation parameter space. Because the Fisher information matrices of $\boldsymbol{Y} = (y_1,\cdots,y_n)^T$ for both $\boldsymbol{\theta}$ and $\boldsymbol{\mu}$ are $\sigma^{-2}\ddot{\boldsymbol{b}}(\boldsymbol{\theta}) = \sigma^{-2}\boldsymbol{I}_n$, the Fisher information inner products in both expectation parameter space and natural parameter space are just the ordinary inner product in Euclidean space, that is $\langle \boldsymbol{a},\boldsymbol{b}\rangle = \langle \boldsymbol{a},\boldsymbol{b}\rangle_\theta = \boldsymbol{a}^T\boldsymbol{b}$. This means that the duality of expectation parameter space is just itself, i.e. it is self-dual. In Table 3.1, two sides of the table are identical for model (3.27). The curvatures defined in Sections 3.2 and 3.3 are coincident and are also the same as those defined by Bates and Watts (1980). It is easily seen that normal model (3.27) is only a self-dual model. In fact, if the distribution of y_i is (1.3) and satisfies $\mu_i = \dot{b}(\theta_i) = \theta_i$ $(i = 1,\cdots,n)$, then $b(\theta_i) = (1/2)\theta_i^2 + c_i$ $(i = 1,\cdots,n)$, c_i's are constants. It follows from (1.7) that the moment generating function of y_i is $\exp\{\tau\theta_i + (1/2)\tau^2/\phi\}$, which results in a normal distribution $N(\theta_i,\phi^{-1})$, $\phi = \sigma^{-2}$.

3.4.2 Generalized linear models

In model (2.1), if $\eta_i = g(\mu_i) = \boldsymbol{x}_i^T\boldsymbol{\beta}$, then it is the well-known generalized linear model (McCullagh and Nelder 1989), which may be written as

$$\boldsymbol{\eta} = \boldsymbol{g}(\boldsymbol{\mu}) = \boldsymbol{X}\boldsymbol{\beta}, \tag{3.28}$$

where $\boldsymbol{X}$ is an $n\times p$ nonsingular design matrix. To compute the curvature arrays defined in (3.5) and (3.23), it is enough to find the first two derivatives $\boldsymbol{D}(\boldsymbol{\beta})$ and $\boldsymbol{W}(\boldsymbol{\beta})$ of $\boldsymbol{\mu} = \boldsymbol{\mu}(\boldsymbol{\beta})$ (note that the directional curvatures and the maximum curvatures can be obtained from curvature arrays, see Theorem 3.1 and Lemma 3.1). Now let $\dot{\boldsymbol{G}} = \mathrm{diag}\,(\dot{g}(\mu_1),\cdots,\dot{g}(\mu_n))$ and $\ddot{\boldsymbol{G}}$ denote an $n\times n\times n$ array with the elements $\ddot{G}_{iii} = \ddot{g}(\mu_i)$ $(i = 1,\cdots,n)$ and zeros elsewhere. By taking the first two derivatives with respect to $\boldsymbol{\beta}$ in (3.28) and using (A.8), we get

$$\frac{\partial \boldsymbol{g}(\boldsymbol{\mu})}{\partial \boldsymbol{\mu}^T}\frac{\partial \boldsymbol{\mu}}{\partial \boldsymbol{\beta}^T} = \boldsymbol{X},$$

and

$$\left(\frac{\partial \boldsymbol{\mu}}{\partial \boldsymbol{\beta}^T}\right)^T\left(\frac{\partial^2 \boldsymbol{g}(\boldsymbol{\mu})}{\partial \boldsymbol{\mu}\partial \boldsymbol{\mu}^T}\right)\left(\frac{\partial \boldsymbol{\mu}}{\partial \boldsymbol{\beta}^T}\right) + \left[\frac{\partial \boldsymbol{g}(\boldsymbol{\mu})}{\partial \boldsymbol{\mu}^T}\right]\left[\frac{\partial^2 \boldsymbol{\mu}}{\partial \boldsymbol{\beta}\partial \boldsymbol{\beta}^T}\right] = 0.$$

From these equations we have

$$\boldsymbol{D} = \dot{\boldsymbol{G}}^{-1}\boldsymbol{X}, \qquad \boldsymbol{W} = -[\dot{\boldsymbol{G}}^{-1}][\boldsymbol{X}^T\dot{\boldsymbol{G}}^{-1}\ddot{\boldsymbol{G}}\dot{\boldsymbol{G}}^{-1}\boldsymbol{X}]. \tag{3.29}$$

These equations show that when $\boldsymbol{X}$ is given, $\boldsymbol{D}$ and $\boldsymbol{W}$ are determined by the link function $\boldsymbol{g}(\boldsymbol{\mu})$. This is quite reasonable. If the link function $\boldsymbol{g}(\boldsymbol{\mu})$ is given, then we can find the QR decomposition of $\boldsymbol{D} = \dot{\boldsymbol{G}}^{-1}\boldsymbol{X}$, and curvature arrays $\boldsymbol{A}^I, \boldsymbol{A}^P$ and $\boldsymbol{A}^I_\theta, \boldsymbol{A}^P_\theta$ can be obtained from equations (3.5), (3.24) and (3.25), respectively.

Now let us consider a common situation in which $\boldsymbol{g}(\boldsymbol{\mu})$ is a canonical link satisfying

$$\boldsymbol{\theta} = \boldsymbol{g}(\boldsymbol{\mu}) = \boldsymbol{X}\boldsymbol{\beta}.$$

Then we have $\dot{\boldsymbol{G}} = \partial\boldsymbol{\theta}/\partial\boldsymbol{\mu}^T, \ddot{\boldsymbol{G}} = \partial^2\boldsymbol{\theta}/\partial\boldsymbol{\mu}\partial\boldsymbol{\mu}^T$. It follows from (1.13) that

$$\dot{\boldsymbol{G}} = \boldsymbol{V}^{-1}, \qquad \ddot{\boldsymbol{G}} = -[\boldsymbol{V}^{-1}][\boldsymbol{V}^{-1}\boldsymbol{S}\boldsymbol{V}^{-1}],$$

where $\boldsymbol{S} = \boldsymbol{b}^{(3)}(\boldsymbol{\theta})$ is given in (1.12). Therefore $\boldsymbol{D}$ and $\boldsymbol{W}$ in (3.29) can be reduced as

$$\boldsymbol{D} = \boldsymbol{V}\boldsymbol{X}, \qquad \boldsymbol{W} = \boldsymbol{\Gamma} = \boldsymbol{X}^T\boldsymbol{S}\boldsymbol{X}, \tag{3.30}$$

where $\boldsymbol{\Gamma}$ is given in (2.7). From these formulas, we can easily obtain the curvature array. In particular, (3.30) results in the following reasonable conclusion:

For generalized linear models with canonical link, both the intrinsic curvature array $\boldsymbol{A}^I_\theta$ and the parameter-effects curvature array $\boldsymbol{A}^P_\theta$ vanish.

This result is easily obtained from (3.30) and Theorem 3.3. But this conclusion may not be true for non-canonical link, since $\boldsymbol{\theta}$ may not be a linear function of $\boldsymbol{\beta}$. Furthermore, even though for a canonical link, it is not necessary to have zero curvature in expectation parameter space because $\boldsymbol{\mu} = \boldsymbol{g}^{-1}(\boldsymbol{X}\boldsymbol{\beta})$ is not necessarily linear in $\boldsymbol{\beta}$. We may compute $\boldsymbol{A}^I$ and $\boldsymbol{A}^P$ based on (3.30) as follows. Suppose that the QR decomposition of $\boldsymbol{D}_\theta = \boldsymbol{X}$ under the inner product (3.22) is given by

$$\boldsymbol{X} = (\boldsymbol{Q}_\theta, \boldsymbol{N}_\theta)\begin{pmatrix}\boldsymbol{R}_\theta \\ 0\end{pmatrix} = \boldsymbol{Q}_\theta\boldsymbol{R}_\theta,$$

where $\boldsymbol{Q}_\theta$ and $\boldsymbol{N}_\theta$ satisfy $\boldsymbol{Q}_\theta^T\boldsymbol{V}\boldsymbol{Q}_\theta = \boldsymbol{I}_p$, $\boldsymbol{Q}_\theta^T\boldsymbol{V}\boldsymbol{N}_\theta = \boldsymbol{0}$ and $\boldsymbol{N}_\theta^T\boldsymbol{V}\boldsymbol{N}_\theta = \boldsymbol{I}_{n-p}$, respectively. Then the QR decomposition of $\boldsymbol{D}$ under the inner product (3.3) can be obtained from (3.30)

$$\boldsymbol{D} = \boldsymbol{V}\boldsymbol{X} = \boldsymbol{V}\boldsymbol{Q}_\theta\boldsymbol{R}_\theta = \boldsymbol{Q}\boldsymbol{R},$$

where $\boldsymbol{Q} = \boldsymbol{V}\boldsymbol{Q}_\theta, \boldsymbol{R} = \boldsymbol{R}_\theta$ and $\boldsymbol{N} = \boldsymbol{V}\boldsymbol{N}_\theta$ because $\boldsymbol{Q}^T\boldsymbol{V}^{-1}\boldsymbol{Q} = \boldsymbol{I}_p$, $\boldsymbol{Q}^T\boldsymbol{V}^{-1}\boldsymbol{N} = \boldsymbol{0}$, and $\boldsymbol{N}^T\boldsymbol{V}^{-1}\boldsymbol{N} = \boldsymbol{I}_{n-p}$ obviously hold. Hence it follows from (3.5) and (3.30) that

$$\boldsymbol{A}^I = [\boldsymbol{N}_\theta^T][\boldsymbol{Q}_\theta^T\boldsymbol{S}\boldsymbol{Q}_\theta], \quad \boldsymbol{A}^P = [\boldsymbol{Q}_\theta^T][\boldsymbol{Q}_\theta^T\boldsymbol{S}\boldsymbol{Q}_\theta]. \tag{3.31}$$

This result shows that the curvature array in expectation parameter space for generalized linear models with canonical link are determined by the design matrix $\boldsymbol{X}$ and the skewness of $\boldsymbol{Y}$ given in (1.12). If $E(e_i^3) = S_{iii} = 0$ $(i = 1, \cdots, n)$, then $\boldsymbol{A}^I = \boldsymbol{0}$ and $\boldsymbol{A}^P = \boldsymbol{0}$. For logistic regression, it can be obtained that $S_{iii} = n_i\mu_i(1-\mu_i)(1-2\mu_i)$, so it is not necessary to have $\boldsymbol{A}^I = \boldsymbol{0}$ or $\boldsymbol{A}^P = \boldsymbol{0}$. Equation (3.31) also shows that when $\boldsymbol{S}$ is known, the curvature arrays are determined by the QR decomposition of the design matrix $\boldsymbol{X}$.

3.4.3 One-parameter model and Efron curvature

One-parameter model is a commonly encountered situation where parameter $\boldsymbol{\beta}$ is a scalar in (2.2), that is $p = 1$. Efron (1975; 1978) studied the one-parameter curved exponential families in detail (the statistical inference of his study is based on the iid observations). For our model (2.2), the curvatures defined in Sections 3.2 and 3.3 can be considerably simplified when $\boldsymbol{\beta}$ is a scalar.

We first consider the curvature arrays. If β is a scalar in (2.2), then all the derivatives of $\boldsymbol{\mu}(\beta)$ are n-vectors. In particular, $\boldsymbol{D}(\beta) = \dot{\boldsymbol{\mu}}_\beta = d\boldsymbol{\mu}/d\beta$, $\boldsymbol{W}(\beta) = \ddot{\boldsymbol{\mu}}_\beta = d^2\boldsymbol{\mu}/d\beta^2$. The QR decomposition of $\boldsymbol{D} = \dot{\boldsymbol{\mu}}_\beta$ is simply

$$\dot{\boldsymbol{\mu}}_\beta = \frac{\dot{\boldsymbol{\mu}}_\beta}{\|\dot{\boldsymbol{\mu}}_\beta\|}\|\dot{\boldsymbol{\mu}}_\beta\| = \boldsymbol{Q}R,$$

where $\boldsymbol{Q} = \dot{\boldsymbol{\mu}}_\beta/\|\dot{\boldsymbol{\mu}}_\beta\|$ is a unit vector; $R = \|\dot{\boldsymbol{\mu}}_\beta\|$ is a scalar and $L = R^{-1} = \|\dot{\boldsymbol{\mu}}_\beta\|^{-1}$. Hence the normal space has the dimension of $n-1$ and $\boldsymbol{N}$ is an $n \times (n-1)$ matrix. Then the curvature arrays defined in (3.5) become

$$\boldsymbol{A}^I = \frac{\boldsymbol{N}^T\boldsymbol{V}^{-1}\ddot{\boldsymbol{\mu}}_\beta}{\|\dot{\boldsymbol{\mu}}_\beta\|^2}, \qquad A^P = \frac{\dot{\boldsymbol{\mu}}_\beta^T\boldsymbol{V}^{-1}\ddot{\boldsymbol{\mu}}_\beta}{\|\dot{\boldsymbol{\mu}}_\beta\|^3}, \tag{3.32}$$

where $\boldsymbol{A}^I$ is an $(n-1)$-vector and A^P is a scalar.

Expression (3.32) is quite similar to those of directional curvatures. In fact, if $\boldsymbol{\beta}$ is a scalar in (3.9) to (3.11), then the direction vector $\boldsymbol{h}$ is also a scalar. Therefore K_h^N and K_h^P do not depend on the "direction" h and the directional curvatures are the same as the maximum curvatures. It is easily seen that both (3.10) and (3.11) are simply

$$K_h^N = K^N = \frac{||\boldsymbol{N}\boldsymbol{N}^T\boldsymbol{V}^{-1}\ddot{\boldsymbol{\mu}}_\beta||}{||\dot{\boldsymbol{\mu}}_\beta||^2} = ||\boldsymbol{N}\boldsymbol{A}^I||, \tag{3.33}$$

and

$$K_h^P = K^P = \frac{\dot{\boldsymbol{\mu}}_\beta^T\boldsymbol{V}^{-1}\ddot{\boldsymbol{\mu}}_\beta}{||\dot{\boldsymbol{\mu}}_\beta||^3} = ||\boldsymbol{Q}A^P||.$$

The geometric meanings of (3.32) and (3.33) are almost the same. Here K^N represents the norm of the projection of the "normed acceleration" onto the normal space at β while $\boldsymbol{A}^I$ is the coordinate of this projection under the orthonormal basis formed by the columns of $\boldsymbol{N}$. Both K^N and $\boldsymbol{A}^I$ indicate the nonlinearity of the model.

Similarly, we can get the dualities of the above curvatures in natural parameter space. In particular, the dualities of (3.32) and (3.33) are namely

$$\boldsymbol{A}_\theta^I = \frac{\boldsymbol{N}_\theta^T\boldsymbol{V}\ddot{\boldsymbol{\theta}}_\beta}{||\dot{\boldsymbol{\theta}}_\beta||_\theta^2}, \qquad \boldsymbol{A}_\theta^P = \frac{\dot{\boldsymbol{\theta}}_\beta^T\boldsymbol{V}\ddot{\boldsymbol{\theta}}_\beta}{||\dot{\boldsymbol{\theta}}_\beta||_\theta^3};$$

and

$$K_\theta^N = \frac{||\boldsymbol{N}_\theta\boldsymbol{N}_\theta^T\boldsymbol{V}\ddot{\boldsymbol{\theta}}_\beta||_\theta}{||\dot{\boldsymbol{\theta}}_\beta||_\theta^2} = ||\boldsymbol{N}_\theta\boldsymbol{A}_\theta^I||_\theta, \tag{3.34}$$

where both $\dot{\boldsymbol{\theta}}_\beta$ and $\ddot{\boldsymbol{\theta}}_\beta$ are derivative vectors and the inner product is defined in (3.22).

It is interesting to note that this normal curvature is exactly consistent with the well-known Efron's (1975) curvature γ_β, detailed as follows. The general definition of Efron curvature for one-parameter (say, β) families is defined as

$$\begin{aligned}\gamma_\beta &= \frac{\det\{\mathrm{Cov}(\dot{l}_\beta, \ddot{l}_\beta)\}}{\{\mathrm{E}(\dot{l}_\beta^2)\}^3} \\ &= \left[\frac{\nu_{02}}{\nu_{20}^2} - \frac{\nu_{11}^2}{\nu_{20}^3}\right]^{\frac{1}{2}},\end{aligned} \tag{3.35}$$

where $\nu_{20} = \mathrm{Var}(\dot{l}_\beta)$, $\nu_{11} = \mathrm{Cov}(\dot{l}_\beta, \ddot{l}_\beta)$ and $\nu_{02} = \mathrm{Var}(\ddot{l}_\beta)$ (Efron 1975, p.1196). Then we have the following theorem.

Theorem 3.5

In natural parameter space, if β is a scalar, then the intrinsic curvature K_θ^N defined in (3.34) is namely the Efron curvature (except a constant) and satisfies

$$\left(K_\theta^N\right)^2 = \phi\gamma_\beta^2 = \frac{(\ddot{\boldsymbol{\theta}}_\beta)_N^T(\ddot{\boldsymbol{\theta}}_\beta)_N}{(\dot{\boldsymbol{\theta}}_\beta^T\boldsymbol{V}\dot{\boldsymbol{\theta}}_\beta)^2}, \tag{3.36}$$

where $(\ddot{\boldsymbol{\theta}}_\beta)_N = \boldsymbol{N}_\theta^T\boldsymbol{V}\ddot{\boldsymbol{\theta}}_\beta$.

Proof. For the one-parameter model, it follows from Lemma 2.3 that

$$\dot{l}_\beta = \phi\dot{\boldsymbol{\theta}}_\beta^T\boldsymbol{e}, \qquad \ddot{l}_\beta = \phi(\ddot{\boldsymbol{\theta}}_\beta^T\boldsymbol{e} - \dot{\boldsymbol{\theta}}_\beta^T\boldsymbol{V}\dot{\boldsymbol{\theta}}_\beta),$$

where $\boldsymbol{e} = \boldsymbol{Y} - \boldsymbol{\mu}(\beta)$. Then it follows from (1.8) that

$$\nu_{20} = \phi\dot{\boldsymbol{\theta}}_\beta^T\boldsymbol{V}\dot{\boldsymbol{\theta}}_\beta = \phi\|\dot{\boldsymbol{\theta}}_\beta\|_\theta^2, \qquad \nu_{02} = \phi\ddot{\boldsymbol{\theta}}_\beta^T\boldsymbol{V}\ddot{\boldsymbol{\theta}}_\beta,$$

$$\nu_{11}^2 = \phi^2(\dot{\boldsymbol{\theta}}_\beta^T\boldsymbol{V}\ddot{\boldsymbol{\theta}}_\beta)^2 = \phi^2(\ddot{\boldsymbol{\theta}}_\beta^T\boldsymbol{V}\dot{\boldsymbol{\theta}}_\beta\dot{\boldsymbol{\theta}}_\beta^T\boldsymbol{V}\ddot{\boldsymbol{\theta}}_\beta).$$

Substitutions of these results into (3.35) give

$$\begin{aligned}\gamma_\beta^2 &= \frac{\ddot{\boldsymbol{\theta}}_\beta^T\boldsymbol{V}\ddot{\boldsymbol{\theta}}_\beta}{\phi(\dot{\boldsymbol{\theta}}_\beta^T\boldsymbol{V}\dot{\boldsymbol{\theta}}_\beta)^2} - \frac{\ddot{\boldsymbol{\theta}}_\beta^T\boldsymbol{V}\dot{\boldsymbol{\theta}}_\beta\dot{\boldsymbol{\theta}}_\beta^T\boldsymbol{V}\ddot{\boldsymbol{\theta}}_\beta}{\phi(\dot{\boldsymbol{\theta}}_\beta^T\boldsymbol{V}\dot{\boldsymbol{\theta}}_\beta)^3} \\ &= \frac{\ddot{\boldsymbol{\theta}}_\beta^T\boldsymbol{V}\ddot{\boldsymbol{\theta}}_\beta - \ddot{\boldsymbol{\theta}}_\beta^T\boldsymbol{V}[\dot{\boldsymbol{\theta}}_\beta(\dot{\boldsymbol{\theta}}_\beta^T\boldsymbol{V}\dot{\boldsymbol{\theta}}_\beta)^{-1}\dot{\boldsymbol{\theta}}_\beta^T\boldsymbol{V}]\ddot{\boldsymbol{\theta}}_\beta}{\phi\|\dot{\boldsymbol{\theta}}_\beta\|_\theta^4}.\end{aligned}$$

Note that $\dot{\boldsymbol{\theta}}_\beta(\dot{\boldsymbol{\theta}}_\beta^T\boldsymbol{V}\dot{\boldsymbol{\theta}}_\beta)^{-1}\dot{\boldsymbol{\theta}}_\beta\boldsymbol{V} = (\boldsymbol{P}_T)_\theta = \boldsymbol{I}_n - (\boldsymbol{P}_N)_\theta = \boldsymbol{I}_n - \boldsymbol{N}_\theta\boldsymbol{N}_\theta^T\boldsymbol{V}$, hence from (3.34) we have

$$\begin{aligned}\phi\gamma_\beta^2 &= \frac{\ddot{\boldsymbol{\theta}}_\beta^T\boldsymbol{V}\{\boldsymbol{I}_n - (\boldsymbol{P}_T)_\theta\}\ddot{\boldsymbol{\theta}}_\beta}{\|\dot{\boldsymbol{\theta}}_\beta\|_\theta^4} \\ &= \frac{\ddot{\boldsymbol{\theta}}_\beta^T\boldsymbol{V}\boldsymbol{N}_\theta\boldsymbol{N}_\theta^T\boldsymbol{V}\ddot{\boldsymbol{\theta}}_\beta}{\|\dot{\boldsymbol{\theta}}_\beta\|_\theta^4} \\ &= (K_\theta^N)^2.\end{aligned}$$

∎

It should be stressed that here the Efron curvature given in (3.36) is obtained from the likelihood function of the whole observations $\boldsymbol{Y} = (y_1, \cdots, y_n)^T$, but not from one of the components of $\boldsymbol{Y}$. In general, if

$y_1, \cdots, y_n$ were iid observations, then the Efron curvature $\tilde{\gamma}_\beta$ corresponding to y_1 would be

$$\tilde{\gamma}_\beta = \sqrt{n}\gamma_\beta.$$

This can be verified from (3.35) directly (see also Efron 1975, p.1198, (6.2)). However, $\tilde{\gamma}_\beta = \sqrt{n}\gamma_\beta$ is also quite meaningful in exponential family nonlinear model, even though that is the non-iid case. In this case, $\tilde{\gamma}_\beta$ may indicate the "average" Efron curvature of the components of $\boldsymbol{Y} = (y_1, \cdots, y_n)^T$. In fact it follows from (3.36) that $\tilde{\gamma}_\beta^2$ can be represented as

$$\tilde{\gamma}_\beta^2 = \frac{n^{-1}(\ddot{\boldsymbol{\theta}}_\beta)_N^T(\ddot{\boldsymbol{\theta}}_\beta)_N}{\phi(n^{-1}\dot{\boldsymbol{\theta}}_\beta^T \boldsymbol{V}\dot{\boldsymbol{\theta}}_\beta)^2}. \tag{3.37}$$

By the Assumption C given in Section 4.1 in the next chapter, this expression is of the order of $O(1)$ as $n \to \infty$, as expected. So $\tilde{\gamma}_\beta$ is definable. The limit of $\tilde{\gamma}_\beta^2$ as $n \to \infty$ will be denoted by $\bar{\gamma}_\beta^2$. In Chapter 4, We will use this curvature to illustrate some asymptotic results of maximum likelihood estimator.

Chapter 4

Some Second Order Asymptotics

It was Efron (1975) who first studied the second order asymptotic theory of estimation by using differential geometric methods. His pioneer work caused the great progress of differential geometric statistics in the 1980s. Efron (1975) presented a differential geometric framework for one-parameter curved exponential families and applied this framework to study the squared error loss and the information loss of the estimate in these families. He first pointed out the important role of the statistical curvature in the asymptotic theory of estimation. In his succeeding papers (Efron 1978; Efron and Hinkley 1978), he studied the relationship among the observed information, the expected information and the statistical curvature for the same families. They proved that the relative difference between the observed information and the expected information times the square root of the sample size is asymptotically normal and the asymptotic variance is just the square of Efron's curvature.

It was Amari (1982a,b; 1985) who first introduced the Riemannian geometric framework to multi-parameter curved exponential families and applied this framework to study several asymptotic properties for these families. He obtained good results similar to those of Efron.

This chapter discusses analogous asymptotic problems mentioned above in our exponential family nonlinear models by using the MBW geometric framework. Section 4.1 derives stochastic expansions related to the maximum likelihood estimator as a basic tool to investigate the asymptotics. Sections 4.2 and 4.3 study the variance and the information loss of the maximum likelihood estimator. The relationship between the observed information and the expected information is investigated in Section 4.4.

It should be emphasized that the asymptotics studied in this chapter have some essential differences from those studied by Efron (1975) and Amari (1985). In fact, we study the asymptotics for quite general nonlinear regression models, in which the observations are independent but not necessarily identically distributed, while Efron and Amari studied curved exponential families with iid observations. On the other hand, the geometric structure corresponding to the regression model is defined for the whole observations $\boldsymbol{Y} = (y_1, \cdots, y_n)$, which varies as the sample size n increases. While for the iid case studied by Efron and Amari, the geometric structure is actually based on one of iid observations (Amari 1985), which is independent of the sample size. Therefore, the asymptotics in iid case are much simpler than regression case, in which more severe regularity conditions are needed (see Assumption C in Section 4.1).

Finally, we notice a fact which will be used very often in later discussions. Equation (2.10) shows that the likelihood equation of $\boldsymbol{\beta}$ does not involve the dispersion parameter $\phi = \sigma^{-2}$, so we may assume σ^2 is known when we investigate the statistical behavior of $\hat{\boldsymbol{\beta}}$. In this chapter, we shall always admit this assumption.

4.1 Stochastic Expansions

In exponential family nonlinear models, the maximum likelihood estimator $\hat{\boldsymbol{\beta}}(\boldsymbol{Y})$ of $\boldsymbol{\beta}$ may be approximately expressed as a function of the "random error" $\boldsymbol{e} = \boldsymbol{Y} - \boldsymbol{\mu}(\boldsymbol{\beta})$ up to some order, where $\boldsymbol{\beta}$ is the true value of the parameter to be estimated. This expression is called the stochastic expansion of $\hat{\boldsymbol{\beta}}$ (Amari 1982a,b; 1985). Similar terminology can be used for other statistics, such as $\boldsymbol{\mu}(\hat{\boldsymbol{\beta}})$, $\hat{e} = \boldsymbol{Y} - \boldsymbol{\mu}(\hat{\boldsymbol{\beta}})$ and so on. The stochastic expansion provides a powerful tool to investigate the asymptotic properties related to the maximum likelihood estimator. Many authors have applied this kind of expansions (e.g. Efron 1975; Clarke 1980; Amari 1982a,b; 1985; Cook and Tsai 1985; Gallant 1987; Morton 1987; Seber and Wild 1989; Wei 1989; 1991; 1996).

To get the stochastic expansions of the estimate and related statistics, some regularity conditions are needed. We now first introduce some notation. In what follows, we use different index letters to denote different components of matrices and arrays. They are denoted by $\boldsymbol{D} = (D_{ia})$, $\boldsymbol{W} = (W_{iab})$, $\boldsymbol{Q} = (Q_{ia})$, $\boldsymbol{N} = (N_{i\kappa})$, $\boldsymbol{V} = (V_{ij})$, $\boldsymbol{V}^{-1} = (V^{ij})$, $\boldsymbol{D}_\theta = (\dot{\theta}_{ia})$, $\boldsymbol{W}_\theta = (\ddot{\theta}_{iab})$ and so forth; where $i, j = 1, \cdots, n$; $a, b = 1, \cdots, p$; $\kappa, \lambda = 1, \cdots, n-p$. These quantities are usually evaluated at the true value $\boldsymbol{\beta}$ that will be mostly dropped for short. For the quantities evaluated at another point $\boldsymbol{\beta}'$ in the neighborhood of $\boldsymbol{\beta}$, we use $D'_{ia} = D_{ia}(\boldsymbol{\beta}')$, $W'_{iab} = W_{iab}(\boldsymbol{\beta}')$, $N'_{i\kappa} = N_{ik}(\boldsymbol{\beta}')$, etc. to distinguish them. The derivatives of $\mu_i(\boldsymbol{\beta}), \theta_i(\boldsymbol{\beta})$ and $N_{i\kappa}(\boldsymbol{\beta})$ with respect to $\boldsymbol{\beta}$ are denoted by adding subscripts, say $A = abc$, then $\mu_{iA} = \partial^3\mu_i/\partial\beta_a\partial\beta_b\partial\beta_c$; if $B = ab$, then

$\theta_{iB} = \partial^2\theta_i/\partial\beta_a\partial\beta_b$, $(N_{i\kappa})_B = \partial^2 N_{i\kappa}/\partial\beta_a\partial\beta_b$.

Before presenting the regularity conditions, we give some description about them. First of all, (2.11) shows that $n^{-1}\boldsymbol{J}_\beta(\boldsymbol{Y}) = n^{-1}\phi\boldsymbol{D}^T\boldsymbol{V}^{-1}\boldsymbol{D} = n^{-1}\phi\boldsymbol{D}_\theta^T\boldsymbol{V}\boldsymbol{D}_\theta$, which can be regarded as the average Fisher information for $\boldsymbol{\beta}$ containing in the observations $y_1, \cdots, y_n$, so it is very natural to assume that the limit of $n^{-1}\boldsymbol{D}^T\boldsymbol{V}^{-1}\boldsymbol{D}$ exists and tends to a positive definite matrix as $n \to \infty$. Most literature has adopted this fundamental assumption (e.g. Jennrich 1969; McCullagh 1983; Fahrmeir and Kaufmann 1985; Gallant 1987; Morton 1987; Wei 1989; 1991). Based on the existence of $n^{-1}\boldsymbol{D}^T\boldsymbol{V}^{-1}\boldsymbol{D} = n^{-1}\boldsymbol{D}_\theta^T\boldsymbol{V}\boldsymbol{D}_\theta$ as $n \to \infty$, we may further assume $n^{-1}\boldsymbol{D}^T\boldsymbol{D}$, $n^{-1}\boldsymbol{D}_\theta^T\boldsymbol{D}_\theta$ and its derivatives (such as $n^{-1}[\boldsymbol{D}^T][\boldsymbol{W}]$, $n^{-1}[\boldsymbol{D}_\theta^T][\boldsymbol{W}_\theta]$) are bounded for any n (note that $\boldsymbol{V} = \text{diag}(V_i)$ and $V_i = \ddot{b}(\theta_i)$, with $\theta_i = \theta(f(\boldsymbol{x}_i, \boldsymbol{\beta}))$ are always bounded in the compact subset $\mathcal{X}$). These assumptions sound as if μ_{iA}, θ_{iA} have the order of $n^{\frac{1}{2}}$ when one takes summations for them with $i = 1, \cdots, n$. Further, since we shall discuss the problems in some neighborhood of the maximum likelihood estimator $\hat{\boldsymbol{\beta}}$ or the true value $\boldsymbol{\beta}$, we may assume $\mu'_{iA} = \mu_{iA}(\boldsymbol{\beta}')$ is also likely to have order of $n^{\frac{1}{2}}$ when $\boldsymbol{\beta}'$ is another point in the neighborhood of $\boldsymbol{\beta}$. Similarly, since $\boldsymbol{N}^T\boldsymbol{V}^{-1}\boldsymbol{N} = \boldsymbol{I}_{n-p}$, $\boldsymbol{Q}^T\boldsymbol{V}^{-1}\boldsymbol{Q} = \boldsymbol{I}_p$, we may assume the derivatives of $\boldsymbol{N}^T\boldsymbol{N}$, $\boldsymbol{Q}^T\boldsymbol{Q}$ are finite. Thus $(N_{i\kappa})_A$ and $(Q_{ia})_B$ are likely to have order 1 when one takes summations for them with $i = 1, \cdots, n$.

Assumption C

(a) The matrices $\boldsymbol{D}(\boldsymbol{\beta})$ and $\boldsymbol{D}_\theta(\boldsymbol{\beta})$ are of full rank in column in $\mathcal{B}$. $\boldsymbol{N}(\boldsymbol{\beta})$ is differentiable up to the third order.

(b) There exists a positive definite and continuous matrix $\boldsymbol{K}(\boldsymbol{\beta})$ such that

$$\frac{1}{n}\boldsymbol{D}^T(\boldsymbol{\beta})\boldsymbol{V}^{-1}(\boldsymbol{\beta})\boldsymbol{D}(\boldsymbol{\beta}) = \frac{1}{n}\boldsymbol{R}^T(\boldsymbol{\beta})\boldsymbol{R}(\boldsymbol{\beta}) \to \boldsymbol{K}(\boldsymbol{\beta}) \qquad (4.1)$$

uniformly in $\overline{N}(\delta)$ as $n \to \infty$, where $\overline{N}(\delta)$ is a neighborhood of the true value $\boldsymbol{\beta}$ with radius δ.

(c) As n tends to infinite, the following summations are always uniformly bounded in $\overline{N}(\delta)$ for any n:

$$\frac{1}{n}\sum_{i=1}^n \mu_{iA}\mu'_{iB}, \quad \frac{1}{\sqrt{n}}\sum_{i=1}^n \mu_{iA}(N'_{i\kappa})_B, \quad \sum_{i=1}^n (N_{i\kappa})_A(N'_{i\lambda})_B,$$

where $\mu_{iA} = \mu_{iA}(\boldsymbol{\beta})$, $\mu'_{iB} = \mu_{iB}(\boldsymbol{\beta}')$, $\boldsymbol{\beta}'$ is another point in the neighborhood of $\boldsymbol{\beta}$; and A, B are any group of subscripts of $\boldsymbol{\beta}$ not exceeding three index letters. $(N_{i\kappa})_A$ and $(N'_{i\lambda})_B$ are similarly defined. ∎

These assumptions are quite similar to those given by Jennrich (1969), Morton (1987) and Wei (1989; 1991) for nonlinear regression models. The assumptions indicate that the derivatives of μ_i, such as $\boldsymbol{D} = (\mu_{ia})$, $\boldsymbol{W} = (\mu_{iab})$ seem to have the order of $n^{\frac{1}{2}}$ when one takes summations for them with $i = 1, \cdots, n$ and $n \to \infty$. Assumption (4.1) also shows that as $n \to \infty$, the matrix $\boldsymbol{R}$ seems to have the order of $n^{\frac{1}{2}}$ and $\boldsymbol{L} = \boldsymbol{R}^{-1}$ seems to have the order of $n^{-\frac{1}{2}}$. In particular, equation (3.5) in Definition 3.1 shows that the curvature arrays A^I and A^P seem to have the order of $n^{-\frac{1}{2}}$. Similar statements are valid for the quantities in natural parameter space such as $\boldsymbol{D}_\theta = (\dot{\theta}_{ia})$, $\boldsymbol{W}_\theta = (\ddot{\theta}_{iab})$, $\boldsymbol{A}_\theta^I$, $\boldsymbol{A}_\theta^P$ and so on.

Note that most of the above assumptions are not needed in iid case studied by Efron (1975) and Amari (1985).

This chapter will discuss some asymptotic properties related to the maximum likelihood estimator in terms of the curvatures. To this end, the consistency and the asymptotic normality of the maximum likelihood estimator must be first considered as a common problem. Jennrich (1969), Malinvaud (1970) and Wu (1981) discussed the consistency and the asymptotic normality of least squares estimator in nonlinear regression models. Fahrmeir and Kaufmann (1985) studied similar problems for generalized linear models. In the following text, we give a theorem, which investigates similar problems for exponential family nonlinear models. The proof of this theorem is rather complicated and is deferred to Appendix B, which is for readers who are interested in the large sample theory.

Theorem 4.1

If Assumptions A to C hold in exponential family nonlinear model (2.2), then there exists a sequence $\{\hat{\boldsymbol{\beta}}_n\}$ of random variables with $\dot{\boldsymbol{l}}(\hat{\boldsymbol{\beta}}_n) = \boldsymbol{0}$ and it satisfies

(a) $P_{\boldsymbol{\beta}}(\hat{\boldsymbol{\beta}}_n \to \boldsymbol{\beta}) = 1$ (strong consistency);

(b) $\sqrt{n}(\hat{\boldsymbol{\beta}}_n - \boldsymbol{\beta}) \xrightarrow{L} N(\boldsymbol{0},\ \sigma^2 \boldsymbol{K}^{-1}(\boldsymbol{\beta}))$ (asymptotic normality). ∎

From this theorem, we may write $\Delta\boldsymbol{\beta} = \hat{\boldsymbol{\beta}}_n - \boldsymbol{\beta} = O_p(n^{-\frac{1}{2}})$ that will be used frequently in later discussions. In what follows, we still denote $\hat{\boldsymbol{\beta}}_n$ by the shortened version $\hat{\boldsymbol{\beta}}$ in most cases.

To get the stochastic expansion of $\hat{\boldsymbol{\beta}}$, we need two lemmas.

Lemma 4.1

Suppose that the Assumptions A to C hold for model (2.2). If an $n \times 1$

vector $\boldsymbol{a}$ satisfies $n^{-1}\boldsymbol{a}^T\boldsymbol{V}^{-1}\boldsymbol{a} \to a_0$ as $n \to \infty$, then

$$(\sqrt{n})^{-1}\boldsymbol{a}^T\boldsymbol{V}^{-1}\boldsymbol{e} \xrightarrow{L} N(0,\ \sigma^2 a_0).$$

In particular, if $\boldsymbol{b}^T\boldsymbol{V}^{-1}\boldsymbol{b} \to b_0$ as $n \to \infty$, then

$$\boldsymbol{b}^T\boldsymbol{V}^{-1}\boldsymbol{e} \xrightarrow{L} N(0,\ \sigma^2 b_0).$$ ∎

This lemma is a direct extension of Jennrich's (1969) Theorem 5. The proof of this lemma is given in Appendix B together with Theorem 4.1.

Lemma 4.2

For model (2.2), if Assumptions A to C hold, then $\boldsymbol{N}(\hat{\boldsymbol{\beta}})$ can be expanded as

$$\boldsymbol{N}(\hat{\boldsymbol{\beta}}) = \boldsymbol{N}(\boldsymbol{\beta}) + \{\boldsymbol{F}(\boldsymbol{\beta}) \triangle \boldsymbol{\beta}\}^T + \boldsymbol{\gamma}', \tag{4.2}$$

where $\Delta\boldsymbol{\beta} = (\Delta\beta_a)$, $\boldsymbol{\gamma}' = (\gamma'_{i\kappa})$ with

$$\gamma'_{i\kappa} = \sum_{a=1}^{p}\sum_{b=1}^{p}(N'_{i\kappa})_{ab} \triangle \beta_a \triangle \beta_b$$

for some $N'_{i\kappa} = N_{i\kappa}(\boldsymbol{\beta}')$, $\boldsymbol{\beta}' = t'\boldsymbol{\beta} + (1-t')\hat{\boldsymbol{\beta}}$ and $0 \le t' \le 1$; $\boldsymbol{F}(\boldsymbol{\beta})$ is an $(n-p) \times n \times p$ array with entry $F_{\kappa ia} = (N_{i\kappa})_a = \partial N_{i\kappa}/\partial\beta_a$ and satisfies

$$\boldsymbol{D}^T\boldsymbol{V}^{-1}\boldsymbol{F} = -\boldsymbol{R}^T\boldsymbol{A}^I_\theta\boldsymbol{R}. \tag{4.3}$$

Proof. The Taylor series expansions of $N_{i\kappa}(\hat{\boldsymbol{\beta}})$ at $\boldsymbol{\beta}$ ($i = 1, \cdots, n$; $\kappa = 1, \cdots, n-p$) give (4.2) directly. To prove (4.3), we write $\boldsymbol{D}^T\boldsymbol{V}^{-1}\boldsymbol{N} = \boldsymbol{0}$ as

$$\sum_{i=1}^{n} D_{ia}(\boldsymbol{\beta})V_i^{-1}(\boldsymbol{\beta})N_{i\kappa}(\boldsymbol{\beta}) = 0$$

$$(a = 1, \cdots, p; \quad \kappa = 1, \cdots, n-p).$$

These equations hold for any $\boldsymbol{\beta}$, so differentiating these equations in β_b ($b = 1, \cdots, p$) gives

$$\sum_{i=1}^{n}\left\{W_{iab}V_i^{-1}N_{i\kappa} + D_{ia}F_{\kappa ib}V_i^{-1} - D_{ia}N_{i\kappa}V_i^{-2}(\partial V_i/\partial\beta_b)\right\} = 0.$$

Since $V_i = \ddot{b}(\theta_i)$, $\mu_i = \dot{b}(\theta_i)$, we have

$$\frac{\partial V_i}{\partial \beta_b} = \frac{\partial V_i}{\partial \theta_i}\frac{\partial \theta_i}{\partial \mu_i}\frac{\partial \mu_i}{\partial \beta_b} = S_i V_i^{-1} D_{ib},$$

where $S_i = b^{(3)}(\theta_i)$. Combining the above equations and using matrix notation yield

$$[\boldsymbol{N}^T\boldsymbol{V}^{-1}][\boldsymbol{W}] + \boldsymbol{D}^T\boldsymbol{V}^{-1}\boldsymbol{F} - [\boldsymbol{N}^T\boldsymbol{V}^{-1}][\boldsymbol{D}^T\boldsymbol{V}^{-1}\boldsymbol{S}\boldsymbol{V}^{-1}\boldsymbol{D}] = \boldsymbol{0}.$$

Using (3.24), this equation reduces to

$$\boldsymbol{D}^T\boldsymbol{V}^{-1}\boldsymbol{F} = -[\boldsymbol{N}^T\boldsymbol{V}^{-1}][\boldsymbol{W} - \boldsymbol{\Gamma}] = -\boldsymbol{R}^T\boldsymbol{A}_\theta^I\boldsymbol{R},$$

which follows (4.3). ∎

To get the stochastic expansion of $\hat{\boldsymbol{\beta}}$, it is very natural to start from the likelihood equation of $\hat{\boldsymbol{\beta}}$. Equation (3.1) shows that the residual vector $\hat{e} = \boldsymbol{Y} - \boldsymbol{\mu}(\hat{\boldsymbol{\beta}})$ is in normal space $T'_{\hat{\beta}}$ of the solution locus at $\hat{\boldsymbol{\beta}}$. Hence $\hat{e}$ may be written as

$$\hat{e} = \boldsymbol{N}(\hat{\boldsymbol{\beta}})\hat{\boldsymbol{\nu}}. \tag{4.4}$$

Here the $(n-p)\times 1$ vector $\hat{\boldsymbol{\nu}}$ is the coordinate of $\hat{e}$ in normal space and the columns of $\boldsymbol{N}(\hat{\boldsymbol{\beta}})$ are taken as an orthonormal basis in this space. The statistic $\hat{\boldsymbol{\nu}}$ will play an important role in our later discussions. Equation (4.4) shows that the observed vector $\boldsymbol{Y}$ can be written as

$$\boldsymbol{Y} = \boldsymbol{\mu}(\hat{\boldsymbol{\beta}}) + \boldsymbol{N}(\hat{\boldsymbol{\beta}})\hat{\boldsymbol{\nu}}, \tag{4.5}$$

which indicates that $(\hat{\boldsymbol{\beta}}, \hat{\boldsymbol{\nu}})$ is a sufficient statistic. It will be seen from (4.11) that each component of $\hat{\boldsymbol{\nu}}$ is an asymptotic ancillary statistic (an ancillary statistic is defined as a statistic based on $\boldsymbol{Y}$ with a distribution independent of $\boldsymbol{\beta}$; see Cox and Hinkley 1974, p.34 for details). The decomposition of the observed vector $\boldsymbol{Y}$ in (4.5) is quite similar to the "ancillary family associated with the estimator" introduced by Amari (1985, p.118; see also Kass 1989).

Theorem 4.2

If Assumptions A to C hold in (2.2), then the maximum likelihood estimator $\hat{\boldsymbol{\beta}}$ can be expanded as

$$\Delta\boldsymbol{\beta} = \boldsymbol{L}\boldsymbol{\tau} + \boldsymbol{L}\{[\boldsymbol{\lambda}^T][\boldsymbol{A}_\theta^I]\boldsymbol{\tau} - \frac{1}{2}\boldsymbol{\tau}^T\boldsymbol{A}^P\boldsymbol{\tau}\} + O_p(n^{-3/2}), \tag{4.6}$$

where $\Delta\boldsymbol{\beta} = \hat{\boldsymbol{\beta}} - \boldsymbol{\beta}$, and

$$\boldsymbol{\tau} = \boldsymbol{Q}^T\boldsymbol{V}^{-1}\boldsymbol{e}, \qquad \boldsymbol{\lambda} = \boldsymbol{N}^T\boldsymbol{V}^{-1}\boldsymbol{e}$$

are uncorrelated. The elements of $\boldsymbol{\tau}$ and $\boldsymbol{\lambda}$ have asymptotically normal distributions of $N(0,\sigma^2)$. All the matrices and arrays in (4.6) such as $\boldsymbol{L}$, $\boldsymbol{A}_\theta^I$, $\boldsymbol{Q}$ etc. are evaluated at $\boldsymbol{\beta}$.

Proof. From (4.4) we may write

$$e = Y - \mu(\beta) = \hat{e} + \Delta\mu = N(\hat{\beta})\hat{\nu} + \Delta\mu, \tag{4.7}$$

where $\Delta\mu = \mu(\hat{\beta}) - \mu(\beta)$. Then it follows from Lemma 4.2 and the Taylor series expansions of $N(\hat{\beta})$ and $\Delta\mu$ that

$$e = N\hat{\nu} + [\hat{\nu}^T][F]\Delta\beta + D\Delta\beta + \frac{1}{2}(\Delta\beta)^T W(\Delta\beta) + \gamma'\hat{\nu} + \gamma'', \tag{4.8}$$

where $\gamma'' = (\gamma_i'')$ with the element

$$\gamma_i'' = \sum_{a=1}^{p}\sum_{b=1}^{p}\sum_{c=1}^{p}\mu_{iabc}(\beta'')\Delta\beta_a\Delta\beta_b\Delta\beta_c,$$

$\beta'' = t''\beta + (1-t'')\hat{\beta}$ and $0 \le t'' \le 1$. Multiplying (4.8) by D^TV^{-1} and N^TV^{-1} respectively yields

$$\begin{aligned} D^TV^{-1}e &= [\hat{\nu}^T][D^TV^{-1}F]\Delta\beta + (D^TV^{-1}D)\Delta\beta + \\ &\quad \frac{1}{2}(\Delta\beta)^T[D^TV^{-1}][W](\Delta\beta) + \alpha_1, \\ N^TV^{-1}e &= \hat{\nu} + [\hat{\nu}^T][N^TV^{-1}F]\Delta\beta + \\ &\quad \frac{1}{2}(\Delta\beta)^T[N^TV^{-1}][W](\Delta\beta) + \alpha_2, \end{aligned}$$

where

$$\alpha_1 = D^TV^{-1}(\gamma'\hat{\nu} + \gamma''), \qquad \alpha_2 = N^TV^{-1}(\gamma'\hat{\nu} + \gamma'').$$

It follows from (3.4), (3.5) and Lemma 4.2 that

$$\begin{aligned} \Delta\beta &= (D^TV^{-1}D)^{-1}\Big\{R^T\tau + [\hat{\nu}^T][R^TA_\theta^IR]\Delta\beta - \\ &\quad \frac{1}{2}(\Delta\beta)^TR^T[R^T][A^P]R(\Delta\beta) + \alpha_1\Big\}, \end{aligned} \tag{4.9}$$

$$\hat{\nu} = \lambda - [\hat{\nu}^T][N^TV^{-1}F]\Delta\beta - \frac{1}{2}(\Delta\beta)^T(R^TA^IR)(\Delta\beta) + \alpha_2. \tag{4.10}$$

Let $\tau = (\tau_a)$, $\tau_a = Q_a^TV^{-1}e$, where Q_a is the a-th column of Q. Since $Q_a^TV^{-1}Q_a = 1$, it follows from Lemma 4.1 that $\tau_a \xrightarrow{L} N(0,\sigma^2)$, so does $\lambda_\kappa \xrightarrow{L} N(0,\sigma^2)$, $\lambda_\kappa = O_p(1)$, where $\lambda = (\lambda_\kappa) = N^TV^{-1}e$. Moreover, it follows from (4.4) that $\hat{\nu} = (N^TV^{-1}e)_{\hat{\beta}} = (\lambda)_{\hat{\beta}}$, so we can write $\hat{\nu} = (\hat{\nu}_\kappa), \hat{\nu}_\kappa = O_p(1)$ by the continuity of $N^T(\beta)V^{-1}(\beta)e(\beta)$ at β. Therefore, in (4.9) and (4.10), it follows from $\Delta\beta = O_p(n^{-\frac{1}{2}})$, $\hat{\nu}_\kappa = O_p(1)$ and Assumption

C that each component of $\boldsymbol{\alpha}_1$ has the order of $O_p(n^{-\frac{1}{2}})$, each component of $\boldsymbol{\alpha}_2$ has the order of $O_p(n^{-1})$ and $(\boldsymbol{D}^T\boldsymbol{V}^{-1}\boldsymbol{D})^{-1} = O(n^{-1})$. Then from (4.9), (4.10) and Assumption C we get

$$\Delta\boldsymbol{\beta} = \boldsymbol{L\tau} + O_p(n^{-1}), \qquad \hat{\boldsymbol{\nu}} = \boldsymbol{\lambda} + O_p(n^{-\frac{1}{2}}). \tag{4.11}$$

Substituting these equations into the right-hand side of (4.9) and putting the higher order terms together with $\boldsymbol{\alpha}_1$, we get equation (4.6). It is easily seen that $\mathrm{Cov}\,(\boldsymbol{\tau}, \boldsymbol{\lambda}) = \boldsymbol{Q}^T\boldsymbol{V}^{-1}\boldsymbol{N} = \boldsymbol{0}$, so $\boldsymbol{\tau}$ and $\boldsymbol{\lambda}$ are uncorrelated. ∎

Corollary 1

The stochastic expansion of $\hat{\boldsymbol{\nu}}$ can be expressed as

$$\hat{\boldsymbol{\nu}} = \boldsymbol{\lambda} - [\boldsymbol{\lambda}^T][\boldsymbol{N}^T\boldsymbol{V}^{-1}\boldsymbol{F}]\boldsymbol{L\tau} - \frac{1}{2}\boldsymbol{\tau}^T\boldsymbol{A}^I\boldsymbol{\tau} + \boldsymbol{a}, \tag{4.12}$$

where $\boldsymbol{a} = (a_\kappa)$ with $a_\kappa = O_p(n^{-1})$.

Proof. Substituting (4.11) into (4.10) gives this result. ∎

Equations (4.11) and (4.12) show that the asymptotic distribution of each component $\hat{\nu}_\kappa$ of $\hat{\boldsymbol{\nu}}$ is independent of the parameter $\boldsymbol{\beta}$, so it is an asymptotic ancillary statistic. Amari (1982b) have obtained a similar result for curved exponential families.

Corollary 2

The stochastic expansions of $\Delta\boldsymbol{\mu} = \boldsymbol{\mu}(\hat{\boldsymbol{\beta}}) - \boldsymbol{\mu}(\boldsymbol{\beta})$ and $\hat{\mathbf{e}}$ can be expressed as

$$\Delta\boldsymbol{\mu} = \boldsymbol{Q\tau} + \boldsymbol{Q}[\boldsymbol{\lambda}^T][\boldsymbol{A}_\theta^I]\boldsymbol{\tau} + \frac{1}{2}\boldsymbol{N}(\boldsymbol{\tau}^T\boldsymbol{A}^I\boldsymbol{\tau}) + \boldsymbol{b}, \tag{4.13}$$

$$\hat{\mathbf{e}} = \boldsymbol{N\lambda} - \boldsymbol{Q}[\boldsymbol{\lambda}^T][\boldsymbol{A}_\theta^I]\boldsymbol{\tau} - \frac{1}{2}\boldsymbol{N}(\boldsymbol{\tau}^T\boldsymbol{A}^I\boldsymbol{\tau}) + \boldsymbol{b}, \tag{4.14}$$

where $\boldsymbol{b} = (b_i)$ with $b_i = O_p(n^{-1})$.

Proof. As shown in (4.8), $\Delta\boldsymbol{\mu}$ can be written as

$$\Delta\boldsymbol{\mu} = \boldsymbol{D}\Delta\boldsymbol{\beta} + \frac{1}{2}(\Delta\boldsymbol{\beta})^T\boldsymbol{W}(\Delta\boldsymbol{\beta}) + \boldsymbol{\gamma}''.$$

Substituting (4.6) and (4.11) respectively into the first term and the second term of the right-hand side of this equation and putting the higher order terms together with $\boldsymbol{\gamma}''$, we get

$$\Delta\boldsymbol{\mu} = \boldsymbol{Q\tau} + \boldsymbol{Q}[\boldsymbol{\lambda}^T][\boldsymbol{A}_\theta^I]\boldsymbol{\tau} - \frac{1}{2}\boldsymbol{Q}(\boldsymbol{\tau}^T\boldsymbol{A}^P\boldsymbol{\tau}) + \frac{1}{2}\boldsymbol{\tau}^T\boldsymbol{U\tau} + \boldsymbol{b}.$$

It follows from (3.7) that

$$\begin{aligned}\frac{1}{2}\boldsymbol{\tau}^T\boldsymbol{U}\boldsymbol{\tau} - \frac{1}{2}\boldsymbol{Q}(\boldsymbol{\tau}^T\boldsymbol{A}^P\boldsymbol{\tau}) &= \frac{1}{2}\boldsymbol{\tau}^T\{\boldsymbol{U} - [\boldsymbol{Q}][\boldsymbol{A}^P]\}\boldsymbol{\tau} \\ &= \frac{1}{2}\boldsymbol{\tau}^T[\boldsymbol{N}][\boldsymbol{A}^I]\boldsymbol{\tau} \\ &= \frac{1}{2}\boldsymbol{N}(\boldsymbol{\tau}^T\boldsymbol{A}^I\boldsymbol{\tau}).\end{aligned}$$

Substituting this result into the right-hand side of the above $\Delta\boldsymbol{\mu}$ yields (4.13). To prove (4.14), we use (4.7) and (4.12), which result in

$$\begin{aligned}\hat{e} &= e - \Delta\boldsymbol{\mu} \\ &= (e - \boldsymbol{Q}\boldsymbol{\tau}) - \boldsymbol{Q}[\boldsymbol{\lambda}^T][\boldsymbol{A}^I]\boldsymbol{\tau} - \frac{1}{2}\boldsymbol{N}(\boldsymbol{\tau}^T\boldsymbol{A}^I\boldsymbol{\tau}) + b.\end{aligned}$$

Then (4.14) can be obtained from $\boldsymbol{e} = (\boldsymbol{Q}\boldsymbol{Q}^T\boldsymbol{V}^{-1} + \boldsymbol{N}\boldsymbol{N}^T\boldsymbol{V}^{-1})e = \boldsymbol{Q}\boldsymbol{\tau} + \boldsymbol{N}\boldsymbol{\lambda}$, and $\boldsymbol{e} - \boldsymbol{Q}\boldsymbol{\tau} = \boldsymbol{N}\boldsymbol{\lambda}$. ∎

In summary, we have obtained four stochastic expansions for the maximum likelihood estimator $\hat{\boldsymbol{\beta}}$ and related statistics $\hat{e}$, $\Delta\boldsymbol{\mu}$ and $\hat{\boldsymbol{\nu}}$. Equation (4.6) shows that the stochastic expansion of $\hat{\boldsymbol{\beta}}$ depends on the intrinsic curvature $\boldsymbol{A}_\theta^I$ in natural parameter space and the parameter-effects curvature $\boldsymbol{A}^P$ in expectation parameter space. This is consistent with the statement of Pazman (1991). Equations (4.12) to (4.14) show that the stochastic expansions of $\hat{\boldsymbol{\nu}}$, $\Delta\boldsymbol{\mu}$ and $\hat{e}$ depend only on the intrinsic curvatures $\boldsymbol{A}_\theta^I$ and $\boldsymbol{A}^I$, so they are all invariant under parameterizations. These conclusions are quite reasonable because $\Delta\boldsymbol{\mu}$ and $\hat{e}$ are invariant under parameterizations. In fact, expressions $\Delta\boldsymbol{\mu} = \boldsymbol{\mu}(\hat{\boldsymbol{\beta}}) - \boldsymbol{\mu}(\boldsymbol{\beta})$ and $\hat{e} = \boldsymbol{Y} - \boldsymbol{\mu}(\hat{\boldsymbol{\beta}})$ show that $\Delta\boldsymbol{\mu}$ and $\hat{e}$ depend on $\boldsymbol{\beta}$ only in terms of $\boldsymbol{\mu} = \boldsymbol{\mu}(\boldsymbol{\beta})$ and hence are independent of the parameterizations of $\boldsymbol{\beta}$.

In the literature, there are some analogous expansions as shown earlier for certain specific models. For nonlinear quasi-likelihood models, McCullagh (1983) obtained an expansion of $\hat{\boldsymbol{\beta}}$ as

$$\Delta\boldsymbol{\beta} = (\boldsymbol{D}^T\boldsymbol{V}^{-1}\boldsymbol{D})^{-1}\boldsymbol{D}^T\boldsymbol{V}^{-1}\boldsymbol{e} + O_p(n^{-1}).$$

This formula actually coincides with our equation (4.11) because we have $(\boldsymbol{D}^T\boldsymbol{V}^{-1}\ \boldsymbol{D})^{-1} = (\boldsymbol{R}^T\boldsymbol{R})^{-1} = \boldsymbol{L}\boldsymbol{L}^T$, $\boldsymbol{D}^T\boldsymbol{V}^{-1}\boldsymbol{e} = \boldsymbol{R}^T\boldsymbol{Q}^T\boldsymbol{V}^{-1}\boldsymbol{e} = \boldsymbol{R}^T\boldsymbol{\tau}$. Gallant (1987, p.16) obtained a similar formula for nonlinear regression models, where $\boldsymbol{V} = \boldsymbol{I}_n$ and $\Delta\boldsymbol{\beta} = (\boldsymbol{D}^T\boldsymbol{D})^{-1}\boldsymbol{D}^T e + O_p(n^{-1})$. Again for normal nonlinear regression models, since $\boldsymbol{V} = \boldsymbol{I}_n$, $\boldsymbol{A}_\theta^I = \boldsymbol{A}^I$, the expansion (4.6) reduces to

$$\Delta\boldsymbol{\beta} = \boldsymbol{L}\boldsymbol{\tau} + \boldsymbol{L}\{[\boldsymbol{\lambda}^T][\boldsymbol{A}^I]\boldsymbol{\tau} - \frac{1}{2}\boldsymbol{\tau}^T\boldsymbol{A}^P\boldsymbol{\tau}\} + O_p(n^{-3/2})$$

with $\boldsymbol{\tau} = \boldsymbol{Q}^T\boldsymbol{e}$, $\boldsymbol{\lambda} = \boldsymbol{N}^T\boldsymbol{e}$. This formula has been given by Seber and Wild (1989), Wei (1989; 1991) and implicitly given by Clarke (1980). The expansion of $\hat{e}$ for nonlinear regression case has been given by Cook and Tsai (1985), Seber and Wild (1989) and Wei (1989).

These expansions are quite useful to investigate the asymptotic properties related to the maximum likelihood estimator $\hat{\boldsymbol{\beta}}$ because $\boldsymbol{\tau}$ and $\boldsymbol{\lambda}$ in expansions are uncorrelated and the components of them are asymptotically normally distributed. In the rest of this chapter, we shall use these stochastic expansions to study some asymptotics related to the curvatures.

4.2 Approximate Bias and Variance

The bias and the variance of an estimator are basic indicators to assess the statistical behavior for that estimator. It is well known that the least squares estimator in normal linear regression models is a uniformly minimum variance unbiased estimator. However, it may not be true for other estimators and other models. In nonlinear regression models, since the existence of nonlinearity, the least squares estimator is biased and can not achieve the usual Cramér-Rao lower bound. The high order terms of bias and the variance are naturally determined by the curvatures of the model (e.g. Bates and Watts 1980; Clarke 1980; Cook, Tsai and Wei 1986; Seber and Wild 1989; Wei 1989; 1991). For exponential family nonlinear models, the nonlinearity of the model also plays an important role in the expressions of bias and variance of an estimator. This section will derive several approximations of biases and variances for the maximum likelihood estimator and related statistics in terms of the curvatures. It will be shown that the existence of nonlinearity of the model is the main reason that causes the bias and the larger variance. The nonlinearity also causes the inefficiency of other statistics, such as the fitted value $\hat{\boldsymbol{\mu}}$ and the residual $\hat{\boldsymbol{e}}$.

We will apply the stochastic expansions introduced in Section 4.1 to derive biases and variances of $\hat{\boldsymbol{\beta}}$, $\hat{\boldsymbol{\mu}}$ and $\hat{\boldsymbol{e}}$. To do so, we need some lemmas.

Lemma 4.3

Let $\boldsymbol{\tau} = \boldsymbol{Q}^T\boldsymbol{V}^{-1}\boldsymbol{e} = (\tau_a)$, $\boldsymbol{Q} = (Q_{ia})$, then for model (2.2), the first four moments of $\boldsymbol{\tau}$ are given by

$$\mathrm{E}(\tau_a) = 0, \qquad \mathrm{E}(\tau_a\tau_b) = \sigma^2\delta_{ab},$$

$$\mathrm{E}(\tau_a\tau_b\tau_c) = \sigma^4\Gamma^P_{abc}, \tag{4.15}$$

$$\mathrm{E}(\tau_a\tau_b\tau_c\tau_d) = \sigma^4(\delta_{ab}\delta_{cd} + \delta_{ac}\delta_{bd} + \delta_{ad}\delta_{bc}) + \sigma^6\Delta_{abcd}, \tag{4.16}$$

$$\Delta_{abcd} = \sum_{i=1}^{n} Q_{ia}Q_{ib}Q_{ic}Q_{id}V_i^{-4}\Delta_i,$$

where δ_{ab} is the Kronecker delta: $\delta_{ab} = 1$ if $a = b$, otherwise $\delta_{ab} = 0$; $\boldsymbol{\Gamma}^P = (\Gamma^P_{abc})$ is given in (3.25) and $\Delta_i = b^{(4)}(\theta_i)$.

Proof. The moments of $\boldsymbol{e}$ given in (1.9) to (1.11) can be applied to model (2.2). In this case, since $y_1, \cdots, y_n$ are mutually independent, the quantities in (1.9) to (1.11) can be simplified as $V_{ij} = V_i\delta_{ij}$; $S_{ij\kappa} = S_i$ if $i = j = \kappa$, otherwise $S_{ij\kappa} = 0$; $\Delta_{ij\kappa l} = \Delta_i$ if $i = j = \kappa = l$, otherwise $\Delta_{ij\kappa l} = 0$. By using these results and $\tau_a = \sum_i Q_{ia}V_i^{-1}e_i$, $\boldsymbol{Q}^T\boldsymbol{V}^{-1}\boldsymbol{Q} = \boldsymbol{I}_p$, we get

$$\begin{aligned} \mathrm{E}(\tau_a\tau_b) &= \mathrm{E}\Big\{(\sum_{i=1}^n Q_{ia}V_i^{-1}e_i)(\sum_{j=1}^n Q_{jb}V_j^{-1}e_j)\Big\} \\ &= \sum_{i=1}^n \sigma^2 Q_{ia}Q_{ib}V_i^{-1} = \sigma^2\delta_{ab}. \end{aligned}$$

Similarly, it follows from (1.10) that

$$\mathrm{E}(\tau_a\tau_b\tau_c) = \sigma^4 \sum_{i=1}^n Q_{ia}Q_{ib}Q_{ic}V_i^{-3}S_i. \tag{4.17}$$

On the other hand, equation (3.25) gives

$$\begin{aligned} \boldsymbol{\Gamma}^P &= [\boldsymbol{Q}^T\boldsymbol{V}^{-1}][\boldsymbol{L}^T\boldsymbol{\Gamma}\boldsymbol{L}] \\ &= [\boldsymbol{Q}^T\boldsymbol{V}^{-1}][\boldsymbol{Q}^T\boldsymbol{V}^{-1}\boldsymbol{S}\boldsymbol{V}^{-1}\boldsymbol{Q}]. \end{aligned}$$

Hence the component of $\boldsymbol{\Gamma}^P$ at (a, b, c) is just the right-hand side of (4.17) divided by σ^4 and we get (4.15). Equation (4.17) also shows that Γ^P_{abc} is symmetric with respect to indices a, b and c. Finally, we have

$$\mathrm{E}(\tau_a\tau_b\tau_c\tau_d) = \sum_{i,j}\sum_{\kappa,l}\Big\{Q_{ia}Q_{jb}Q_{kc}Q_{ld}V_i^{-1}V_j^{-1}V_\kappa^{-1}V_l^{-1}\mathrm{E}(e_ie_je_\kappa e_l)\Big\}.$$

Substituting (1.11) into this equation and using the results such as

$$\begin{aligned} &\sum_{i,j}\sum_{\kappa,l}\Big\{Q_{ia}Q_{jb}Q_{\kappa c}Q_{ld}V_i^{-1}V_j^{-1}V_\kappa^{-1}V_l^{-1}(\delta_{ij}V_i\delta_{\kappa l}V_\kappa)\Big\} \\ &= \Big(\sum_{i=1}^n Q_{ia}Q_{ib}V_i^{-1}\Big)\Big(\sum_{\kappa=1}^n Q_{\kappa c}Q_{\kappa d}V_\kappa^{-1}\Big) = \delta_{ab}\delta_{cd}, \end{aligned}$$

we can get (4.16). ∎

Lemma 4.4

For model (2.2), if $\boldsymbol{B}, \boldsymbol{C}$ are respectively $n \times p \times p$ and $m \times p \times p$ arrays and each face of $\boldsymbol{C}$ is symmetric, then

$$\mathrm{E}(\boldsymbol{\tau}^T \boldsymbol{B} \boldsymbol{\tau}) = \sigma^2 \mathrm{tr}\,[\boldsymbol{B}], \tag{4.18}$$

$$\mathrm{Cov}\,(\boldsymbol{\tau}^T \boldsymbol{B} \boldsymbol{\tau},\ \boldsymbol{\tau}^T \boldsymbol{C} \boldsymbol{\tau}) = (\alpha_{ij}) \tag{4.19}$$

with

$$\alpha_{ij} = 2\sigma^4 \mathrm{tr}\,(\boldsymbol{B}_i \boldsymbol{C}_j) + \sum_{a,b} \sum_{c,d} \sigma^6 B_{iab} C_{jcd} \Delta_{abcd},$$

where $\boldsymbol{B}_i$ and $\boldsymbol{C}_j$ are the i-th face of $\boldsymbol{B}$ and the j-th face of $\boldsymbol{C}$, respectively.

Proof. From $\mathrm{E}(\boldsymbol{\tau}\boldsymbol{\tau}^T) = \sigma^2 I_p$, it is easy to get $\mathrm{E}(\boldsymbol{\tau}^T \boldsymbol{B}_i \boldsymbol{\tau}) = \sigma^2 \mathrm{tr}\,(\boldsymbol{B}_i)$ which gives (4.18). We now prove (4.19). It follows from (4.16) that

$$\begin{aligned}
\alpha_{ij} &= \mathrm{Cov}\,(\boldsymbol{\tau}^T \boldsymbol{B}_i \boldsymbol{\tau},\ \boldsymbol{\tau}^T \boldsymbol{C}_j \boldsymbol{\tau}) \\
&= \Big\{\sum_{a,b} \sum_{c,d} B_{iab} C_{jcd} \mathrm{E}(\tau_a \tau_b \tau_c \tau_d)\Big\} - \\
&\quad \Big\{\sum_{a,b} B_{iab} \mathrm{E}(\tau_a \tau_b)\Big\}\Big\{\sum_{c,d} C_{jcd} \mathrm{E}(\tau_c \tau_d)\Big\} \\
&= \Big\{\sum_{a,b} \sum_{c,d} \sigma^4 B_{iab} C_{jcd} (\delta_{ab}\delta_{cd} + \delta_{ac}\delta_{bd} + \delta_{ad}\delta_{bc})\Big\} + \\
&\quad \Big\{\sum_{a,b} \sum_{c,d} \sigma^6 B_{iab} C_{jcd} \Delta_{abcd}\Big\} - \Big\{\sum_{a,b} \sum_{c,d} \sigma^4 B_{iab} C_{jcd} \delta_{ab} \delta_{cd}\Big\} \\
&= \Big\{\sum_{a,b} 2\sigma^4 B_{iab} C_{jab}\Big\} + \Big\{\sum_{a,b} \sum_{c,d} \sigma^6 B_{iab} C_{jcd} \Delta_{abcd}\Big\}.
\end{aligned}$$

This results in (4.19). ∎

From stochastic expansions (4.6), (4.13), (4.14), and Lemma 4.4, we can easily obtain the approximate biases of $\hat{\boldsymbol{\beta}}$, $\hat{\boldsymbol{\mu}}$ and $\hat{\boldsymbol{e}}$. The results are summarized as follows:

$$\begin{aligned}
\mathrm{bias}\,(\hat{\boldsymbol{\beta}}) &\approx -\frac{1}{2}\sigma^2 \boldsymbol{L} \mathrm{tr}\,[\boldsymbol{A}^P], \\
\mathrm{bias}\,(\hat{\boldsymbol{\mu}}) &\approx \frac{1}{2}\sigma^2 \boldsymbol{N} \mathrm{tr}\,[\boldsymbol{A}^I], \\
\mathrm{E}(\hat{\boldsymbol{e}}) &\approx -\frac{1}{2}\sigma^2 \boldsymbol{N} \mathrm{tr}\,[\boldsymbol{A}^I],
\end{aligned}$$

where $\mathrm{bias}\,(\hat{\boldsymbol{\beta}}) = \mathrm{E}(\hat{\boldsymbol{\beta}} - \boldsymbol{\beta})$, and $\mathrm{bias}\,(\hat{\boldsymbol{\mu}}) = \mathrm{E}\{\boldsymbol{\mu}(\hat{\boldsymbol{\beta}}) - \boldsymbol{\mu}(\boldsymbol{\beta})\}$. These results show that the approximate bias of $\hat{\boldsymbol{\beta}}$ depends only on the parameter-effects curvature; the approximate biases of $\hat{\mu}$ and $\hat{e}$ depend only on the intrinsic curvature and are invariant under parameter transformations. The above results include several special cases which have appeared in the literature (e.g. Bates and Watts 1980; Cook and Tsai 1985; Cook, Tsai and Wei 1986; Seber and Wild 1989; Wei 1989 for nonlinear regression models; Cordeiro and McCullagh 1991 for generalized linear models; and Efron 1975; Amari 1982b; 1985 for curved exponential families).

The accurate second order expressions of variances of $\hat{\boldsymbol{\beta}}$, $\hat{\mu}$ and $\hat{e}$ are quite complicated (Clarke 1980; Seber and Wild 1989, p.184). We prefer to derive a little rough but rather intuitive expressions for them in terms of curvatures.

Theorem 4.3

For model (2.2), if Assumptions A to C hold, then the approximate variance of $\hat{\boldsymbol{\beta}}$ can be represented as

$$\mathrm{Var}\,(\hat{\boldsymbol{\beta}}) \approx \boldsymbol{J}_{\beta}^{-1}(\boldsymbol{Y}) + \sigma^4 \boldsymbol{L}\boldsymbol{V}_{\theta I}\boldsymbol{L}^T + \frac{1}{2}\sigma^4 \boldsymbol{L}(\boldsymbol{V}_{P\Gamma} + \frac{1}{2}\sigma^2 \boldsymbol{V}_{P\Delta})\boldsymbol{L}^T, \quad (4.20)$$

where

$$\boldsymbol{V}_{\theta I} = \sum_{\kappa=1}^{n-p} (\boldsymbol{A}_{\theta\kappa}^I)^2. \quad (4.21)$$

$\boldsymbol{A}_{\theta\kappa}^I$ is the κ-th face of $\boldsymbol{A}_\theta^I$, and the components of $\boldsymbol{V}_{P\Gamma}$ and $\boldsymbol{V}_{P\Delta}$ at (a,b) are respectively

$$(\boldsymbol{V}_{P\Gamma})_{ab} = \mathrm{tr}\,(\boldsymbol{A}_a^P \boldsymbol{A}_b^P - \boldsymbol{A}_a^P \boldsymbol{\Gamma}_b^P - \boldsymbol{A}_b^P \boldsymbol{\Gamma}_a^P),$$

$$(\boldsymbol{V}_{P\Delta})_{ab} = \sum_{c,d}\sum_{e,f} A_{acd}^P A_{bef}^P \Delta_{cdef}.$$

Proof. We shall use Theorem 4.2, Lemmas 4.3, 4.4 and the following equality to get an approximation of $\mathrm{Var}(\hat{\boldsymbol{\beta}})$:

$$\mathrm{Var}(\hat{\boldsymbol{\beta}}) = \mathrm{E}\Big\{\mathrm{Var}(\hat{\boldsymbol{\beta}}|\boldsymbol{\tau})\Big\} + \mathrm{Var}\Big\{\mathrm{E}(\hat{\boldsymbol{\beta}}|\boldsymbol{\tau})\Big\}.$$

It follows from Theorem 4.2 that

$$\begin{aligned}
\mathrm{E}\Big\{\mathrm{Var}(\hat{\boldsymbol{\beta}}|\boldsymbol{\tau})\Big\} &\approx \boldsymbol{L}\mathrm{E}\Big\{\mathrm{Var}([\boldsymbol{\lambda}^T][\boldsymbol{A}_\theta^I]\boldsymbol{\tau}|\boldsymbol{\tau})\Big\}\boldsymbol{L}^T \\
&= \boldsymbol{L}\mathrm{E}\Big\{\sum_{\kappa=1}^{n-p}\mathrm{Var}(\lambda_\kappa \boldsymbol{A}_{\theta\kappa}^I\boldsymbol{\tau}|\boldsymbol{\tau})\Big\}\boldsymbol{L}^T \\
&= \boldsymbol{L}\mathrm{E}\Big\{\sum_{\kappa=1}^{n-p}\sigma^2 \boldsymbol{A}_{\theta\kappa}^I\boldsymbol{\tau}\boldsymbol{\tau}^T\boldsymbol{A}_{\theta\kappa}^I\Big\}\boldsymbol{L}^T
\end{aligned}$$

$$
\begin{aligned}
&= L\Big\{\sum_{\kappa=1}^{n-p} \sigma^4 (A^I_{\theta\kappa})^2\Big\} L^T \\
&= \sigma^4 L V_{\theta I} L^T,
\end{aligned}
$$

which is the second term of (4.20). It follows again from Theorem 4.2 that

$$
\begin{aligned}
\mathrm{Var}\Big\{\mathrm{E}(\hat{\beta}|\tau)\Big\} &\approx \mathrm{Var}\Big\{L\tau - \frac{1}{2}L(\tau^T A^P \tau)\Big\} \\
&= \mathrm{Var}(L\tau) - \frac{1}{2}L\ \mathrm{Cov}(\tau,\ \tau^T A^P \tau) L^T - \\
&\quad \frac{1}{2}L\ \mathrm{Cov}(\tau^T A^P \tau,\ \tau) L^T + \frac{1}{4} L \mathrm{Var}(\tau^T A^P \tau) L^T \\
&= \alpha_1 + L(\alpha_2 + \alpha_3 + \alpha_4) L^T,
\end{aligned}
$$

where α_i is the i-th term ($i = 1, \cdots, 4$) of this expression. It follows from (2.11) and (3.4) that

$$
\alpha_1 = \sigma^2 L L^T = J_\beta^{-1}(Y).
$$

This is the first term of (4.20). It follows from (4.15) of Lemma 4.3 that the component of α_2 at (a, b) is

$$
\begin{aligned}
(\alpha_2)_{ab} &= -\frac{1}{2}\mathrm{E}\Big\{\sum_{c=1}^{p}\sum_{d=1}^{p} A^P_{bcd}\tau_a\tau_c\tau_d\Big\} \\
&= -\frac{1}{2}\sigma^4 \sum_{c=1}^{p}\sum_{d=1}^{p} A^P_{bcd}\Gamma^P_{acd} \\
&= -\frac{1}{2}\sigma^4 \mathrm{tr}(A^P_b \Gamma^P_a).
\end{aligned}
$$

Similarly, we obtain

$$
(\alpha_3)_{ab} = -\frac{1}{2}\sigma^4 \mathrm{tr}(A^P_a \Gamma^P_b).
$$

It follows from Lemma 4.4 that

$$
(\alpha_4)_{ab} = \frac{1}{2}\sigma^4 \mathrm{tr}(A^P_a A^P_b) + \frac{1}{4}\sigma^6 \sum_{c,d}\sum_{e,f} A^P_{acd} A^P_{bef} \Delta_{cdef}.
$$

Combining the expressions of α_2, α_3 and α_4, we obtain the last term of (4.20) and the theorem is proved. ∎

The approximation of variance $\mathrm{Var}(\hat{\beta})$ given in (4.20) may be not accurate enough (Clarke 1980; Seber and Wild 1989, p.184) but is quite intuitive and reasonable. In expression (4.20), the variance $\mathrm{Var}(\hat{\beta})$ is decomposed into a sum of three terms. The first term is the usual Cramér-Rao lower bound which has the order of $O(n^{-1})$ and the rest of the terms have the order of $O(n^{-2})$. The second term depends only on the intrinsic curvature and hence is invariant under parameterizations. As we shall see in Theorem 4.4, this term is closely connected with the information loss of the maximum likelihood estimator. The third term of (4.20) depends on the parameter-effects curvature and some high order moments of $\boldsymbol{Y}$. This term strongly depends on the manner of parameterizations of the model.

Efron (1975) first made a decomposition of variance for an estimator having the form like (4.20) in one-parameter curved exponential families. Amari (1982a; 1985) extended Efron's work to multi-parameter families. The decomposition and its geometric interpretation given in Theorem 4.3 are very similar to Efron and Amari's well-known results for curved exponential families. But here we deal with the general regression models. As a direct deduction of Theorem 4.3, we have the following corollary.

Corollary 1

For normal nonlinear regression models, we have

$$\mathrm{Var}(\hat{\boldsymbol{\beta}}) \approx \boldsymbol{J}_{\beta}^{-1}(\boldsymbol{Y}) + \sigma^4 \boldsymbol{L}(\boldsymbol{V}_I + \frac{1}{2}\boldsymbol{V}_P)\boldsymbol{L}^T, \tag{4.22}$$

where

$$\boldsymbol{V}_I = \sum_{\kappa=1}^{n-p} (\boldsymbol{A}_{\kappa}^I)^2, \quad (\boldsymbol{V}_P)_{ab} = \mathrm{tr}(\boldsymbol{A}_a^P \boldsymbol{A}_b^P),$$

and $\boldsymbol{A}_{\kappa}^I$ is the κ-th face of $\boldsymbol{A}^I$.

Proof. For normal nonlinear regression models, we have $\boldsymbol{V} = \boldsymbol{I}_n$, $\boldsymbol{A}_{\theta}^I = \boldsymbol{A}^I$, and $S_i = 0$ and $\Delta_i = 0$, which result in $\boldsymbol{\Gamma} = \boldsymbol{0}$ and $\Delta_{abcd} = 0$. Therefore (4.20) reduces to (4.22). ∎

The expression (4.22) has been obtained by Seber and Wild (1989) and Wei (1989; 1991).

By similar derivations to Theorem 4.3, we can find the approximate variances for $\hat{\boldsymbol{\mu}}$ and $\hat{\boldsymbol{e}}$ as follows, but the details are omitted here.

Corollary 2

The approximations of variances of $\hat{\boldsymbol{\mu}}$ and $\hat{\boldsymbol{e}}$ are respectively given by

$$\begin{aligned}\mathrm{Var}(\hat{\boldsymbol{\mu}}) \approx\ & \sigma^2 \boldsymbol{Q}\boldsymbol{Q}^T + \sigma^4 \boldsymbol{Q}\boldsymbol{V}_{\theta I}\boldsymbol{Q}^T + \\ & \frac{1}{2}\sigma^4 \left\{\boldsymbol{Q}\boldsymbol{V}_{\Gamma I}\boldsymbol{N}^T + \boldsymbol{N}\boldsymbol{V}_{I\Gamma}\boldsymbol{Q}^T + \boldsymbol{N}\boldsymbol{V}_{II}\boldsymbol{N}^T\right\} + \\ & \frac{1}{4}\sigma^6 \boldsymbol{N}\boldsymbol{V}_{I\Delta}\boldsymbol{N}^T,\end{aligned}$$

$$\begin{aligned}\mathrm{Var}(\hat{\boldsymbol{e}}) &\approx \sigma^2 \boldsymbol{N}\boldsymbol{N}^T + \sigma^4 \boldsymbol{Q}\boldsymbol{V}_{\theta I}\boldsymbol{Q}^T + \\ &\quad \frac{1}{2}\sigma^4 \boldsymbol{N}\boldsymbol{V}_{II}\boldsymbol{N}^T + \frac{1}{4}\sigma^6 \boldsymbol{N}\boldsymbol{V}_{I\triangle}\boldsymbol{N}^T,\end{aligned}$$

where

$$(\boldsymbol{V}_{\Gamma I})_{ab} = \mathrm{tr}(\boldsymbol{\Gamma}_a^P \boldsymbol{A}_b^I), \quad (\boldsymbol{V}_{I\Gamma})_{ab} = \mathrm{tr}(\boldsymbol{A}_a^I \boldsymbol{\Gamma}_b^P), \quad (\boldsymbol{V}_{II})_{ab} = \mathrm{tr}(\boldsymbol{A}_a^I \boldsymbol{A}_b^I),$$

$$(\boldsymbol{V}_{I\triangle})_{ab} = \sum_{c,d}\sum_{e,f} A_{acd}^I A_{bef}^I \triangle_{cdef}.$$

In particular, if the model is a normal nonlinear regression, then the above expressions reduce to

$$\begin{aligned}\mathrm{Var}(\hat{\boldsymbol{\mu}}) &\approx \sigma^2 \boldsymbol{Q}\boldsymbol{Q}^T + \sigma^4 \boldsymbol{Q}\boldsymbol{V}_I\boldsymbol{Q}^T + \frac{1}{2}\sigma^4 \boldsymbol{N}\boldsymbol{V}_{II}\boldsymbol{N}^T, \\ \mathrm{Var}(\hat{\boldsymbol{e}}) &\approx \sigma^2 \boldsymbol{N}\boldsymbol{N}^T + \sigma^4 \boldsymbol{Q}\boldsymbol{V}_I\boldsymbol{Q}^T + \frac{1}{2}\sigma^4 \boldsymbol{N}\boldsymbol{V}_{II}\boldsymbol{N}^T.\end{aligned}$$ ∎

The last two results have been given by Seber and Wild (1989) and Wei (1989; 1991).

In summary, we have the approximations of biases and variances for $\hat{\boldsymbol{\beta}}$, $\hat{\boldsymbol{\mu}}$ and $\hat{\boldsymbol{e}}$ in terms of curvatures. These results show that all these quantities are dominated by the nonlinearity of the model, as expected. It should be noticed that there are some differences between the behaviors of $\hat{\boldsymbol{\beta}}$ and $\hat{\boldsymbol{\mu}}$, $\hat{\boldsymbol{e}}$. Here bias($\hat{\boldsymbol{\mu}}$), E($\hat{\boldsymbol{e}}$), Var($\hat{\boldsymbol{\mu}}$) and Var($\hat{\boldsymbol{e}}$) depend only on the intrinsic curvature and hence are invariant under parameter transformations. By contrast, bias($\hat{\boldsymbol{\beta}}$) depends only on the parameter-effects curvature while Var($\hat{\boldsymbol{\beta}}$) depends on both intrinsic curvature and parameter-effects curvature.

From Theorem 3.3, we may make some parameter transformations to diminish the bias and the variance of $\hat{\boldsymbol{\beta}}$. Now let $\gamma = \gamma(\boldsymbol{\beta})$ be a one-to-one transformation from $\boldsymbol{\beta}$ to γ and $\hat{\gamma} = \gamma(\hat{\boldsymbol{\beta}})$ (see Subsection 3.2.3). If $\gamma(\boldsymbol{\beta})$ satisfies equation (3.20), then the parameter-effects curvature array $\boldsymbol{A}_\gamma^P = \boldsymbol{0}$. Therefore under this parameter, the bias of $\hat{\gamma}$ will approximately vanish and the variance of $\hat{\gamma}$ in (4.20) only has the first two terms. For the general transformation $\gamma(\boldsymbol{\beta})$, we have the following corollary.

Corollary 3

Under the parameter transformation $\gamma = \gamma(\boldsymbol{\beta})$, the bias of $\hat{\gamma} = \gamma(\hat{\boldsymbol{\beta}})$ can be represented as

$$\mathrm{bias}(\hat{\gamma}) \approx \dot{\gamma}_\beta \mathrm{bias}(\hat{\boldsymbol{\beta}}) + \frac{1}{2}\sigma^2 \mathrm{tr}[\ddot{\gamma}_\beta (\boldsymbol{D}^T\boldsymbol{D})^{-1}]. \tag{4.23}$$

Proof. It follows from (4.20) and Theorem 3.3 that

$$\begin{aligned}\text{bias}(\hat{\gamma}) &\approx -\frac{1}{2}\sigma^2 \boldsymbol{L}_\gamma \text{tr}[\boldsymbol{A}_\gamma^P] \\ &= -\frac{1}{2}\sigma^2 \dot{\gamma}_\beta \boldsymbol{L}\text{tr}[\boldsymbol{A}^P + \boldsymbol{A}^G] \\ &= \dot{\gamma}_\beta \text{bias}(\hat{\boldsymbol{\beta}}) - \frac{1}{2}\sigma^2 \dot{\gamma}_\beta \boldsymbol{L}\text{tr}[\boldsymbol{A}^G].\end{aligned}$$

It follows from (3.19) and Section A.2 of Appendix A that

$$\begin{aligned}-\frac{\sigma^2}{2}\dot{\gamma}_\beta \boldsymbol{L}\text{tr}[\boldsymbol{A}^G] &= \frac{1}{2}\sigma^2 \dot{\gamma}_\beta \boldsymbol{L}\text{tr}\Big[[\boldsymbol{R}]\boldsymbol{L}^T[\dot{\gamma}_\beta^{-1}][\ddot{\gamma}_\beta]\boldsymbol{L}\Big] \\ &= \frac{1}{2}\sigma^2 \text{tr}[\boldsymbol{L}^T \ddot{\gamma}_\beta \boldsymbol{L}] \\ &= \frac{1}{2}\sigma^2 \text{tr}[\ddot{\gamma}_\beta \boldsymbol{L}\boldsymbol{L}^T] \\ &= \frac{1}{2}\sigma^2 \text{tr}[\ddot{\gamma}_\beta (\boldsymbol{D}^T\boldsymbol{D})^{-1}].\end{aligned}$$

Combining these results leads to (4.23). ∎

In equation (3.20), we give transformations which produce zero curvature $\boldsymbol{A}_\gamma^P = \boldsymbol{0}$, then we have bias$(\hat{\gamma}) \approx \boldsymbol{0}$. However, it is not easy to get these transformations.

Under the parameter transformation $\gamma = \gamma(\boldsymbol{\beta})$, the expression of Var$(\hat{\gamma})$ is quite complicated, so the details are omitted here.

4.3 Information Loss

The information loss of an estimator is an important measure to assess the efficiency for that estimator. The problem goes back to Fisher (1925) and Rao (1961; 1962; 1963). Efron (1975) first obtained an elegant expression and pointed out the essential connection between information loss and statistical curvature. Efron's (1975) work was restricted within one-parameter curved exponential families at that time. Amari (1982a) extended Efron's result to multi-parameter curved exponential families by using a Riemannian geometric framework. Since then, many authors have been concerned with this problem (e.g. Amari 1985; Amari *et al.* 1987; Kass 1989; Okamoto, Amari and Takeuchi 1991). However, all the discussions mentioned earlier were still restricted within iid observations from curved exponential families. Hosoya (1988; 1990) investigated the information amount of estimation for very general models (such as time series, etc.) in a general form by using Edgeworth expansions. His expressions, of course, are rather complicated. For a non-iid regression case, Wei (1989; 1991) discussed this problem for the least squares estimator in normal nonlinear regression models and obtained a rather simple and intuitive expression by using the BW geometric framework. In this section, we deal with the information loss of the maximum likelihood estimator for exponential family nonlinear models which include many commonly encountered regression models as their special cases.

For model (2.2), the information loss of the maximum likelihood estimator $\hat{\boldsymbol{\beta}}$ is defined as

$$\triangle \boldsymbol{J}(\hat{\boldsymbol{\beta}}) = \boldsymbol{J}_\beta(\boldsymbol{Y}) - \boldsymbol{J}(\hat{\boldsymbol{\beta}}),$$

where $\boldsymbol{J}_\beta(\boldsymbol{Y})$ and $\boldsymbol{J}(\hat{\boldsymbol{\beta}})$ are respectively Fisher information matrices of the observed vector $\boldsymbol{Y}$ and the maximum likelihood estimator $\hat{\boldsymbol{\beta}}$ for the parameter $\boldsymbol{\beta}$. By Assumption C, $\boldsymbol{J}_\beta(\boldsymbol{Y})$ has the order of $O(n)$. We may find an approximation of $\triangle \boldsymbol{J}(\hat{\boldsymbol{\beta}})$ up to the order of $O(1)$. Now we first give a lemma.

Lemma 4.5

If Assumptions A and B hold in model (2.2), then the information loss of $\hat{\boldsymbol{\beta}}$ can be represented as

$$\triangle \boldsymbol{J}(\hat{\boldsymbol{\beta}}) = \mathrm{E}_\beta\Big\{\mathrm{Var}_\beta\Big(\frac{\partial l}{\partial \boldsymbol{\beta}}\Big|\hat{\boldsymbol{\beta}}\Big)\Big\}, \tag{4.24}$$

where $l = l(\boldsymbol{\beta})$ is given in (2.5) and $\partial l/\partial \boldsymbol{\beta} = \dot{\boldsymbol{l}}(\boldsymbol{\beta})$ is given in (2.8).

Proof. Let the log-likelihood of $\hat{\boldsymbol{\beta}}$ be $\tilde{l}(\boldsymbol{\beta})$, then by Rao (1973, p.330), $\tilde{l}(\boldsymbol{\beta})$ satisfies

$$\frac{\partial \tilde{l}}{\partial \boldsymbol{\beta}} = \mathrm{E}_\beta\Big(\frac{\partial l}{\partial \boldsymbol{\beta}}\Big|\hat{\boldsymbol{\beta}}\Big),$$

which leads to

$$\boldsymbol{J}(\hat{\boldsymbol{\beta}}) = \mathrm{Var}_\beta\Big(\frac{\partial \tilde{l}}{\partial \boldsymbol{\beta}}\Big) = \mathrm{Var}_\beta\Big\{\mathrm{E}_\beta\Big(\frac{\partial l}{\partial \boldsymbol{\beta}}\Big|\hat{\boldsymbol{\beta}}\Big)\Big\}.$$

On the other hand, $\boldsymbol{J}_\beta(\boldsymbol{Y}) = \mathrm{Var}_\beta(\partial l/\partial \boldsymbol{\beta})$ can be expressed as

$$\begin{aligned}
\boldsymbol{J}_\beta(\boldsymbol{Y}) &= \mathrm{E}_\beta\Big\{\mathrm{Var}_\beta\Big(\frac{\partial l}{\partial \boldsymbol{\beta}}\Big|\hat{\boldsymbol{\beta}}\Big)\Big\} + \mathrm{Var}_\beta\Big\{\mathrm{E}_\beta\Big(\frac{\partial l}{\partial \boldsymbol{\beta}}\Big|\hat{\boldsymbol{\beta}}\Big)\Big\} \\
&= \mathrm{E}_\beta\Big\{\mathrm{Var}_\beta\Big(\frac{\partial l}{\partial \boldsymbol{\beta}}\Big|\hat{\boldsymbol{\beta}}\Big)\Big\} + \boldsymbol{J}(\hat{\boldsymbol{\beta}}).
\end{aligned}$$

This yields (4.24). ∎

Using this lemma, we get the following theorem.

Theorem 4.4

For model (2.2), if Assumptions A to C hold, then the information loss of the maximum likelihood estimator $\hat{\boldsymbol{\beta}}$ can be approximately represented as

$$\Delta \boldsymbol{J}(\hat{\boldsymbol{\beta}}) \approx \boldsymbol{R}^T \boldsymbol{V}_{\theta I} \boldsymbol{R}, \tag{4.25}$$

where $\boldsymbol{V}_{\theta I}$ is given in (4.21).

Proof. In order to use formula (4.24), we first derive a stochastic expansion of $\partial l/\partial \boldsymbol{\beta}$. From equations (2.8) and (4.7), $\partial l/\partial \boldsymbol{\beta}$ can be written as

$$\begin{aligned}\frac{\partial l}{\partial \boldsymbol{\beta}} &= \sigma^{-2} \boldsymbol{D}^T(\boldsymbol{\beta}) \boldsymbol{V}^{-1}(\boldsymbol{\beta}) \boldsymbol{e}(\boldsymbol{\beta}) \\ &= \sigma^{-2} \boldsymbol{D}^T \boldsymbol{V}^{-1} \{\Delta \boldsymbol{\mu} + \boldsymbol{N}(\hat{\boldsymbol{\beta}}) \hat{\boldsymbol{\nu}}\}.\end{aligned}$$

It follows from Lemma 4.2 that

$$\begin{aligned}\frac{\partial l}{\partial \boldsymbol{\beta}} &= \sigma^{-2} \boldsymbol{D}^T \boldsymbol{V}^{-1} \Big\{\Delta \boldsymbol{\mu} + \boldsymbol{N} \hat{\boldsymbol{\nu}} + [\hat{\boldsymbol{\nu}}^T][\boldsymbol{F}] \Delta \boldsymbol{\beta} + \boldsymbol{\gamma}' \hat{\boldsymbol{\nu}}\Big\} \\ &= \sigma^{-2} \Big\{\boldsymbol{D}^T \boldsymbol{V}^{-1} \Delta \boldsymbol{\mu} + [\hat{\boldsymbol{\nu}}^T][\boldsymbol{D}^T \boldsymbol{V}^{-1} \boldsymbol{F}] \Delta \boldsymbol{\beta} + \boldsymbol{\alpha}_1'\Big\},\end{aligned}$$

where $\boldsymbol{\alpha}_1' = \boldsymbol{D}^T \boldsymbol{V}^{-1} \boldsymbol{\gamma}' \hat{\boldsymbol{\nu}}$ is of the order of $O_p(n^{-\frac{1}{2}})$ (see the proof of Theorem 4.2). It follows from Assumption C, Lemma 4.2 and (4.11) that

$$\frac{\partial l}{\partial \boldsymbol{\beta}} = \sigma^{-2} \Big\{\boldsymbol{D}^T \boldsymbol{V}^{-1} \Delta \boldsymbol{\mu} - [\boldsymbol{\lambda}^T][\boldsymbol{R}^T \boldsymbol{A}_\theta^I \boldsymbol{R}] \Delta \boldsymbol{\beta}\Big\} + O_p(n^{-\frac{1}{2}}).$$

Then from Theorem 4.2 we get

$$\begin{aligned}\mathrm{Var}_\beta\Big(\frac{\partial l}{\partial \boldsymbol{\beta}}\Big|\hat{\boldsymbol{\beta}}\Big) &\approx \sigma^{-4} \mathrm{Var}_\beta \Big\{\boldsymbol{R}^T [\boldsymbol{\lambda}^T][\boldsymbol{A}_\theta^I] \boldsymbol{R} \Delta \boldsymbol{\beta} | \hat{\boldsymbol{\beta}}\Big\} \\ &\approx \sigma^{-4} \sum_{\kappa=1}^{n-p} \mathrm{Var}_\beta \Big\{\lambda_\kappa \boldsymbol{R}^T \boldsymbol{A}_{\theta\kappa}^I \boldsymbol{R} \Delta \boldsymbol{\beta} | \hat{\boldsymbol{\beta}}\Big\} \\ &\approx \sigma^{-2} \sum_{\kappa=1}^{n-p} \Big(\boldsymbol{R}^T \boldsymbol{A}_{\theta\kappa}^I \boldsymbol{R} \Delta \boldsymbol{\beta} \Delta \boldsymbol{\beta}^T \boldsymbol{R}^T \boldsymbol{A}_{\theta\kappa}^I \boldsymbol{R}\Big).\end{aligned}$$

It follows from (4.11) that $\mathrm{E}_\beta(\Delta \boldsymbol{\beta} \Delta \boldsymbol{\beta}^T) \approx \sigma^2 \boldsymbol{L} \boldsymbol{L}^T$. Therefore from (4.24) we get

$$\begin{aligned}\Delta \boldsymbol{J}(\hat{\boldsymbol{\beta}}) &= \mathrm{E}_\beta \Big\{\mathrm{Var}_\beta\Big(\frac{\partial l}{\partial \boldsymbol{\beta}}\Big|\hat{\boldsymbol{\beta}}\Big)\Big\} \\ &\approx \sum_{\kappa=1}^{n-p} \Big\{\boldsymbol{R}^T (\boldsymbol{A}_{\theta\kappa}^I)^2 \boldsymbol{R}\Big\} \\ &= \boldsymbol{R}^T \boldsymbol{V}_{\theta I} \boldsymbol{R},\end{aligned}$$

which leads to (4.25). ∎

From Assumption C and its succeeding description, the approximate expression given in (4.25) is of the order of $O(1)$ while $\boldsymbol{J}_\beta(\boldsymbol{Y})$ is of the order of $O(n)$. This theorem also shows that the approximate information loss of $\hat{\boldsymbol{\beta}}$ depends only on the intrinsic curvature, so it is invariant under parameterizations. This result is consistent with the well-known results given by Efron (1975) and Amari (1982a). Note that there is one more term in Amari's expression, but this term is determined by the estimate and vanishes for the maximum likelihood estimator. But the models studied here are essentially different from the models studied by Efron (1975) and Amari (1982a; 1985). In fact, the former is actually a general regression model and the latter is based on iid observations.

Corollary 1

The approximate variance of $\hat{\boldsymbol{\beta}}$ can be written as

$$\begin{aligned}\mathrm{Var}(\hat{\boldsymbol{\beta}}) &\approx \boldsymbol{J}^{-1}(\boldsymbol{Y}) + \sigma^4 \boldsymbol{J}^{-1}(\boldsymbol{Y})\Delta \boldsymbol{J}(\hat{\boldsymbol{\beta}})\boldsymbol{J}^{-1}(\boldsymbol{Y}) + \\ &\quad \frac{1}{2}\sigma^4 \boldsymbol{L}(\boldsymbol{V}_{P\Gamma} + \frac{1}{2}\sigma^2 \boldsymbol{V}_{P\Delta})\boldsymbol{L}^T.\end{aligned}$$

Proof. Since $J(\boldsymbol{Y}) = \sigma^{-2}\boldsymbol{D}^T\boldsymbol{V}^{-1}\boldsymbol{D} = \sigma^{-2}\boldsymbol{R}^T\boldsymbol{R},\ \boldsymbol{L} = \boldsymbol{R}^{-1}$, we have

$$\Delta \boldsymbol{J}(\hat{\boldsymbol{\beta}}) \approx \sigma^4 \boldsymbol{J}(\boldsymbol{Y})(\boldsymbol{L}\boldsymbol{V}_{\theta I}\boldsymbol{L}^T)\boldsymbol{J}(\boldsymbol{Y}).$$

From this and (4.20), we get the result. ∎

This result shows that the variance of an estimator sounds as if it is "proportional to" the information loss of that estimator.

Corollary 2

For normal nonlinear regression models, we have

$$\Delta \boldsymbol{J}(\hat{\boldsymbol{\beta}}) \approx \boldsymbol{R}^T\boldsymbol{V}_I\boldsymbol{R} = \sigma^4 \boldsymbol{J}(\boldsymbol{Y})(\boldsymbol{L}\boldsymbol{V}_I\boldsymbol{L}^T)\boldsymbol{J}(\boldsymbol{Y}),$$

$$\mathrm{Var}(\hat{\boldsymbol{\beta}}) \approx \boldsymbol{J}^{-1}(\boldsymbol{Y}) + \sigma^4 \boldsymbol{J}^{-1}(\boldsymbol{Y})\Delta \boldsymbol{J}(\hat{\boldsymbol{\beta}})\boldsymbol{J}^{-1}(\boldsymbol{Y}) + \frac{1}{2}\sigma^4 \boldsymbol{L}\boldsymbol{V}_P\boldsymbol{L}^T,$$

where $\boldsymbol{V}_I$ and $\boldsymbol{V}_P$ are given in (4.22) (Wei 1989; 1991). ∎

The relative information loss of an estimate has been discussed by Kass (1989). For model (2.2) the relative information loss of $\hat{\boldsymbol{\beta}}$ can be defined as

$$\Delta \boldsymbol{J}_R(\hat{\boldsymbol{\beta}}) = \boldsymbol{J}^{-1}(\boldsymbol{Y})\Delta \boldsymbol{J}(\hat{\boldsymbol{\beta}}).$$

Corollary 3

The trace of the relative information loss of $\hat{\boldsymbol{\beta}}$ can be represented as

$$\text{tr}\{\triangle \boldsymbol{J}_R(\hat{\boldsymbol{\beta}})\} \approx \sigma^2 \sum_{\kappa=1}^{n-p}\sum_{a=1}^{p}\sum_{b=1}^{p}(\boldsymbol{A}_\theta^I)^2_{\kappa ab},$$

where $(\boldsymbol{A}_\theta^I)_{\kappa ab}$ is the element of $\boldsymbol{A}_\theta^I$ at (κ, a, b).

Proof. It follows from (2.11) and (4.25) that

$$\begin{aligned}\text{tr}\,\{\triangle \boldsymbol{J}_R(\hat{\boldsymbol{\beta}})\} &\approx \text{tr}\left\{\sigma^2(\boldsymbol{D}^T\boldsymbol{V}^{-1}\boldsymbol{D})^{-1}\boldsymbol{R}^T\boldsymbol{V}_{\theta I}\boldsymbol{R}\right\}\\ &= \sigma^2\text{tr}(\boldsymbol{L}\boldsymbol{L}^T\boldsymbol{R}^T\boldsymbol{V}_{\theta I}\boldsymbol{R})\\ &= \sigma^2\text{tr}(\boldsymbol{V}_{\theta I})\\ &= \sigma^2\sum_{\kappa=1}^{n-p}\text{tr}(\boldsymbol{A}_{\theta\kappa}^I)^2,\end{aligned}$$

which leads to the desired result. ∎

Kass (1989) presented a similar result for one-parameter curved exponential families based on iid observations. In the following, we give an expression of information loss in terms of the Efron's curvature.

Corollary 4

If $\boldsymbol{\beta}$ is a scalar, then we have

$$\triangle \boldsymbol{J}(\hat{\boldsymbol{\beta}}) = \{n^{-1}\boldsymbol{J}_\beta(\boldsymbol{Y})\}\tilde{\gamma}_\beta^2 \to \phi\boldsymbol{K}(\boldsymbol{\beta})\bar{\gamma}_\beta^2 \quad (\text{as } n \to \infty), \tag{4.26}$$

where $\boldsymbol{K}(\boldsymbol{\beta})$ is given in (4.1), $\tilde{\gamma}_\beta^2$ is the Efron curvature given in (3.37) and $\bar{\gamma}_\beta^2$ is the limit of $\tilde{\gamma}_\beta^2$ (as $n \to \infty$), which is assumed to exist.

Proof. When $\boldsymbol{\beta}$ is a scalar, $\boldsymbol{A}_\theta^I$ is given in Subsection 3.4.3 and it may be written as $\boldsymbol{A}_\theta^I = (\ddot{\boldsymbol{\theta}}_\beta)_N(\dot{\boldsymbol{\theta}}_\beta^T\boldsymbol{V}\dot{\boldsymbol{\theta}}_\beta)^{-1}$ with $(\ddot{\boldsymbol{\theta}}_\beta)_N = \boldsymbol{N}_\theta^T\boldsymbol{V}\ddot{\boldsymbol{\theta}}_\beta$. Hence $\boldsymbol{V}_{\theta I}$ given in (4.21) becomes

$$\begin{aligned}\boldsymbol{V}_{\theta I} &= \frac{(\ddot{\boldsymbol{\theta}}_\beta)_N^T(\ddot{\boldsymbol{\theta}}_\beta)_N}{(\dot{\boldsymbol{\theta}}_\beta^T\boldsymbol{V}\dot{\boldsymbol{\theta}}_\beta)^2}\\ &= \phi\gamma_\beta^2 = n^{-1}\phi\tilde{\gamma}_\beta^2,\end{aligned}$$

obtained from (3.36) and (3.37). On the other hand, if $\boldsymbol{\beta}$ is a scalar, it follows from Subsection 3.4.3 that $R = R^T = \|\dot{\boldsymbol{\mu}}_\beta\| = \|\dot{\boldsymbol{\theta}}_\beta\|_\theta$, $L = L^T = \|\dot{\boldsymbol{\theta}}_\beta\|_\theta^{-1}$ and it follows from (2.11) that $\boldsymbol{J}_\beta(\boldsymbol{Y}) = \phi\|\dot{\boldsymbol{\mu}}_\beta\|^2 = \phi\|\dot{\boldsymbol{\theta}}_\beta\|_\theta^2$. Therefore (4.25) becomes

$$\triangle \boldsymbol{J}(\hat{\beta}) = n^{-1}\boldsymbol{J}_\beta(\boldsymbol{Y})\tilde{\gamma}_\beta^2,$$

which leads to the desired result. ∎

Expression (4.26) shown in Corollary 4 is the same as Efron's (1975) well-known expression (9.1), but here we treat the non-iid regression case. The problem of information loss seems a successful example of differential geometric method in statistical inference. If one adopts the classical approach to derive the information loss, the expression is much more complicated and hard to explain, as pointed out by Efron (1975; see also Hosoya 1990; Okamoto, Amari and Takeuchi 1991).

4.4 Observed Information and Fisher Information

In the literature, minus the second order derivative of the log-likelihood function at the estimated value is usually called the observed information while the Fisher information is sometimes called the expected information (Efron and Hinkley 1978). The relationship between the observed information and the Fisher information is a common problem in statistical inference (e.g. Pierce 1975; Efron 1978; Efron and Hinkley 1978; Amari 1982b; 1985; McCullagh 1983; Skovgaard 1985; Wei and Zhao 1987; Wei 1988; 1991; Kass 1989).

For exponential family nonlinear models (2.2), let

$$\boldsymbol{\Lambda} = \sqrt{n}\{-\ddot{\boldsymbol{l}}(\boldsymbol{\beta})\boldsymbol{J}_{\beta}^{-1}(\boldsymbol{Y}) - \boldsymbol{I}_p\}_{\beta=\hat{\beta}}, \tag{4.27}$$

which may be rewritten as

$$\boldsymbol{\Lambda} = \sqrt{n}\{-\ddot{\boldsymbol{l}}(\hat{\boldsymbol{\beta}}) - \boldsymbol{J}_{\hat{\beta}}(\boldsymbol{Y})\}\boldsymbol{J}_{\hat{\beta}}^{-1}(\boldsymbol{Y}),$$

that is just the standardized version of the observed information at $\hat{\boldsymbol{\beta}}$ (Skovgaard 1985). Pierce (1975) explicitly introduced this expression (but without $\sqrt{n}$) and pointed out that $-\ddot{\boldsymbol{l}}(\hat{\boldsymbol{\beta}}) = \boldsymbol{J}_{\hat{\beta}}(\boldsymbol{Y}) + O_p(n^{\frac{1}{2}})$ may hold (note that $\boldsymbol{J}_{\beta}(\boldsymbol{Y}) = O(n)$). He also indicated the importance of Fisher's (1925) argument that suggests the use of $-\ddot{\boldsymbol{l}}(\hat{\boldsymbol{\beta}})$ to get ancillary statistic. Efron and Hinkley (1978) discussed this problem in detail for some specific one-parameter families based on the iid observations. They argued that $(\hat{\beta}, -\ddot{l}(\hat{\beta}))$ acts like a sufficient statistic and showed that Λ in some one-parameter case is asymptotically normally distributed as $N(0, \gamma_{\beta}^2)$ where γ_{β} is Efron curvature. They further defined a statistic Q as

$$Q = \frac{1 - \{-\ddot{l}(\hat{\beta})/J_{\hat{\beta}}(\boldsymbol{Y})\}}{\gamma_{\beta}},$$

which is an asymptotic ancillary statistic because $\sqrt{n}Q \xrightarrow{L} N(0,1)$ as $n \to \infty$, while Q is the function of $(\hat{\beta}, \ddot{l}(\hat{\beta}))$ linear in $\ddot{l}(\hat{\beta})$ for $\hat{\beta}$ fixed (Efron and Hinkley 1978, p.470). Since then, several authors have investigated this problem and a lot of progress has been made. Amari (1982b) studied the asymptotic ancillary statistic and conditional inference for multi-parameter curved exponential families from the geometric point of view. Wei and Zhao (1987) and Wei (1988) discussed the asymptotic normality of $\boldsymbol{\Lambda}$ based on its vectorization and trace in multi-parameter curved exponential families. Skovgaard (1985) investigated the second order asymptotic ancillary for very general models by using Edgeworth expansions. His results are more precise but, of course, rather complicated. However, all the discussions mentioned earlier were still restricted within iid observations. For the non-iid case, Wei (1989; 1991) studied the asymptotic ancillary and conditional inference for a special case: normal nonlinear regression models, by using the BW geometric framework. This section will investigate some analogous problems in exponential family nonlinear models which include several important regression models as their special cases.

For the distribution of $\boldsymbol{\Lambda}$, as pointed out by Efron and Hinkley (1978, p.482), there are certain problems of definition with the case of multi-parameter families. Indeed, since $\boldsymbol{\Lambda}$ given in (4.27) is a $p \times p$ matrix, we can not directly study its asymptotic distribution. As an alternative, we shall investigate the asymptotic behaviors of vectorization and trace of $\boldsymbol{\Lambda}$, which are consistent with one-parameter case when $p = 1$.

Theorem 4.5

Suppose that Assumptions A to C hold for model (2.2), and the following limits exist for any indices a, b, c and $d = 1, \cdots, p$

$$\lim_{n\to\infty} \frac{1}{n} (\ddot{\boldsymbol{\theta}}_N)_{ab}^T (\ddot{\boldsymbol{\theta}}_N)_{cd}, \tag{4.28}$$

where $(\ddot{\boldsymbol{\theta}}_N)_{ab} = \boldsymbol{N}_\theta^T \boldsymbol{V} \ddot{\boldsymbol{\theta}}_{ab}$ and $\ddot{\boldsymbol{\theta}}_{ab} = \partial^2 \boldsymbol{\theta} / \partial\beta_a \partial\beta_b$ is an n-vector and $(\ddot{\boldsymbol{\theta}}_N)_{cd}$ is similarly defined. Then $\mathrm{Vec}(\boldsymbol{\Lambda})$ is asymptotically normally distributed as $N(\boldsymbol{0}, \sigma^2 \boldsymbol{\Sigma})$ with

$$\boldsymbol{\Sigma} = (\boldsymbol{K}^{-1} \otimes \boldsymbol{I}_p) \boldsymbol{\Omega} (\boldsymbol{K}^{-1} \otimes \boldsymbol{I}_p), \tag{4.29}$$

$$\boldsymbol{\Omega} = \lim_{n\to\infty} \frac{1}{n} \left\{ \mathrm{Vec}\, [\boldsymbol{R}^T \boldsymbol{A}_\theta^I \boldsymbol{R}] \right\} \left\{ \mathrm{Vec}\, [\boldsymbol{R}^T \boldsymbol{A}_\theta^I \boldsymbol{R}] \right\}^T, \tag{4.30}$$

where $\otimes$ denotes the Kronecker product and $\mathrm{Vec}[\cdot]$ denotes the vectorization of an array (see Section A.2).

Proof. $-\ddot{l}(\boldsymbol{\beta})$ and $\boldsymbol{J}_\beta(\boldsymbol{Y})$ are respectively given in (2.9) and (2.11), so $\boldsymbol{\Lambda}$ may be written as

$$\boldsymbol{\Lambda} \;=\; -\sqrt{n} [\hat{\boldsymbol{e}}^T] [\hat{\boldsymbol{W}}_\theta] (\hat{\boldsymbol{D}}^T \hat{\boldsymbol{V}}^{-1} \hat{\boldsymbol{D}})^{-1}$$

$$= -\frac{1}{\sqrt{n}}[\hat{\boldsymbol{\nu}}^T][[\hat{\boldsymbol{N}}^T][\hat{\boldsymbol{W}}_\theta]]\left(\frac{\hat{\boldsymbol{D}}^T\hat{\boldsymbol{V}}^{-1}\hat{\boldsymbol{D}}}{n}\right)^{-1},$$

where $\hat{\boldsymbol{D}} = \boldsymbol{D}(\hat{\boldsymbol{\beta}})$, $\hat{\boldsymbol{W}}_\theta = \boldsymbol{W}_\theta(\hat{\boldsymbol{\beta}})$, $\hat{\boldsymbol{N}} = \boldsymbol{N}(\hat{\boldsymbol{\beta}})$ and $\hat{\boldsymbol{V}} = \boldsymbol{V}(\hat{\boldsymbol{\beta}})$ whose elements may be replaced by the elements of $\boldsymbol{D}, \boldsymbol{W}_\theta, \boldsymbol{N}$ and $\boldsymbol{V}$ respectively because $\hat{\boldsymbol{\beta}} - \boldsymbol{\beta} = O_p(n^{-\frac{1}{2}})$ and the Taylor series expansions can be used for each of their elements. Thus we have

$$\begin{aligned}
\boldsymbol{\Lambda} &= -\frac{1}{\sqrt{n}}[\hat{\boldsymbol{\nu}}^T][[\boldsymbol{N}_\theta^T\boldsymbol{V}][\boldsymbol{W}_\theta]]\left(\frac{\boldsymbol{D}^T\boldsymbol{V}^{-1}\boldsymbol{D}}{n}\right)^{-1} + O_p(n^{-\frac{1}{2}}) && (4.31)\\
&= -\frac{1}{\sqrt{n}}[\hat{\boldsymbol{\nu}}^T][[\boldsymbol{N}_\theta^T\boldsymbol{V}][\boldsymbol{W}_\theta]]\boldsymbol{K}^{-1} + O_p(n^{-\frac{1}{2}}), && (4.32)
\end{aligned}$$

where $\boldsymbol{N} = \boldsymbol{V}\boldsymbol{N}_\theta$ and the assumption (4.1) are used. It follows from (3.23) and (4.11) that

$$\boldsymbol{\Lambda} = -\frac{1}{\sqrt{n}}[\boldsymbol{\lambda}^T][\boldsymbol{R}^T\boldsymbol{A}_\theta^I\boldsymbol{R}]\boldsymbol{K}^{-1} + O_p(n^{-\frac{1}{2}}).$$

Taking vectorization of $\boldsymbol{\Lambda}$ and using formulas (11) and (12) in Section A.2, we obtain

$$\begin{aligned}
\text{Vec}\,(\boldsymbol{\Lambda}) &= -\frac{1}{\sqrt{n}}(\boldsymbol{K}^{-1}\otimes\boldsymbol{I}_p)\text{Vec}\left\{[\boldsymbol{\lambda}^T][\boldsymbol{R}^T\boldsymbol{A}_\theta^I\boldsymbol{R}]\right\} + O_p(n^{-\frac{1}{2}})\\
&= -(\boldsymbol{K}^{-1}\otimes\boldsymbol{I}_p)\frac{1}{\sqrt{n}}\left\{\text{Vec}\,[\boldsymbol{R}^T\boldsymbol{A}_\theta^I\boldsymbol{R}]\right\}\boldsymbol{\lambda} + O_p(n^{-\frac{1}{2}}).
\end{aligned}$$

By Theorem 4.1, each element of $\boldsymbol{\lambda}$ has asymptotic distribution $N(0,\sigma^2)$, hence Vec$(\boldsymbol{\Lambda})$ is also asymptotically normally distributed with $N(\boldsymbol{0},\sigma^2\boldsymbol{\Sigma})$ and $\boldsymbol{\Sigma}$ is given in (4.29). To see the existence of the limit (4.30), we write (4.30) as

$$\begin{aligned}
\boldsymbol{\Omega} &= \lim_{n\to\infty}\frac{1}{n}\left\{\text{Vec}\left[[\boldsymbol{N}_\theta^T\boldsymbol{V}][\boldsymbol{W}_\theta]\right]\right\}\left\{\text{Vec}\left[[\boldsymbol{N}_\theta^T\boldsymbol{V}][\boldsymbol{W}_\theta]\right]\right\}^T\\
&= \lim_{n\to\infty}\frac{1}{n}\left\{\text{Vec}\,[\ddot{\boldsymbol{\theta}}_N]\right\}\left\{\text{Vec}\,[\ddot{\boldsymbol{\theta}}_N]\right\}^T,
\end{aligned}$$

where $\ddot{\boldsymbol{\theta}}_N = [\boldsymbol{N}_\theta^T\boldsymbol{V}][\boldsymbol{W}_\theta]$ is an $(n-p)\times p\times p$ array. Now for any integers $1\le s\le p^2$ and $1\le t\le p^2$, s and t can be uniquely decomposed as $s=(b-1)p+a$ and $t=(d-1)p+c$ respectively with $1\le a,b,c,d\le p$. It is easily seen that the element of $\boldsymbol{\Omega}$ at (s,t) is just

$$\Omega_{st} = \lim_{n\to\infty}\frac{1}{n}(\ddot{\boldsymbol{\theta}}_N)_{ab}^T(\ddot{\boldsymbol{\theta}}_N)_{cd},$$

which exists by the assumption (4.28) and hence we complete the proof of the theorem. ∎

Theorem 4.6

For model (2.2), if Assumption A to C hold and the following limit exists:

$$\lim_{n\to\infty}\frac{1}{n}\Big\{\mathrm{tr}\,[\ddot{\boldsymbol{\theta}}_N\boldsymbol{K}^{-1}]\Big\}^T\Big\{\mathrm{tr}\,[\ddot{\boldsymbol{\theta}}_N\boldsymbol{K}^{-1}]\Big\} = \lim_{n\to\infty} n\Big\{\mathrm{tr}\,[\boldsymbol{A}_\theta^I]\Big\}^T\Big\{\mathrm{tr}\,[\boldsymbol{A}_\theta^I]\Big\} = a^2, \tag{4.33}$$

where $\ddot{\boldsymbol{\theta}}_N = [\boldsymbol{N}_\theta^T\boldsymbol{V}][\boldsymbol{W}_\theta]$, then $\mathrm{tr}(\boldsymbol{\Lambda})$ is asymptotically normally distributed as $N(0,\ \sigma^2 a^2)$.

Proof. It follows from (4.32) that

$$\begin{aligned}\mathrm{tr}\,(\boldsymbol{\Lambda}) &= -\frac{1}{\sqrt{n}}\mathrm{tr}\left\{[\boldsymbol{\lambda}^T][\ddot{\boldsymbol{\theta}}_N]\boldsymbol{K}^{-1}\right\} + O_p(n^{-\frac{1}{2}})\\ &= -\frac{1}{\sqrt{n}}\Big\{\mathrm{tr}[\ddot{\boldsymbol{\theta}}_N\boldsymbol{K}^{-1}]\Big\}^T\boldsymbol{\lambda} + O_p(n^{-\frac{1}{2}}),\end{aligned}$$

where $\mathrm{tr}[\cdot]$ is defined in Section A.2. From the first equation of (4.33), we get $\mathrm{tr}(\boldsymbol{\Lambda}) \xrightarrow{L} N(0, \sigma^2 a^2)$ as $n \to \infty$. On the other hand, since $(\boldsymbol{D}^T\boldsymbol{V}^{-1}\boldsymbol{D})^{-1} = \boldsymbol{L}\boldsymbol{L}^T$, it follows from (4.32) that

$$\begin{aligned}\mathrm{tr}\,(\boldsymbol{\Lambda}) &= -\sqrt{n}\mathrm{tr}\left\{[\boldsymbol{\lambda}^T][\boldsymbol{R}^T\boldsymbol{A}_\theta^I\boldsymbol{R}]\boldsymbol{L}\boldsymbol{L}^T\right\} + O_p(n^{-\frac{1}{2}})\\ &= -\sqrt{n}\Big\{\mathrm{tr}\,[\boldsymbol{R}^T\boldsymbol{A}_\theta^I\boldsymbol{L}^T]\Big\}^T\boldsymbol{\lambda} + O_p(n^{-\frac{1}{2}})\\ &= -\sqrt{n}\Big\{\mathrm{tr}\,[\boldsymbol{A}_\theta^I]\Big\}^T\boldsymbol{\lambda} + O_p(n^{-\frac{1}{2}}).\end{aligned}$$

From the second equation of (4.33), we get the desired result. ∎

Note that the assumption (4.28) is actually analogous to Assumption C, but a little bit stronger, while the assumption (4.33) is actually analogous to (4.28). In fact, as pointed out in the succeeding paragraph of Assumption C, $\ddot{\boldsymbol{\theta}}_{ab}$ seems to have the order of $n^{\frac{1}{2}}$ and $\boldsymbol{A}_\theta^I$ seems to have the order of $n^{-\frac{1}{2}}$, so the assumption (4.33) is not surprising.

From Theorems 4.5 and 4.6, we can get several corollaries which are summarized as follows.

(1) The asymptotic distribution of $\mathrm{Vec}(\boldsymbol{\Lambda})$ and $\mathrm{tr}(\boldsymbol{\Lambda})$ depend only on the intrinsic curvature $\boldsymbol{A}_\theta^I$, so they are invariant under parameter transformations. ∎

(2) The statistics $\boldsymbol{\Sigma}^{-1/2}\mathrm{Vec}(\boldsymbol{\Lambda})$ and $a^{-1}\mathrm{tr}(\boldsymbol{\Lambda})$ are asymptotic ancillary statistics because their asymptotic distributions are standard normal and hence independent of the parameter β. Similar results have been studied by Efron and Hinkley (1978), Amari (1982b), Wei and Zhao (1987) and Wei (1988) for curved exponential families. From (4.5), we can see that $(\hat{\beta}, \hat{\nu})$ is a sufficient statistic, while (4.32) shows that Λ is a function of $\hat{\nu}$ in asymptotic sense. Therefore $(\hat{\beta}, \Lambda)$ acts like an asymptotic sufficient statistic. Furthermore, as we pointed out after (4.27), since Λ is just the standardized version of the observed information, one may regard $(\hat{\beta}, \ddot{l}(\hat{\beta}))$ as an asymptotic sufficient statistic. Skovgaard (1985) discussed this in detail. ∎

(3) For normal nonlinear regression model, which is the special case of (2.2), we have $\boldsymbol{V} = \boldsymbol{I}_n, \boldsymbol{\theta} = \boldsymbol{\mu}$, and $\boldsymbol{A}_\theta^I = \boldsymbol{A}^I$, then Theorems 4.5 and 4.6 lead to the results of Wei (1991). ∎

(4) It follows from Theorems 4.5 and 4.6 that

$$\Lambda = \sqrt{n}\Big\{-\ddot{l}(\hat{\beta}) - \boldsymbol{J}_{\hat{\beta}}(\boldsymbol{Y})\Big\}\boldsymbol{J}_{\hat{\beta}}^{-1}(\boldsymbol{Y}) = O_p(1).$$

Since $\boldsymbol{J}_{\hat{\beta}}^{-1}(\boldsymbol{Y}) = O_p(n^{-1})$ (see Assumption C), we get $-\ddot{l}(\hat{\beta}) - \boldsymbol{J}_{\hat{\beta}}(\boldsymbol{Y}) = O_p(n^{\frac{1}{2}})$ or $-\ddot{l}^{-1}(\hat{\beta}) = \boldsymbol{J}_{\hat{\beta}}^{-1}(\boldsymbol{Y}) + O_p(n^{-3/2})$. Similar results to these for iid observations have been mentioned by Pierce (1975), Efron and Hinkley (1978), Skovgaard (1985) and Kass (1989). McCullagh (1983) stated that these results may hold for nonlinear models based on quasi-likelihood functions. ∎

(5) If β is a scalar, then we have

$$\Lambda \xrightarrow{L} N(0,\ \overline{\gamma}_\beta^2), \qquad n \to \infty, \tag{4.34}$$

where $\overline{\gamma}_\beta^2$ is the average Efron curvature given in (3.37) and (4.26).

Proof. When β is a scalar, then $p = 1$ and $\mathrm{Vec}(\Lambda) = \mathrm{tr}(\Lambda) = \Lambda$ is also a scalar. Hence Theorems 4.5 and 4.6 lead to the same result and Λ is asymptotically normally distributed as $N(0, \sigma^2 a^2)$. Here a^2 is determined by (4.29) or (4.33) with the same value

$$a^2 = \lim_{n\to\infty} \frac{1}{n}(\ddot{\boldsymbol{\theta}}_\beta)_N^T(\ddot{\boldsymbol{\theta}}_\beta)_N \boldsymbol{K}^{-2},$$

where $(\ddot{\boldsymbol{\theta}}_\beta)_N = \boldsymbol{N}_\theta^T \boldsymbol{V}\ddot{\boldsymbol{\theta}}_\beta$, $\ddot{\boldsymbol{\theta}}_\beta = \partial^2\boldsymbol{\theta}/\partial\beta^2$ and the matrix $\boldsymbol{K}$ given in (4.1) is reduced to

$$\boldsymbol{K} = \lim_{n\to\infty} \frac{1}{n}\dot{\boldsymbol{\theta}}_\beta^T \boldsymbol{V}\dot{\boldsymbol{\theta}}_\beta.$$

Thus we get

$$\begin{aligned} a^2 &= \lim_{n\to\infty} \frac{n^{-1}(\ddot{\boldsymbol{\theta}}_\beta)_N^T(\ddot{\boldsymbol{\theta}}_\beta)_N}{(n^{-1}\dot{\boldsymbol{\theta}}_\beta^T \boldsymbol{V}\dot{\boldsymbol{\theta}}_\beta)^2} \\ &= \lim_{n\to\infty} \phi\tilde{\gamma}_\beta^2 = \sigma^{-2}\overline{\gamma}_\beta^2, \end{aligned}$$

which follows from (3.36), (3.37) and (4.26). Then we have $\sigma^2 a^2 = \overline{\gamma}_\beta^2$ that leads to (4.34). ∎

Equation (4.34) exactly coincides with the Lemma 2 of Efron and Hinkley (1978), but here we treat the non-iid regression case.

Finally, we briefly investigate the approximate conditional moments of $\hat{\boldsymbol{\beta}}$. It follows from (4.9) and (4.11) that

$$\Delta\boldsymbol{\beta} = \boldsymbol{L\tau} + \boldsymbol{L}\{[\hat{\boldsymbol{\nu}}][\boldsymbol{A}_\theta^I]\boldsymbol{\tau} - \frac{1}{2}\boldsymbol{\tau}^T\boldsymbol{A}^P\boldsymbol{\tau}\} + O_p(n^{-3/2}).$$

From this expansion we have $\mathrm{E}(\Delta\boldsymbol{\beta}|\hat{\boldsymbol{\nu}}) \approx -\frac{1}{2}\sigma^2\boldsymbol{L}\mathrm{tr}[\boldsymbol{A}^P]$ which is independent of $\hat{\boldsymbol{\nu}}$ (so is the third moment, see Skovgaard 1985). The conditional variance of $\hat{\boldsymbol{\beta}}$ given $\hat{\boldsymbol{\nu}}$ may be roughly denoted by

$$\mathrm{Var}_\beta(\hat{\boldsymbol{\beta}}|\hat{\boldsymbol{\nu}}) \approx \sigma^2\boldsymbol{L}\boldsymbol{L}^T + \sigma^2\boldsymbol{L}([\hat{\boldsymbol{\nu}}^T][\boldsymbol{A}_\theta^I])\boldsymbol{L}^T + \sigma^2\boldsymbol{L}([\hat{\boldsymbol{\nu}}^T][\boldsymbol{A}_\theta^I])^2\boldsymbol{L} + \cdots.$$

On the other hand, it follows from (2.9) and (4.4) that

$$\begin{aligned} -\ddot{l}^{-1}(\hat{\boldsymbol{\beta}}) &= \sigma^2\{\boldsymbol{R}^T\boldsymbol{R} - \boldsymbol{R}^T[\hat{\boldsymbol{\nu}}^T][\boldsymbol{A}_\theta^I]\boldsymbol{R}\}_{\hat{\beta}}^{-1} \\ &= \sigma^2\hat{\boldsymbol{L}}\{\boldsymbol{I}_p - [\hat{\boldsymbol{\nu}}^T][\hat{\boldsymbol{A}}_\theta^I]\}^{-1}\hat{\boldsymbol{L}}^T, \end{aligned}$$

where $\hat{\boldsymbol{L}} = \boldsymbol{L}(\hat{\boldsymbol{\beta}})$ and $\hat{\boldsymbol{A}}_\theta^I = \boldsymbol{A}_\theta^I(\hat{\boldsymbol{\beta}})$. We may use the expansion $(\boldsymbol{I}_p - \boldsymbol{A})^{-1}$ $= \boldsymbol{I}_p + \boldsymbol{A} + \boldsymbol{A}^2 + \cdots$ for the above equation to obtain

$$-\ddot{l}^{-1}(\hat{\boldsymbol{\beta}}) \approx \sigma^2\hat{\boldsymbol{L}}\{\boldsymbol{I}_p + [\hat{\boldsymbol{\nu}}^T][\boldsymbol{A}_\theta^I] + ([\hat{\boldsymbol{\nu}}^T][\boldsymbol{A}_\theta^I])^2 + \cdots\}_{\beta=\hat{\beta}}\hat{\boldsymbol{L}}^T.$$

Combining the above equations, we may roughly have

$$\left\{\mathrm{Var}_\beta(\hat{\boldsymbol{\beta}}|\hat{\boldsymbol{\nu}})\right\}_{\beta=\hat{\beta}} = -\ddot{l}^{-1}(\hat{\boldsymbol{\beta}}) + r_n. \tag{4.35}$$

As pointed out by Skovgaard (1985, p.543), the remaining term r_n is not necessary to have higher order. Equation (4.35) has been investigated by Efron and Hinkley (1978), Amari (1982b), Wei (1991) and discussed by Skovgaard (1985) in detail.

Chapter 5

Confidence Regions

Hamilton, Watts and Bates (1982) studied confidence regions of parameters in normal nonlinear regression models by using the geometric framework of Bates and Watts (1980). They obtained quadratic approximations in terms of curvatures for likelihood regions, from which the ellipsoids corresponding to the inference regions can be obtained in tangent space (see also Beale 1960; Bates and Watts 1981). Hamilton (1986) extended their results to the confidence regions of parameter subsets, for which the regions based on the likelihood ratio and the score statistic are obtained from the geometric viewpoint. Cook and Tsai (1990) studied confidence regions for exponential family nonlinear models by using linear mappings. They obtained good results based on the cubic approximations of the likelihood. Wei (1994) derived confidence regions for embedded models in regular parametric families by using modified BW geometric framework. In this chapter, we extend the results of Hamilton, Watts and Bates (1982) and Hamilton (1986) from normal nonlinear regression to exponential family nonlinear models. We use a relatively simple and unified method to derive confidence regions for parameters and parameter subsets based on the modified BW geometric framework introduced in Chapter 3. Our results show that the extent to which the tangent space inference ellipsoid differs from the linear approximate sphere gives a direct indication of the effects of intrinsic nonlinearity on the inference. In Section 5.1, we first briefly review some basis of multi-parameter confidence regions and then describe our basic idea of constructing the inference regions used in this chapter. Sections 5.2 and 5.3, respectively derive the inference regions for parameters and parameter subsets from the geometric point of view. The method introduced in this chapter can also be extended to more general class of models (see Chapter 7 for details).

5.1 Introduction

Now we first introduce some basis and notation for multi-parameter confidence regions corresponding to a parametric family including exponential family nonlinear models. For such a family, such as (2.2), the confidence region $C(\boldsymbol{Y})$ of parameters $\boldsymbol{\beta}$ with a confidence level α can be defined by

$$P_{\beta}\{\boldsymbol{\beta}: \ \boldsymbol{\beta} \in C(\boldsymbol{Y})\} \geq 1 - \alpha, \tag{5.1}$$

where $C(\boldsymbol{Y})$ is a region in parameter space $\mathcal{B}$. A commonly used, and usually preferable region, is based on the likelihood ratio statistic (2.15). Since (2.15) holds for any $\boldsymbol{\beta}_0$, the likelihood region of $\boldsymbol{\beta}$ can be represented as (Cox and Hinkley 1974, p.314)

$$C(\boldsymbol{Y}) = \{\boldsymbol{\beta}: \ \sigma^2 LR(\boldsymbol{\beta}) \leq \rho^2(\alpha)\}, \tag{5.2}$$

where

$$LR(\boldsymbol{\beta}) = 2\{l(\hat{\boldsymbol{\beta}}) - l(\boldsymbol{\beta})\} \tag{5.3}$$

with $l(\boldsymbol{\beta}) = l(\mu(\boldsymbol{\beta}), \boldsymbol{Y})$. The quantity $\rho^2(\alpha)$ is independent of the parameter $\boldsymbol{\beta}$ and designated to the nominal level of the region. Note that the likelihood ratio statistic $LR(\boldsymbol{\beta})$ is actually equal to the product of $\phi = \sigma^{-2}$ and the deviance (see the equation after (2.15); and also (5.13) in Section 5.2). So if σ^2 is known and $\rho^2(\alpha) = \sigma^2\chi^2(p, \alpha)$, where $\chi^2(p, \alpha)$ is the upper α percentile of a chi-squared distribution with p degree of freedom, then the expression (5.2) corresponds to the usual likelihood region for $\boldsymbol{\beta}$. When σ^2 is unknown, it is appropriate to set $\rho^2(\alpha) = \tilde{\sigma}^2 pF(p, n-p, \alpha)$ where $\tilde{\sigma}^2$ is an appropriate estimator of σ^2 and may asymptotically have chi-squared distribution $(n-p)^{-1}\sigma^2\chi^2(n-p)$ (Jorgensen 1987; Cook and Tsai 1990; see also Section 2.2). In this case, the expression (5.2) is no longer a strict likelihood region in general but can be regarded as a quasi-likelihood region, as discussed by McCullagh and Nelder (1989; see also Cook and Tsai 1990). In this chapter, we shall not pay more attention to the distinction of these two cases and just call (5.2) the likelihood region, because these two cases are quite similar in most situations.

For a general parametric family, such as exponential family nonlinear models, the likelihood region $C(\boldsymbol{Y})$ denoted by (5.2) is usually complicated in parameter space $\mathcal{B}$ and is difficult to display because the log-likelihood function $l(\boldsymbol{\beta})$ strongly depends on several functions such as $b(\cdot)$, $c(\cdot)$, $s(\cdot, \cdot)$, $g(\cdot)$ and $f(\cdot; \cdot)$ (see (2.1) and (2.2)). So, as pointed out by Hamilton, Watts and Bates (1982), there is considerable value in obtaining adequate approximate inference regions which can be easily presented and summarized. To make the region (5.2) easy to understand and to deal with, one usually performs a transformation from $\boldsymbol{\beta}$ to a set of new parameters, say, $\boldsymbol{t} = \boldsymbol{t}(\boldsymbol{\beta})$, so that $\mathcal{B}$ maps to a space $\mathcal{T}$ and $\{\boldsymbol{\beta} \in C(\boldsymbol{Y})\}$ maps to a region

$\{\boldsymbol{t} = \boldsymbol{t}(\boldsymbol{\beta}) \in K(\boldsymbol{Y})\}$ in $\mathcal{T}$ satisfying

$$P_\beta\{\boldsymbol{\beta}: \ \boldsymbol{t}(\boldsymbol{\beta}) \in K(\boldsymbol{Y})\} = P_\beta\{\boldsymbol{\beta}: \ \boldsymbol{\beta} \in C(\boldsymbol{Y})\} \geq 1 - \alpha. \qquad (5.4)$$

We may expect that the region $K(\boldsymbol{Y})$ in $\mathcal{T}$ is much simpler than $C(\boldsymbol{Y})$ in $\mathcal{B}$. Indeed, for normal nonlinear regression models, Hamilton, Watts and Bates (1982) have successfully presented a transformation from parameter space to tangent space to get a good approximate projection that is an ellipsoid in terms of the curvature measure of the model. We can use a similar idea to exponential family nonlinear models by using our modified BW geometric framework. Since the tangent space is a linear one, it is natural to make a transformation from parameter space to tangent space. We now introduce such a (nonlinear) transformation of $\boldsymbol{\beta}$ by means of the solution locus, which may be regarded as an extension of Hamilton, Watts and Bates (1982). In parameter space, the points $\boldsymbol{\beta}$ and $\hat{\boldsymbol{\beta}}$ respectively map to vectors $\boldsymbol{\mu}(\boldsymbol{\beta})$ and $\boldsymbol{\mu}(\hat{\boldsymbol{\beta}})$ on the solution locus in expectation parameter space. The projection of $\boldsymbol{\mu}(\boldsymbol{\beta}) - \boldsymbol{\mu}(\hat{\boldsymbol{\beta}})$ onto tangent space at $\hat{\boldsymbol{\beta}}$ may be written as $\boldsymbol{t}(\boldsymbol{\beta}) = \boldsymbol{Q}\boldsymbol{Q}^T\boldsymbol{V}^{-1}$ $\{\boldsymbol{\mu}(\boldsymbol{\beta}) - \boldsymbol{\mu}(\hat{\boldsymbol{\beta}})\}$, where both $\boldsymbol{Q}$ and $\boldsymbol{V}$ are evaluated at $\hat{\boldsymbol{\beta}}$ (in this chapter, all the quantities such as $\boldsymbol{Q}$, $\boldsymbol{V}$, $\boldsymbol{D}$, $\boldsymbol{R}$ and so on are all evaluated at $\hat{\boldsymbol{\beta}}$ which will be omitted in most cases). If the columns of $\boldsymbol{Q}$ are taken as an orthonormal basis for the tangent space $T_{\hat{\beta}}$ at $\hat{\boldsymbol{\beta}}$, then the coordinates of the projection $\boldsymbol{t}(\boldsymbol{\beta})$ in tangent space are just

$$\boldsymbol{u}(\boldsymbol{\beta}) = \boldsymbol{Q}^T\boldsymbol{V}^{-1}\{\boldsymbol{\mu}(\boldsymbol{\beta}) - \boldsymbol{\mu}(\hat{\boldsymbol{\beta}})\}. \qquad (5.5)$$

Here, as a new parameter, $\boldsymbol{u} = \boldsymbol{u}(\boldsymbol{\beta})$ represents a nonlinear mapping from parameter space $\mathcal{B}$ to the tangent space $T_{\hat{\beta}}$. For normal nonlinear regression models, since $\boldsymbol{V} = \boldsymbol{I}_n$, the transformation (5.5) becomes $\boldsymbol{u}(\boldsymbol{\beta}) = \boldsymbol{Q}^T\{\boldsymbol{\mu}(\boldsymbol{\beta})$ $-\boldsymbol{\mu}(\hat{\boldsymbol{\beta}})\}$ (Hamilton, Watts and Bates 1982).

The transformation (5.5) has several advantages, such as: (a) it maps $\mathcal{B}$ to a linear space $T_{\hat{\beta}}$ which is much easier to handle than $\mathcal{B}$; (b) this transformation connects the parameter space, the solution locus and the tangent space; in particular, the coordinates $\boldsymbol{u}$ provide a natural reference system for the solution locus and the approximations to it (Hamilton, Watts and Bates 1982). We may construct confidence regions for parameters $\boldsymbol{\beta}$ in terms of the coordinates $\boldsymbol{u} = \boldsymbol{u}(\boldsymbol{\beta})$ by using quadratic approximations. In this chapter, we shall show that the quadratic approximation of $\{\boldsymbol{t}(\boldsymbol{\beta}) \in K(Y)\}$ specified by (5.4) and (5.5) is an ellipsoid in tangent space for exponential family nonlinear models; and (c) under the parameter $\boldsymbol{u}$, the parameter-effects curvature array $\boldsymbol{A}_u^P$ defined in Chapter 3 vanishes at $\hat{\boldsymbol{\beta}}$. The following is a brief proof.

Proof. We can use formula (3.19) of Theorem 3.3 to get $\boldsymbol{A}_u^P$. In fact, from (3.19) we have $\boldsymbol{A}_u^P = \boldsymbol{A}^P + \boldsymbol{A}^G$, and

$$\boldsymbol{A}^G = -[\boldsymbol{R}][\boldsymbol{L}^T[\dot{\boldsymbol{u}}_\beta^{-1}][\ddot{\boldsymbol{u}}_\beta]\boldsymbol{L}].$$

It follows from (5.5) that

$$\begin{aligned}(\dot{\boldsymbol{u}}_\beta)_{\hat{\beta}} &= (\frac{\partial \boldsymbol{u}}{\partial \boldsymbol{\beta}^T})_{\hat{\beta}} = \boldsymbol{Q}^T\boldsymbol{V}^{-1}\boldsymbol{D} = \boldsymbol{R},\\ (\ddot{\boldsymbol{u}}_\beta)_{\hat{\beta}} &= (\frac{\partial^2 \boldsymbol{u}}{\partial \boldsymbol{\beta}\partial \boldsymbol{\beta}^T})_{\hat{\beta}} = [\boldsymbol{Q}^T\boldsymbol{V}^{-1}][\boldsymbol{W}] = \boldsymbol{R}^T\boldsymbol{A}^P\boldsymbol{R}.\end{aligned}$$

Substituting these results into $\boldsymbol{A}^G$ shown earlier, we get $\boldsymbol{A}^G = -\boldsymbol{A}^P$ which leads to $\boldsymbol{A}_u^P = \boldsymbol{0}$. ∎

The transformation (5.5) gives a one-to-one mapping between $\boldsymbol{\beta}$ and $\boldsymbol{u}$ in some neighborhood of $\hat{\boldsymbol{\beta}}$ and $\boldsymbol{\beta} = \hat{\boldsymbol{\beta}}$ just corresponds to $\boldsymbol{u} = \boldsymbol{0}$. In the following, we denote the inverse of $\boldsymbol{u} = \boldsymbol{u}(\boldsymbol{\beta})$ by $\boldsymbol{\beta} = \boldsymbol{\beta}(\boldsymbol{u})$. Then $\boldsymbol{u}(\hat{\boldsymbol{\beta}}) = \boldsymbol{0}$ and $\boldsymbol{\beta}(\boldsymbol{0}) = \hat{\boldsymbol{\beta}}$ hold.

5.2 Likelihood Region in Terms of Curvatures

This section derives the tangent space projection of likelihood region and investigates the effects of curvatures on the inference.

5.2.1 Tangent space projection of likelihood region

We can derive an improved approximate projection of the solution locus likelihood region onto tangent space by using the transformation (5.5). For the sake of simplicity, the log-likelihood $l(\boldsymbol{\beta})$ and the likelihood ratio statistic $LR(\boldsymbol{\beta})$ in (5.2) are often denoted by $l(\boldsymbol{u})$ and $LR(\boldsymbol{u})$ respectively when the transformation $\boldsymbol{\beta} = \boldsymbol{\beta}(\boldsymbol{u})$ is considered. Now we derive a quadratic approximation for $LR(\boldsymbol{\beta})$ in terms of parameter $\boldsymbol{u} = \boldsymbol{u}(\boldsymbol{\beta})$ instead of the original parameter $\boldsymbol{\beta}$. To do so, we need two lemmas.

Lemma 5.1

Suppose that Assumptions A and B hold in model (2.2), then the observed information $-\ddot{l}(\hat{\boldsymbol{\beta}})$ given in (2.9) can be represented as

$$-\ddot{l}(\hat{\boldsymbol{\beta}}) = \sigma^{-2}\boldsymbol{R}^T(\boldsymbol{I}_p - \boldsymbol{B}_\theta)\boldsymbol{R}, \tag{5.6}$$

where

$$\boldsymbol{B}_\theta = \boldsymbol{B} - \boldsymbol{B}^\Gamma = [\hat{\boldsymbol{e}}^T\boldsymbol{N}_\theta][\boldsymbol{A}_\theta^I], \tag{5.7}$$

$$\boldsymbol{B} = [\hat{\boldsymbol{e}}^T\boldsymbol{V}^{-1}\boldsymbol{N}][\boldsymbol{A}^I], \quad \boldsymbol{B}^\Gamma = [\hat{\boldsymbol{e}}^T\boldsymbol{V}^{-1}\boldsymbol{N}][\boldsymbol{\Gamma}^I], \tag{5.8}$$

$\boldsymbol{\Gamma}^I$ is given in (3.24) and $\boldsymbol{I} - \boldsymbol{B}_\theta$ is a positive definite matrix.

Proof. It follows from (2.9) and $\boldsymbol{D} = \boldsymbol{QR}$ that

$$-\ddot{l}(\hat{\boldsymbol{\beta}}) = \sigma^{-2}\{\boldsymbol{R}^T\boldsymbol{R} - [\hat{\mathbf{e}}^T\boldsymbol{V}^{-1}][\boldsymbol{W} - \boldsymbol{\Gamma}]\}.$$

Notice that $\hat{\mathbf{e}}$ is orthogonal to the tangent space at $\hat{\boldsymbol{\beta}}$ (see (3.1)). So at $\hat{\boldsymbol{\beta}}$ we have

$$\hat{\mathbf{e}}^T\boldsymbol{V}^{-1} = \hat{\mathbf{e}}^T\boldsymbol{V}^{-1}(\boldsymbol{N}\boldsymbol{N}^T\boldsymbol{V}^{-1} + \boldsymbol{Q}\boldsymbol{Q}^T\boldsymbol{V}^{-1}) = \hat{\mathbf{e}}^T\boldsymbol{V}^{-1}\boldsymbol{N}\boldsymbol{N}^T\boldsymbol{V}^{-1}.$$

From this we get

$$-\ddot{l}(\hat{\boldsymbol{\beta}}) = \sigma^{-2}\boldsymbol{R}^T\{\boldsymbol{I}_p - [\hat{\mathbf{e}}^T\boldsymbol{V}^{-1}\boldsymbol{N}\boldsymbol{N}^T\boldsymbol{V}^{-1}][\boldsymbol{L}^T(\boldsymbol{W} - \boldsymbol{\Gamma})\boldsymbol{L}]\}\boldsymbol{R}\,.$$

It follows from (3.5), (3.24) and $\boldsymbol{N}_\theta = \boldsymbol{V}^{-1}\boldsymbol{N}$ that

$$\begin{aligned}
-\ddot{l}(\hat{\boldsymbol{\beta}}) &= \sigma^{-2}\boldsymbol{R}^T\{\boldsymbol{I}_p - [\hat{\mathbf{e}}^T\boldsymbol{V}^{-1}\boldsymbol{N}][\boldsymbol{A}^I - \boldsymbol{\Gamma}^I]\}\boldsymbol{R} \\
&= \sigma^{-2}\boldsymbol{R}^T\{\boldsymbol{I}_p - [\hat{\mathbf{e}}^T\boldsymbol{V}^{-1}\boldsymbol{N}][\boldsymbol{A}^I_\theta]\}\boldsymbol{R} \\
&= \sigma^{-2}\boldsymbol{R}^T(\boldsymbol{I}_p - \boldsymbol{B}_\theta)\boldsymbol{R}.
\end{aligned}$$

This leads to (5.6) and (5.7), and (5.8) can be obtained from $\boldsymbol{A}^I_\theta = \boldsymbol{A}^I - \boldsymbol{\Gamma}^I$ (see (3.24)). By Assumption B, $-\ddot{l}(\hat{\boldsymbol{\beta}}) > \mathbf{0}$ then $\boldsymbol{I}_p - \boldsymbol{B}_\theta$ is positive definite since $\boldsymbol{R}$ is nonsingular. ∎

Lemma 5.2

Suppose that Assumptions A and B hold in model (2.2), then the derivatives of the functions $\boldsymbol{u}(\boldsymbol{\beta})$ and $\boldsymbol{\beta}(\boldsymbol{u})$ defined in (5.5) evaluated at $\hat{\boldsymbol{\beta}}$ (i.e. at $\boldsymbol{u} = \mathbf{0}$) are given by

$$\frac{\partial \boldsymbol{u}}{\partial \boldsymbol{\beta}^T} = \boldsymbol{R}, \qquad \frac{\partial^2 \boldsymbol{u}}{\partial \boldsymbol{\beta}\partial \boldsymbol{\beta}^T} = \boldsymbol{R}^T\boldsymbol{A}^P\boldsymbol{R}; \tag{5.9}$$

$$\frac{\partial \boldsymbol{\beta}}{\partial \boldsymbol{u}^T} = \boldsymbol{L}, \qquad \frac{\partial^2 \boldsymbol{\beta}}{\partial \boldsymbol{u}\partial \boldsymbol{u}^T} = -[\boldsymbol{L}][\boldsymbol{A}^P]. \tag{5.10}$$

Proof. It is easy to get (5.9) from (5.5) directly, so we just prove (5.10). Since

$$\frac{\partial \boldsymbol{u}}{\partial \boldsymbol{u}^T} = \frac{\partial \boldsymbol{u}}{\partial \boldsymbol{\beta}^T}\frac{\partial \boldsymbol{\beta}}{\partial \boldsymbol{u}^T} = \boldsymbol{I}_p,$$

we have $\partial\boldsymbol{\beta}/\partial\boldsymbol{u}^T = \boldsymbol{R}^{-1} = \boldsymbol{L}$. It follows from (A.8) in Appendix A that

$$\frac{\partial^2 \boldsymbol{u}}{\partial \boldsymbol{u}\partial \boldsymbol{u}^T} = (\frac{\partial \boldsymbol{\beta}}{\partial \boldsymbol{u}^T})^T(\frac{\partial^2 \boldsymbol{u}}{\partial \boldsymbol{\beta}\partial \boldsymbol{\beta}^T})(\frac{\partial \boldsymbol{\beta}}{\partial \boldsymbol{u}^T}) + [\frac{\partial \boldsymbol{u}}{\partial \boldsymbol{\beta}^T}][\frac{\partial^2 \boldsymbol{\beta}}{\partial \boldsymbol{u}\partial \boldsymbol{u}^T}] = \mathbf{0}$$

when $\beta = \hat{\beta}$. Further, from (5.9) we have

$$L^T(R^T A^P R)L + [R][\frac{\partial^2 \beta}{\partial u \partial u^T}]$$

$$= A^P + [R][\frac{\partial^2 \beta}{\partial u \partial u^T}] = 0$$

that leads to (5.10). ∎

Based on these lemmas, we have the following theorem.

Theorem 5.1

Under notation and assumptions stated in Lemmas 5.1 and 5.2, the approximate tangent space projection of the solution locus likelihood region (5.2) for β with the confidence level $100(1-\alpha)\%$ can be represented as

$$K(Y) = \{\beta : \ u^T(I_p - B_\theta)u \le \rho^2(\alpha); \ u = u(\beta)\}, \tag{5.11}$$

where quantities $\rho^2(\alpha)$, $u(\beta)$ and B_θ are respectively given in (5.2), (5.5) and (5.7).

Proof. By using transformation (5.5), the likelihood ratio statistic $LR(\beta)$ given in (5.2) and (5.3) may be denoted by

$$LR(\beta) = LR(u) = -2\{l(u) - l(0)\}, \tag{5.12}$$

where $l(u)$ and $l(0)$ are actually $l(\beta(u))$ and $l(\beta(0)) = l(\hat{\beta})$, respectively. We also use $\dot{l}(0)$ and $\ddot{l}(0)$ to denote the first two derivatives of $l(u)$ at $u = 0$ for short. These derivatives will be used for the Taylor series expansion of $LR(u)$. It follows from (A.7) and (A.8) that

$$\begin{aligned} \frac{\partial l}{\partial u} &= (\frac{\partial \beta}{\partial u^T})^T(\frac{\partial l}{\partial \beta}), \\ \frac{\partial^2 l}{\partial u \partial u^T} &= (\frac{\partial \beta}{\partial u^T})^T(\frac{\partial^2 l}{\partial \beta \partial \beta^T})(\frac{\partial \beta}{\partial u^T}) + [(\frac{\partial l}{\partial \beta})^T][\frac{\partial^2 \beta}{\partial u \partial u^T}]. \end{aligned}$$

Then it follows from Lemma 5.2 that

$$\dot{l}(0) = L^T \dot{l}(\hat{\beta}) = 0, \quad \ddot{l}(0) = L^T \ddot{l}(\hat{\beta})L.$$

Expanding (5.12) and using the above results, we get

$$\begin{aligned} LR(\beta) &\simeq u^T(\beta)\{-\ddot{l}(0)\}u(\beta) \\ &= u^T(\beta)L^T\{-\ddot{l}(\hat{\beta})\}Lu(\beta). \end{aligned}$$

It follows from (5.6) of Lemma 5.1 that

$$LR(\boldsymbol{\beta}) \simeq \sigma^{-2}\boldsymbol{u}^T(\boldsymbol{\beta})(\boldsymbol{I}_p - \boldsymbol{B}_\theta)\boldsymbol{u}(\boldsymbol{\beta}). \tag{5.13}$$

From this and (5.2), we get (5.11). ∎

By Lemma 5.1, $\boldsymbol{I}_p - \boldsymbol{B}_\theta > \boldsymbol{0}$, so expression (5.11) shows that the approximate tangent space projection of the solution locus likelihood region (5.2) for $\boldsymbol{\beta}$ is an ellipsoid $\boldsymbol{u}^T(\boldsymbol{I}_p - \boldsymbol{B}_\theta)\boldsymbol{u} \leq \rho^2(\alpha)$ in the tangent space $T_{\hat{\beta}}$ at $\hat{\boldsymbol{\beta}}$. Moreover, this ellipsoid (in tangent space) is independent of parameter-effects curvature and depends only on the intrinsic curvature $\boldsymbol{A}_\theta^I$ in natural parameter space, so the region $K(\boldsymbol{Y})$ given in (5.11) is invariant under parameter transformations. The matrix $\boldsymbol{B}_\theta = [\hat{\mathbf{e}}^T\boldsymbol{V}^{-1}\boldsymbol{N}][\boldsymbol{A}_\theta^I]$ is referred to as the "effective residual curvature matrix" because it gives the effective normal curvatures relative to the residual vector $\hat{\mathbf{e}}$ (Hamilton, Watts and Bates 1982).

Theorem 5.1 can be applied to several special cases.

Corollary 1

For normal nonlinear regression model (3.27) we have

$$K(\boldsymbol{Y}) = \{\boldsymbol{\beta} : \ \boldsymbol{u}^T(\boldsymbol{\beta})(\boldsymbol{I}_p - \boldsymbol{B})\boldsymbol{u}(\boldsymbol{\beta}) \leq \rho^2(\alpha)\},$$

where $\boldsymbol{u}(\boldsymbol{\beta}) = \boldsymbol{Q}^T\{\boldsymbol{\mu}(\boldsymbol{\beta}) - \boldsymbol{\mu}(\hat{\boldsymbol{\beta}})\}$ and $\boldsymbol{B} = [\hat{\mathbf{e}}^T\boldsymbol{N}][\boldsymbol{A}^I]$.

Proof. Since the normal nonlinear regression model is the special case of (2.2) with $\boldsymbol{V} = \boldsymbol{I}_n$, $\boldsymbol{A}_\theta^I = \boldsymbol{A}^I$ and $\boldsymbol{\Gamma}^I = \boldsymbol{0}$, from (5.5), (5.7) and (5.11), we get the desired result. ∎

Corollary 2

If $\boldsymbol{B}_\theta = \boldsymbol{0}$ in model (2.2), then (5.11) reduces to

$$K(\boldsymbol{Y}) = \{\boldsymbol{\beta} : \ \|\boldsymbol{P}_T\{\boldsymbol{\mu}(\boldsymbol{\beta}) - \boldsymbol{\mu}(\hat{\boldsymbol{\beta}})\|^2 \leq \rho^2(\alpha)\}, \tag{5.14}$$

where $\boldsymbol{P}_T$ is the projection matrix of the tangent space at $\hat{\boldsymbol{\beta}}$, that is $\boldsymbol{P}_T = \boldsymbol{Q}\boldsymbol{Q}^T\boldsymbol{V}^{-1}$ evaluated at $\hat{\boldsymbol{\beta}}$.

Proof. From (5.11) we have $K(\boldsymbol{Y}) = \{\boldsymbol{\beta} : \ \boldsymbol{u}^T(\boldsymbol{\beta})\boldsymbol{u}(\hat{\boldsymbol{\beta}}) \leq \rho^2(\alpha)\}$ when $\boldsymbol{B}_\theta = \boldsymbol{0}$. It follows from (5.5) that

$$\begin{aligned}
\boldsymbol{u}^T(\boldsymbol{\beta})\boldsymbol{u}(\boldsymbol{\beta}) &= \{\boldsymbol{\mu}(\boldsymbol{\beta}) - \boldsymbol{\mu}(\hat{\boldsymbol{\beta}})\}^T\boldsymbol{V}^{-1}\boldsymbol{Q}\boldsymbol{Q}^T\boldsymbol{V}^{-1}\{\boldsymbol{\mu}(\boldsymbol{\beta}) - \boldsymbol{\mu}(\hat{\boldsymbol{\beta}})\} \\
&= \{\boldsymbol{\mu}(\boldsymbol{\beta}) - \boldsymbol{\mu}(\hat{\boldsymbol{\beta}})\}^T\boldsymbol{V}^{-1}\boldsymbol{Q}\boldsymbol{Q}^T\boldsymbol{V}^{-1}\boldsymbol{Q}\boldsymbol{Q}^T\boldsymbol{V}^{-1}\{\boldsymbol{\mu}(\boldsymbol{\beta}) - \boldsymbol{\mu}(\hat{\boldsymbol{\beta}})\} \\
&= \{\boldsymbol{\mu}(\boldsymbol{\beta}) - \boldsymbol{\mu}(\hat{\boldsymbol{\beta}})\}^T\boldsymbol{P}_T^T\boldsymbol{V}^{-1}\boldsymbol{P}_T\{\boldsymbol{\mu}(\boldsymbol{\beta}) - \boldsymbol{\mu}(\hat{\boldsymbol{\beta}})\} \\
&= \|\boldsymbol{P}_T\{\boldsymbol{\mu}(\boldsymbol{\beta}) - \boldsymbol{\mu}(\hat{\boldsymbol{\beta}})\|^2
\end{aligned}$$

which leads to the desired results. ∎

As we have shown in Section 3.4, $\boldsymbol{A}_\theta^I = \boldsymbol{0}$ holds for generalized linear models with canonical link, so Corollary 2 can be used for this case. In particular, for normal linear regression models, we have $\boldsymbol{\mu} = \boldsymbol{X\beta}$ (see (3.28)), $\boldsymbol{P}_T = \boldsymbol{X}(\boldsymbol{X}^T\boldsymbol{X})^{-1}\boldsymbol{X}^T$ and then (5.14) reduces to

$$K(\boldsymbol{Y}) = \{\boldsymbol{\beta} :\ (\boldsymbol{\beta} - \hat{\boldsymbol{\beta}})^T\boldsymbol{X}^T\boldsymbol{X}(\boldsymbol{\beta} - \hat{\boldsymbol{\beta}}) \leq \rho^2(\alpha)\}.$$

This is just the well-known result for linear regression models (Draper and Smith 1981, p.94).

From Theorem 5.1 and transformation (5.5), we can indirectly calculate the approximate volume $V(C(\boldsymbol{Y}))$ for the likelihood region $C(\boldsymbol{Y})$ given in (5.2) in parameter space. This volume can be used for many purposes such as experimental design (Hamilton and Watts 1985), diagnostics (Cook and Weisberg 1982; Thomas 1990) and so on. It is hard to get $V(C(\boldsymbol{Y}))$ directly in parameter space $\mathcal{B}$, but it is not too difficult to get $V(K(\boldsymbol{Y}))$ in tangent space $T_{\hat{\beta}}$ because $K(\boldsymbol{Y})$ in $T_{\hat{\beta}}$ is an ellipsoid. Then the approximation of $V(C(\boldsymbol{Y}))$ can be obtained by using transformation (5.5). In fact, it is easily seen that

$$\begin{aligned} V(C(\boldsymbol{Y})) &\simeq V(K(\boldsymbol{Y})) = \int_{K(Y)} d\boldsymbol{\beta} \\ &= \int_{\boldsymbol{u}^T(\boldsymbol{I}_p - \boldsymbol{B}_\theta)\boldsymbol{u} \leq \rho^2(\alpha)} \det\left(\frac{\partial\boldsymbol{\beta}}{\partial\boldsymbol{u}^T}\right)d\boldsymbol{u}, \end{aligned}$$

where $\det(\partial\boldsymbol{\beta}/\partial\boldsymbol{u}^T)$ is the determinant of the Jacobian $(\partial\boldsymbol{\beta}/\partial\boldsymbol{u}^T)$, which is assumed to be positive without loss of generality. To compute this integral, we may use the quadratic Taylor series expansion of $\det(\partial\boldsymbol{\beta}/\partial\boldsymbol{u}^T)$ at $\boldsymbol{u} = \boldsymbol{0}$ (see Hamilton and Watts 1985 for details).

In the following subsection, we shall investigate the influence of intrinsic curvature and parameter-effects curvature on the approximate likelihood regions shown in Theorem 5.1 and its corollaries.

5.2.2 Influence of curvature measures

The expression (5.11) shows that if the model is intrinsically linear (i.e. $\boldsymbol{A}_\theta^I = \boldsymbol{0}$), then $K(\boldsymbol{Y})$ reduces to $\{\boldsymbol{u}^T\boldsymbol{u} \leq \rho^2(\alpha)\}$ that is a sphere in tangent space $T_{\hat{\beta}}$ with radius $\rho(\alpha)$. This sphere is denoted by S_α. If $\boldsymbol{A}_\theta^I \neq \boldsymbol{0}$, then $\boldsymbol{B}_\theta \neq \boldsymbol{0}$ and the sphere S_α may be viewed as a linear approximation of likelihood region (5.2). As pointed out by Hamilton, Watts and Bates (1982), the extent to which the tangent space inference ellipsoid $K(\boldsymbol{Y})$ differs from the linear approximate sphere S_α gives a direct indication of the effects of intrinsic nonlinearity on the inference. To see the difference between $K(\boldsymbol{Y})$ and the sphere S_α more clearly, the radius $\rho(\alpha)$ of S_α can be compared with the length of each axis of the inference ellipsoid $K(\boldsymbol{Y})$. Let the spectral decomposition of $\boldsymbol{B}_\theta$ be $\boldsymbol{B}_\theta = \boldsymbol{\Gamma\Lambda\Gamma}^T$, here $\boldsymbol{\Lambda} = \text{diag}(\lambda_1, \cdots, \lambda_p)$ with $\lambda_1 \leq \lambda_2 \leq \cdots \leq \lambda_p < 1$ (since $\boldsymbol{I}_p - \boldsymbol{B}_\theta > 0$) and $\boldsymbol{\Gamma}$ is an orthogonal matrix. Then the ellipsoid given by (5.11) becomes

$$\boldsymbol{u}^T\boldsymbol{\Gamma}(\boldsymbol{I}_p - \boldsymbol{\Lambda})\boldsymbol{\Gamma}^T\boldsymbol{u} \leq \rho^2(\alpha).$$

After making an orthogonal transformation $\boldsymbol{\Gamma}^T\boldsymbol{u} = \boldsymbol{v} = (v_1, \cdots, v_p)^T$, the previous expression reduces to $\boldsymbol{v}^T(\boldsymbol{I}_p - \boldsymbol{\Lambda})\boldsymbol{v} \leq \rho^2(\alpha)$ which is just

$$\sum_{a=1}^{p}(1 - \lambda_a)v_a^2 \leq \rho^2(\alpha).$$

Then the length of the a-th axis of the ellipsoid $K(\boldsymbol{Y})$ is $(1 - \lambda_a)^{-\frac{1}{2}}\rho(\alpha)$, $a = 1, \cdots, p$. The maximum and the minimum axis lengths are respectively $(1 - \lambda_p)^{-\frac{1}{2}}\rho(\alpha)$ and $(1 - \lambda_1)^{-\frac{1}{2}}\rho(\alpha)$. Further, the extreme axis length ratio $\{(1 - \lambda_1)/(1 - \lambda_p)\}^{\frac{1}{2}}$ obtained from λ_1 and λ_p can therefore be used to assess directly the effect of intrinsic nonlinearity on the inference region (Hamilton, Watts and Bates 1982). It is also easily seen from expressions (5.1) and (5.11) that

$$P\{\boldsymbol{u}^T\boldsymbol{u} \leq (1 - \lambda_p)^{-1}\rho^2(\alpha)\} \geq P\{\boldsymbol{u}^T(\boldsymbol{I}_p - \boldsymbol{B}_\theta)\boldsymbol{u} \leq \rho^2(\alpha)\} \geq 1 - \alpha. \tag{5.15}$$

So the sphere

$$K_c(\boldsymbol{Y}) = \{\boldsymbol{u}^T\boldsymbol{u} \leq (1 - \lambda_p)^{-1}\rho^2(\alpha), \; \boldsymbol{u} = \boldsymbol{u}(\boldsymbol{\beta})\}$$

is a conservative sphere region for $\boldsymbol{\beta}$.

Now let us see the influence of parameter-effects curvature on the approximate likelihood region in parameter space. Notice the fact that the ellipsoid $\boldsymbol{u}^T(\boldsymbol{I}_p - \boldsymbol{B}_\theta)\boldsymbol{u} \leq \rho^2(\alpha)$ in tangent space depends only on the intrinsic curvature, but $K(\boldsymbol{Y})$ given by (5.11) in parameter space may be connected with the parameter-effects curvature. To see this, we may use the quadratic approximation of $\boldsymbol{u}(\boldsymbol{\beta})$ given in (5.5) to get a quartic polynomial approximation for $K(\boldsymbol{Y})$ in parameter space.

Theorem 5.2

In parameter space, the region $K(\boldsymbol{Y})$ given in (5.11) can be approximately expressed as

$$K'(\boldsymbol{Y}) = \{\boldsymbol{\beta}: \ (\boldsymbol{\phi} + \frac{1}{2}\boldsymbol{\phi}^T\boldsymbol{A}^P\boldsymbol{\phi})^T(\boldsymbol{I}_p - \boldsymbol{B}_\theta)(\boldsymbol{\phi} + \frac{1}{2}\boldsymbol{\phi}^T\boldsymbol{A}^P\boldsymbol{\phi}) \le \rho^2(\alpha)\}, \tag{5.16}$$

where $\boldsymbol{\phi} = \boldsymbol{R}\Delta\boldsymbol{\beta} = \boldsymbol{R}(\boldsymbol{\beta} - \hat{\boldsymbol{\beta}})$.

Proof. $\boldsymbol{u}(\boldsymbol{\beta})$ given in (5.5) can be expanded as

$$\boldsymbol{u}(\boldsymbol{\beta}) \simeq (\frac{\partial \boldsymbol{u}}{\partial \boldsymbol{\beta}^T})_{\hat{\beta}}\Delta\boldsymbol{\beta} + \frac{1}{2}(\Delta\boldsymbol{\beta})^T(\frac{\partial^2 \boldsymbol{u}}{\partial\boldsymbol{\beta}\partial\boldsymbol{\beta}^T})_{\hat{\beta}}\Delta\boldsymbol{\beta}.$$

It follows from Lemma 5.2 that

$$\begin{aligned} \boldsymbol{u}(\boldsymbol{\beta}) &\simeq \boldsymbol{R}\Delta\boldsymbol{\beta} + \frac{1}{2}(\Delta\boldsymbol{\beta})^T\boldsymbol{R}^T\boldsymbol{A}^P\boldsymbol{R}(\Delta\boldsymbol{\beta}) \\ &= \boldsymbol{\phi} + \frac{1}{2}\boldsymbol{\phi}^T\boldsymbol{A}^P\boldsymbol{\phi}. \end{aligned} \tag{5.17}$$

Substituting this result into (5.11) readily yields (5.16). ∎

Expression (5.16) clearly reveals the relationship between the approximate likelihood region and two kinds of curvature measures. When $\boldsymbol{\phi}^T\boldsymbol{A}^P\boldsymbol{\phi}$ is sufficiently small, the linear approximation of (5.16) is

$$K''(\boldsymbol{Y}) = \{\boldsymbol{\beta}: \ \boldsymbol{\phi}^T(\boldsymbol{I} - \boldsymbol{B}_\theta)\boldsymbol{\phi} \le \rho^2(\alpha)\}. \tag{5.18}$$

This is an ellipsoid in parameter space. To see the validity of this linear approximation, we can compare the length of the quadratic term $\frac{1}{2}\boldsymbol{\phi}^T\boldsymbol{A}^P\boldsymbol{\phi}$ to that of the linear term $\boldsymbol{\phi}$ in (5.17). Now let

$$\delta = \frac{1}{2}\frac{\|\boldsymbol{\phi}^T\boldsymbol{A}^P\boldsymbol{\phi}\|'}{\|\boldsymbol{\phi}\|'}.$$

Here we use the norm under the ordinary inner product, i.e. $\|\boldsymbol{a}\|' = (\boldsymbol{a}^T\boldsymbol{a})^{1/2}$. If δ is sufficiently small, say $\delta \le \delta_0$ for a δ_0, then we may ignore the quadratic term $\frac{1}{2}\boldsymbol{\phi}^T\boldsymbol{A}^P\boldsymbol{\phi}$ in (5.16), and (5.18) can be used in practice. Bates and Watts (1981) suggested that if $\delta > 1/2$ or $\delta > 1/4$, then $\frac{1}{2}\boldsymbol{\phi}^T\boldsymbol{A}^P\boldsymbol{\phi}$ cannot be ignored. δ can be rewritten as

$$\begin{aligned} \delta &= \frac{1}{2}(\|\boldsymbol{\phi}\|')(\|\boldsymbol{d}^T\boldsymbol{A}^P\boldsymbol{d}\|') \\ &= \frac{1}{2}(\|\boldsymbol{\phi}\|')K_d^P \le \frac{1}{2}(\|\boldsymbol{\phi}\|')K^P, \end{aligned} \tag{5.19}$$

where $\boldsymbol{d} = \boldsymbol{\phi}/\|\boldsymbol{\phi}\|'$ is a unit vector, K_d^P is the directional curvature (see (3.14)) and K^P is the maximum curvature (see (3.10)). If $\frac{1}{2}\boldsymbol{\phi}^T\boldsymbol{A}^P\boldsymbol{\phi}$ can be ignored, then $\boldsymbol{u}(\boldsymbol{\beta}) \simeq \boldsymbol{\phi}$ (see (5.17)) and it follows from (5.15) that

$$P\{\boldsymbol{\phi}^T\boldsymbol{\phi} \le (1 - \lambda_p)^{-1}\rho^2(\alpha)\} \ge 1 - \alpha.$$

In this case, (5.19) results in

$$\delta \leq \frac{1}{2} R_\alpha K^P, \quad \text{where } R_\alpha = (1-\lambda_p)^{-\frac{1}{2}} \rho(\alpha)$$

with probability $(1-\alpha)$. Therefore if $\frac{1}{2} R_\alpha K^P \leq \delta_0$ for some small δ_0, then the quadratic term $\frac{1}{2}\boldsymbol{\phi}^T \boldsymbol{A}^P \boldsymbol{\phi}$ in (5.16) can be ignored. This inequality is equivalent to

$$K^P \leq 2\delta_0 R_\alpha^{-1} = \frac{1}{4} R_\alpha^{-1} \quad (\text{say } \delta_0 = \frac{1}{8}).$$

This inequality indicates the usefulness of the linear approximation (5.18) for a $1-\alpha$ confidence region. A large value of K^P indicates unaccepted nonlinearity at the level $1-\alpha$ for the region (5.18) because the quadratic term in (5.16) is comparable to the linear term rather than being a small correction to the linear term (Bates and Watts 1981).

Finally we note that all the previous investigations are based on the expectation parameter space and the associated curvatures. We can also study the approximate likelihood region based on the natural parameter space and associated curvatures. The necessary change is that the transformation (5.5) must be replaced by its duality of the form

$$\boldsymbol{v}(\boldsymbol{\beta}) = \boldsymbol{Q_\theta}^T \boldsymbol{V}\{\boldsymbol{\theta}(\boldsymbol{\beta}) - \boldsymbol{\theta}(\hat{\boldsymbol{\beta}})\}.$$

From this transformation, we can also derive an improved approximate projection of the solution locus likelihood region to get parallel results with Theorems 5.1, 5.2 and their corollaries.

5.3 Confidence Regions for Parameter Subsets

If a subset of parameters is of primary interest as discussed by Hamilton (1986) and Cook and Tsai (1990), often the parameter vector $\boldsymbol{\beta}$ can be partitioned as $\boldsymbol{\beta} = (\boldsymbol{\beta}_1^T, \boldsymbol{\beta}_2^T)^T$, where the last p_2 parameters $\boldsymbol{\beta}_2$ are of interest and the first p_1 parameters $\boldsymbol{\beta}_1$ are treated as nuisance parameters. Further, we respectively partition $\boldsymbol{u}^T = (\boldsymbol{u}_1^T, \boldsymbol{u}_2^T)$, $\boldsymbol{D} = (\boldsymbol{D}_1, \boldsymbol{D}_2)$, $\boldsymbol{R} = (\boldsymbol{R}_{ij})$, $\boldsymbol{B} = (\boldsymbol{B}_{ij})$ and $\boldsymbol{B}_\theta = (\boldsymbol{B}_{\theta ij})$, $i, j = 1, 2$, to conform to the partitioning of $\boldsymbol{\beta}$. Similar partition will also be used for other quantities discussed in this section.

The likelihood region corresponding to $\boldsymbol{\beta}_2$ is quite similar to (5.2) and can be expressed as

$$C_s(\boldsymbol{Y}) = \{\boldsymbol{\beta}_2 : \sigma^2 LR_s(\boldsymbol{\beta}_2) \leq \rho_s^2(\alpha)\} \tag{5.20}$$

with

$$\begin{aligned} LR_s(\boldsymbol{\beta}_2) &= 2\{l_p(\hat{\boldsymbol{\beta}}_2) - l_p(\boldsymbol{\beta}_2)\} \\ &= -2\{l(\tilde{\boldsymbol{\beta}}) - l(\hat{\boldsymbol{\beta}})\} = LR(\tilde{\boldsymbol{\beta}}), \end{aligned} \tag{5.21}$$

where $\tilde{\boldsymbol{\beta}} = (\tilde{\boldsymbol{\beta}}_1^T(\boldsymbol{\beta}_2), \boldsymbol{\beta}_2^T)^T$; $\tilde{\boldsymbol{\beta}}_1 = \tilde{\boldsymbol{\beta}}_1(\boldsymbol{\beta}_2)$ maximizes $l(\boldsymbol{\beta}_1, \boldsymbol{\beta}_2)$ for each value of $\boldsymbol{\beta}_2$ and $l_p(\boldsymbol{\beta}_2) \triangleq l(\tilde{\boldsymbol{\beta}})$ is the profile likelihood for $\boldsymbol{\beta}_2$ (see Section 2.2). $\rho_s^2(\alpha)$ is designated to the nominal level of the region. If σ^2 is known, then $\rho_s^2(\alpha) = \sigma^2\chi^2(p_2, \alpha)$ and if σ^2 is unknown, then $\rho_s^2(\alpha) = \tilde{\sigma}^2 p_2 F(p_2, n-p, \alpha)$ (see Hamilton 1986 and (5.2)).

Usually there are two approaches to reduce $LR_s(\boldsymbol{\beta}_2)$ in (5.21). One is based on the likelihood ratio statistic itself. Another is based on the score statistic, a linear approximation of the likelihood ratio statistic. In the following, we shall introduce two kinds of regions based on these two approaches.

5.3.1 Region based on likelihood ratio statistic

To obtain an improved approximate likelihood region for the parameter subset $\boldsymbol{\beta}_2$ based on (5.20) and (5.21), the transformation (5.5) is still valid. This transformation holds for any $\boldsymbol{\beta}$, so we can take $\boldsymbol{\beta} = \tilde{\boldsymbol{\beta}} = (\tilde{\boldsymbol{\beta}}_1^T(\boldsymbol{\beta}_2), \boldsymbol{\beta}_2^T)^T$ to get

$$\tilde{\boldsymbol{u}} = \boldsymbol{u}(\tilde{\boldsymbol{\beta}}) = \boldsymbol{Q}^T\boldsymbol{V}^{-1}\{\boldsymbol{\mu}(\tilde{\boldsymbol{\beta}}) - \boldsymbol{\mu}(\hat{\boldsymbol{\beta}})\}. \tag{5.22}$$

This transformation maps $\boldsymbol{\mu}(\tilde{\boldsymbol{\beta}}) - \boldsymbol{\mu}(\hat{\boldsymbol{\beta}})$ into tangent space and $\tilde{\boldsymbol{u}} = \boldsymbol{u}(\tilde{\boldsymbol{\beta}})$ is a function of $\boldsymbol{\beta}_2$. Therefore $\tilde{\boldsymbol{u}} = (\tilde{\boldsymbol{u}}_1^T, \tilde{\boldsymbol{u}}_2^T)^T$ only consists of p_2 independent parameters. We may derive a quadratic approximation for $LR_s(\boldsymbol{\beta}_2)$ based on (5.22). This procedure is exactly the same as that shown in Section 5.2 except changing $\boldsymbol{\beta}$ for $\tilde{\boldsymbol{\beta}}$. Therefore the approximation (5.13) still holds for $\boldsymbol{\beta} = \tilde{\boldsymbol{\beta}}$, i.e.

$$LR_s(\boldsymbol{\beta}_2) \simeq \sigma^{-2}\tilde{\boldsymbol{u}}^T(\boldsymbol{I}_p - \boldsymbol{B}_\theta)\tilde{\boldsymbol{u}}, \qquad \tilde{\boldsymbol{u}} = \boldsymbol{u}(\tilde{\boldsymbol{\beta}}). \tag{5.23}$$

However, we must eliminate p_1 dependent parameters in this expression. In fact, $\tilde{\boldsymbol{u}}_1$ and $\tilde{\boldsymbol{u}}_2$ in $\tilde{\boldsymbol{u}} = (\tilde{\boldsymbol{u}}_1^T, \tilde{\boldsymbol{u}}_2^T)^T$ should be subject to (5.22) and not independent. We have the following lemma.

Lemma 5.3

Under the transformation (5.22), the vector $\tilde{\boldsymbol{u}} = (\tilde{\boldsymbol{u}}_1^T, \tilde{\boldsymbol{u}}_2^T)^T$ approximately satisfies

$$\tilde{\boldsymbol{u}}_1 \simeq (\boldsymbol{I}_1 - \boldsymbol{B}_{\theta 11})^{-1}\boldsymbol{B}_{\theta 12}\boldsymbol{u}_2, \qquad \tilde{\boldsymbol{u}}_2 \simeq \boldsymbol{u}_2, \tag{5.24}$$

where $\boldsymbol{I}_1$ is the identity matrix with order of p_1.

Proof. Since $\partial\boldsymbol{\mu}/\partial\boldsymbol{\beta}^T = \boldsymbol{D}(\boldsymbol{\beta})$, the first order approximations of (5.5) and (5.22) are respectively given by $\boldsymbol{u} \simeq \boldsymbol{Q}^T\boldsymbol{V}^{-1}\boldsymbol{D}(\boldsymbol{\beta}-\hat{\boldsymbol{\beta}}) = \boldsymbol{R}(\boldsymbol{\beta}-\hat{\boldsymbol{\beta}})$ and $\tilde{\boldsymbol{u}} \simeq \boldsymbol{R}(\tilde{\boldsymbol{\beta}}-\hat{\boldsymbol{\beta}})$. Since $\tilde{\boldsymbol{\beta}} = (\tilde{\boldsymbol{\beta}}_1^T(\boldsymbol{\beta}_2), \boldsymbol{\beta}_2^T)^T$, the partition of $\tilde{\boldsymbol{u}}$ is given by

$$\begin{aligned}\tilde{\boldsymbol{u}}_1 &\simeq \boldsymbol{R}_{11}(\tilde{\boldsymbol{\beta}}_1-\hat{\boldsymbol{\beta}}_1)+\boldsymbol{R}_{12}(\boldsymbol{\beta}_2-\hat{\boldsymbol{\beta}}_2),\\ \tilde{\boldsymbol{u}}_2 &\simeq \boldsymbol{u}_2 \simeq \boldsymbol{R}_{22}(\boldsymbol{\beta}_2-\hat{\boldsymbol{\beta}}_2).\end{aligned}$$

Then we get the second equation of (5.24). On the other hand, it follows from (5.6) and $\tilde{\boldsymbol{u}} \simeq \boldsymbol{R}(\tilde{\boldsymbol{\beta}}-\hat{\boldsymbol{\beta}})$ that

$$\dot{\boldsymbol{l}}(\tilde{\boldsymbol{\beta}}) \simeq \ddot{\boldsymbol{l}}(\hat{\boldsymbol{\beta}})(\tilde{\boldsymbol{\beta}}-\hat{\boldsymbol{\beta}}) \simeq -\sigma^{-2}\boldsymbol{R}^T(\boldsymbol{I}_p-\boldsymbol{B}_\theta)\tilde{\boldsymbol{u}},$$

which results in an approximation

$$-\sigma^2\boldsymbol{L}^T\dot{\boldsymbol{l}}(\tilde{\boldsymbol{\beta}}) \simeq (\boldsymbol{I}_p-\boldsymbol{B}_\theta)\tilde{\boldsymbol{u}}.$$

Since $\partial l/\partial\boldsymbol{\beta}_1 = \boldsymbol{0}$ at $\tilde{\boldsymbol{\beta}}$, the partition of the above equation gives

$$(\boldsymbol{I}_1-\boldsymbol{B}_{\theta 11})\tilde{\boldsymbol{u}}_1-\boldsymbol{B}_{\theta 12}\tilde{\boldsymbol{u}}_2 \simeq \boldsymbol{0}.$$

Then the first equation of (5.24) can be obtained from this. ∎

From (5.23) and (5.24), we get the following theorem.

Theorem 5.3

Under the notation shown earlier, the approximate tangent space projection of the solution locus likelihood region (5.20) for the parameter subset $\boldsymbol{\beta}_2$ with confidence level $100(1-\alpha)\%$ can be represented as

$$K_s(\boldsymbol{Y}) = \{\boldsymbol{\beta}_2 : \tilde{\boldsymbol{u}}_2^T(\boldsymbol{I}_2-\boldsymbol{T})\tilde{\boldsymbol{u}}_2 \le \rho_s^2(\alpha),\ \tilde{\boldsymbol{u}}_2 = \boldsymbol{u}_2(\tilde{\boldsymbol{\beta}})\}, \tag{5.25}$$

where $\tilde{\boldsymbol{u}}_2 = \boldsymbol{u}_2(\tilde{\boldsymbol{\beta}})$ is the subset of $\boldsymbol{u}$ obtained from (5.22) and $\boldsymbol{T} = \boldsymbol{B}_{\theta 22}+\boldsymbol{B}_{\theta 21}(\boldsymbol{I}_1-\boldsymbol{B}_{\theta 11})^{-1}\boldsymbol{B}_{\theta 12}$. Further, $K_s(\boldsymbol{Y})$ is an ellipsoid in tangent space and is invariant under parameterizations.

Proof. The partitioned expression of (5.23) is

$$\begin{aligned}LR_s(\boldsymbol{\beta}_2) \simeq\ & \sigma^{-2}\tilde{\boldsymbol{u}}_1^T(\boldsymbol{I}_1-\boldsymbol{B}_{\theta 11})\tilde{\boldsymbol{u}}_1-\sigma^{-2}\tilde{\boldsymbol{u}}_1^T\boldsymbol{B}_{\theta 12}\tilde{\boldsymbol{u}}_2-\\ & \sigma^{-2}\tilde{\boldsymbol{u}}_2^T\boldsymbol{B}_{\theta 21}\tilde{\boldsymbol{u}}_1+\sigma^{-2}\tilde{\boldsymbol{u}}_2^T(\boldsymbol{I}_2-\boldsymbol{B}_{\theta 22})\tilde{\boldsymbol{u}}_2.\end{aligned}$$

Substituting the expression (5.24) into this equation yields $LR_s(\boldsymbol{\beta}_2) \simeq \tilde{\boldsymbol{u}}_2^T(\boldsymbol{I}_2-\boldsymbol{T})\tilde{\boldsymbol{u}}_2$ which leads to (5.25). Further, $\boldsymbol{I}_2-\boldsymbol{T} = \boldsymbol{I}_2-\boldsymbol{B}_{\theta 22}-\boldsymbol{B}_{\theta 21}(\boldsymbol{I}_1-\boldsymbol{B}_{\theta 11})^{-1}\boldsymbol{B}_{\theta 12}$ is just the inverse of the lower right corner of $(\boldsymbol{I}-\boldsymbol{B}_\theta)^{-1}$, so

$\boldsymbol{I}_2 - \boldsymbol{T} > 0$. Therefore $K_s(\boldsymbol{Y})$ is an ellipsoid in tangent space. Since $\boldsymbol{T}$ depends only on the intrinsic curvature $\boldsymbol{A}_\theta^I$ (see (5.7)), the region $K_s(\boldsymbol{Y})$ is invariant under parameterizations. ∎

Most of the discussions following Theorem 5.1 are still valid for Theorem 5.3. In particular, for normal nonlinear regression models, we have $\boldsymbol{V} = \boldsymbol{I}_n$, $\boldsymbol{A}_\theta^I = \boldsymbol{A}^I$ and $\boldsymbol{B}_\theta = \boldsymbol{B}$. Then from (5.25), we can get the equation (3.3) of Hamilton (1986), as expected. Besides, if $\boldsymbol{\beta}_2 = \boldsymbol{\beta}$, i.e. $\boldsymbol{\beta}_1 = \boldsymbol{0}$, then $\boldsymbol{B}_{\theta 22} = \boldsymbol{B}_\theta$, $\boldsymbol{B}_{\theta 11} = \boldsymbol{0}$ and (5.25) just reduces to (5.11), as expected.

5.3.2 Region based on score statistic

The score statistic (denoted by SC in this book) has been widely used in the literature in the past two decades; see Cox and Hinkley (1974, p.324) for a comprehensive description. Hamilton (1986) derived a confidence region for parameter subset based on the score statistic in normal nonlinear regression models. This subsection will extend Hamilton's (1986) work to exponential family nonlinear models. The score statistic associated with (5.21) is

$$SC = \{(\frac{\partial l}{\partial \boldsymbol{\beta}_2})^T \boldsymbol{J}^{22} (\frac{\partial l}{\partial \boldsymbol{\beta}_2})\}_{\beta=\tilde{\beta}}, \tag{5.26}$$

where $\boldsymbol{J}^{22}$ is the lower right corner of the partition of $\boldsymbol{J}^{-1}(\boldsymbol{Y}) = (\boldsymbol{J}^{ij}), (i,j = 1,2)$ and $\boldsymbol{J}(\boldsymbol{Y})$ is the Fisher information matrix of $\boldsymbol{Y}$ for $\boldsymbol{\beta}$ given in (2.11). It is well known that SC asymptotically has the $\chi^2(p_2)$ distribution for each $\boldsymbol{\beta}_2$ (Cox and Hinkley 1974, p.324). To get a quadratic approximation of SC in terms of the curvature, we first give a useful lemma.

Lemma 5.4

Let $\tilde{e} = \boldsymbol{e}(\tilde{\boldsymbol{\beta}})$ and

$$\begin{aligned} \boldsymbol{P} &= \boldsymbol{D}(\boldsymbol{D}^T\boldsymbol{V}^{-1}\boldsymbol{D})^{-1}\boldsymbol{D}^T\boldsymbol{V}^{-1}, \\ \boldsymbol{P}_1 &= \boldsymbol{D}_1(\boldsymbol{D}_1^T\boldsymbol{V}^{-1}\boldsymbol{D}_1)^{-1}\boldsymbol{D}_1^T\boldsymbol{V}^{-1}, \end{aligned}$$

then for exponential family nonlinear models, SC in (5.26) can be expressed as

$$SC = \phi \tilde{e}^T \tilde{\boldsymbol{V}}^{-1}(\tilde{\boldsymbol{P}} - \tilde{\boldsymbol{P}}_1)\tilde{e}, \tag{5.27}$$

where $\tilde{\boldsymbol{V}}$, $\tilde{\boldsymbol{P}}$ and $\tilde{\boldsymbol{P}}_1$ are all evaluated at $\tilde{\boldsymbol{\beta}} = (\tilde{\boldsymbol{\beta}}_1^T(\boldsymbol{\beta}_2), \boldsymbol{\beta}_2^T)^T$.

Proof. It follows from (2.8) and (2.11) that the partitions of $\dot{\boldsymbol{l}}(\boldsymbol{\beta})$ and $\boldsymbol{J}(\boldsymbol{Y})$ are

$$\frac{\partial l}{\partial \boldsymbol{\beta}} = \phi \begin{pmatrix} \boldsymbol{D}_1^T\boldsymbol{V}^{-1}\boldsymbol{e} \\ \boldsymbol{D}_2^T\boldsymbol{V}^{-1}\boldsymbol{e} \end{pmatrix},$$

and

$$J(Y) = \phi \begin{pmatrix} D_1^T V^{-1} D_1 & D_1^T V^{-1} D_2 \\ D_2^T V^{-1} D_1 & D_2^T V^{-1} D_2 \end{pmatrix},$$

respectively. Then we have

$$J^{22} = \phi^{-1}(D_2^T V^{-1} P_1' D_2)^{-1}, \qquad P_1' = I_n - P_1.$$

Hence from (5.26), SC can be represented as

$$SC = \{\phi e^T V^{-1} D_2 (D_2^T V^{-1} P_1' D_2)^{-1} D_2^T V^{-1} e\}_{\beta=\tilde{\beta}}.$$

Notice that $(\partial l/\partial \beta_1) = 0$ at $\tilde{\beta} = (\tilde{\beta}_1^T(\beta_2), \beta_2^T)^T$ for any β_2. So we have $D_1^T V^{-1} e = 0$ at $\tilde{\beta}$, which results in $\tilde{P}_1 \tilde{e} = 0$, $\tilde{P}'_1 \tilde{e} = \tilde{e}$. Substituting $\tilde{e} = \tilde{P}'_1 \tilde{e}$ into SC shown above and using $(P_1')^2 = P_1'$ and $(P_1')^T V^{-1} = V^{-1} P_1'$ (see (A.2) of Appendix A) yield

$$\begin{aligned} SC &= \{\phi e^T (P_1')^T V^{-1} D_2 (D_2^T V^{-1} P_1' D_2)^{-1} D_2^T V^{-1} P_1' e\}_{\beta=\tilde{\beta}} \\ &= [\phi e^T V^{-1} P'_1 D_2 \{D_2^T (P_1')^T V^{-1} P_1' D_2\}^{-1} D_2^T (P_1')^T V^{-1} e]_{\beta=\tilde{\beta}} \\ &= \{\phi e^T V^{-1} (P - P_1) e\}_{\beta=\tilde{\beta}}. \end{aligned}$$

Here we use the fact that $P - P_1$ is equal to the projection matrix of $P_1' D_2$ (see Theorem A.3 of Appendix A). ∎

By Lemma 5.4, we can calculate $\tilde{e}^T \tilde{V}^{-1} \tilde{P} \tilde{e}$ and $\tilde{e}^T \tilde{V}^{-1} \tilde{P}_1 \tilde{e}$ in terms of $\tilde{u}$ to get a quadratic approximation of SC. To do so, let $\overline{D} = \partial \mu / \partial u^T = (\overline{D}_1, \overline{D}_2)$, then it follows from (5.9) that $D = \partial \mu / \partial \beta^T = \overline{D} R$ and $D_1 = \overline{D}_1 R_{11}$ at $\hat{\beta}$. Hence $P = P_{\overline{D}}$, $P_1 = P_{\overline{D}_1}$ and we may use $\overline{D}$ instead of D to calculate projection matrices P and P_1. To this aim, we need one more lemma.

Lemma 5.5

If $\beta = \beta(u)$ is determined by (5.5), then the quadratic approximation of $\mu(\beta(u))$ can be expressed as

$$\mu(\beta(u)) \simeq \mu(\hat{\beta}) + Q u + \frac{1}{2} N (u^T A^I u). \tag{5.28}$$

Proof. It follows from Lemma 5.2 and (A.8) that when $u = 0$ (i.e. $\beta = \hat{\beta}$), we have

$$\frac{\partial \mu}{\partial u^T} = \frac{\partial \mu}{\partial \beta^T} \frac{\partial \beta}{\partial u^T} = DL = Q,$$

and

$$\begin{aligned}\frac{\partial^2 \mu}{\partial u \partial u^T} &= (\frac{\partial \beta}{\partial u^T})^T (\frac{\partial^2 \mu}{\partial \beta \partial \beta^T})(\frac{\partial \beta}{\partial u^T}) + [\frac{\partial \mu}{\partial \beta^T}][\frac{\partial^2 \beta}{\partial u \partial u^T}] \\ &= L^T W L + [D][-[L][A^P]] \\ &= U - [Q][A^P] = [N][A^I].\end{aligned}$$

In the above equation, the formula (3.7) is used. Then (5.28) can be obtained by using the second order Taylor series expansion for $\mu(\beta(u))$ at $u = 0$ (i.e. $\beta = \hat{\beta}$). ∎

Hamilton, Watts and Bates (1982) have got similar expansion to (5.28) for normal nonlinear regression models.

By Lemmas 5.4 and 5.5 we can get a quadratic approximation for the score statistic (5.26).

Theorem 5.4

Under the notation shown earlier, the approximate tangent space projection of the solution locus inference region of β_2 based on the score statistic with confidence level $100(1-\alpha)\%$ can be represented as

$$K_{sc}(Y) = \{\beta_2 : \ \tilde{u}_2^T (I_2 - \Phi)^T (I_2 - \Phi)\tilde{u}_2 \leq \rho_s^2(\alpha), \ \tilde{u}_2 = u_2(\tilde{\beta})\}, \quad (5.29)$$

where I_2 is an identity matrix of order p_2 and

$$\Phi = B_{22} + B_{21}(I_1 - B_{\theta 11})^{-1} B_{\theta 12}$$

evaluated at $\hat{\beta}$.

Proof. By Lemma 5.4, we now calculate $e^T V^{-1} P e$ and $e^T V^{-1} P_1 e$ at $\tilde{\beta}$, respectively. The calculations of both are quite similar, so we just calculate $(e^T V^{-1} P e)_{\tilde{\beta}}$. It is easily seen that

$$e^T V^{-1} P e = e^T V^{-1} \overline{D} (\overline{D}^T V^{-1} \overline{D})^{-1} \overline{D}^T V^{-1} e. \quad (5.30)$$

To calculate $e^T V^{-1} \overline{D}$ and $\overline{D}^T V^{-1} \overline{D}$ at $\tilde{\beta}$, we use the expansion (5.28) at $\tilde{u}$ (i.e. $\beta = \tilde{\beta}$) and get

$$\begin{aligned}\mu(\tilde{\beta}) &\simeq \mu(\hat{\beta}) + Q\tilde{u} + \frac{1}{2} N(\tilde{u}^T A^I \tilde{u}), \\ \tilde{e} &= Y - \mu(\tilde{\beta}) = Y - \mu(\hat{\beta}) + \mu(\hat{\beta}) - \mu(\tilde{\beta}) \\ &= \hat{e} - Q\tilde{u} - \frac{1}{2} N(\tilde{u}^T A^I \tilde{u}).\end{aligned}$$

Moreover, taking the derivative in (5.28) with respect to u and evaluating at $\tilde{u}$ yield

$$\overline{D} \simeq Q + NM, \quad \overline{D}_1 \simeq Q_1 + NM_1,$$

where $\boldsymbol{Q} = (\boldsymbol{Q}_1, \boldsymbol{Q}_2)$ and $\boldsymbol{M} = (\boldsymbol{M}_1, \boldsymbol{M}_2) = \boldsymbol{A}^I \tilde{\boldsymbol{u}}$ is an $(n-p) \times p$ matrix. Based on the above expressions, we get

$$\begin{aligned} (\overline{\boldsymbol{D}}^T \boldsymbol{V}^{-1} \overline{\boldsymbol{D}})_{\tilde{\beta}} &\simeq (\boldsymbol{Q} + \boldsymbol{N}\boldsymbol{M})^T \boldsymbol{V}^{-1} (\boldsymbol{Q} + \boldsymbol{N}\boldsymbol{M}) \\ &= (\boldsymbol{I}_p + \boldsymbol{M}^T \boldsymbol{M}), \\ (\boldsymbol{e}^T \boldsymbol{V}^{-1} \overline{\boldsymbol{D}})_{\tilde{\beta}} &\simeq \{\hat{\mathbf{e}} - \boldsymbol{Q}\tilde{\boldsymbol{u}} - \frac{1}{2}\boldsymbol{N}(\tilde{\boldsymbol{u}}^T \boldsymbol{A}^I \tilde{\boldsymbol{u}})\}^T \boldsymbol{V}^{-1} (\boldsymbol{Q} + \boldsymbol{N}\boldsymbol{M}) \\ &= \hat{\mathbf{e}}^T \boldsymbol{V}^{-1} \boldsymbol{N}\boldsymbol{M} - \tilde{\boldsymbol{u}}^T - \frac{1}{2}\tilde{\boldsymbol{u}}^T \boldsymbol{M}^T \boldsymbol{M}, \end{aligned}$$

here we use the formulas $\tilde{\boldsymbol{u}}\boldsymbol{A}^I \tilde{\boldsymbol{u}} = (\boldsymbol{A}^I \tilde{\boldsymbol{u}})\tilde{\boldsymbol{u}} = \boldsymbol{M}\tilde{\boldsymbol{u}}$ (see Section A.2 in Appendix A). Since each face of $\boldsymbol{A}^I$ is symmetric, by using formulas given in Section A.2, we have

$$\hat{\mathbf{e}}^T \boldsymbol{V}^{-1} \boldsymbol{N}\boldsymbol{M} = \tilde{\boldsymbol{u}}^T [\hat{\mathbf{e}}^T \boldsymbol{V}^{-1} \boldsymbol{N}][\boldsymbol{A}^I] = \tilde{\boldsymbol{u}}^T \boldsymbol{B}.$$

Therefore

$$(\boldsymbol{e}^T \boldsymbol{V}^{-1} \overline{\boldsymbol{D}})_{\tilde{\beta}} \simeq -\tilde{\boldsymbol{u}}^T (\boldsymbol{I}_p - \boldsymbol{B} + \frac{1}{2}\boldsymbol{M}^T \boldsymbol{M}).$$

Substituting the above results into (5.30) and neglecting the quartic terms, we get

$$(\boldsymbol{e}^T \boldsymbol{V}^{-1} \boldsymbol{P}\boldsymbol{e})_{\tilde{\beta}} \simeq \tilde{\boldsymbol{u}}^T (\boldsymbol{I}_p - \boldsymbol{B})^2 \tilde{\boldsymbol{u}}.$$

Moreover, since $\overline{\boldsymbol{D}}_1 = (\overline{\boldsymbol{D}}_1, \overline{\boldsymbol{D}}_2)(\boldsymbol{I}_1, \boldsymbol{0})^T$, we get

$$(\boldsymbol{e}^T \boldsymbol{V}^{-1} \boldsymbol{P}_1 \boldsymbol{e})_{\tilde{\beta}} \simeq \tilde{\boldsymbol{u}}^T (\boldsymbol{I}_p - \boldsymbol{B}) \boldsymbol{E}_1 (\boldsymbol{I}_p - \boldsymbol{B}) \tilde{\boldsymbol{u}},$$

where $\boldsymbol{E}_1 = \text{diag}(\boldsymbol{I}_1, \boldsymbol{0})$ is a $p \times p$ matrix. So it follows from Lemma 5.4 that

$$SC \simeq \sigma^{-2} \tilde{\boldsymbol{u}}^T (\boldsymbol{I}_p - \boldsymbol{B})(\boldsymbol{I}_p - \boldsymbol{E}_1)(\boldsymbol{I}_p - \boldsymbol{B}) \tilde{\boldsymbol{u}}.$$

Partitioning this equation and substituting (5.24) into the expression, we get (5.29). ∎

Let us see some special cases of Theorem 5.4. For normal nonlinear regression models, we have $\boldsymbol{V} = \boldsymbol{I}_n$, $\boldsymbol{B}^\Gamma = \boldsymbol{0}$, $\boldsymbol{B}_\theta = \boldsymbol{B}$ and $\boldsymbol{\Phi}$ becomes $\tilde{\boldsymbol{\Phi}} = \boldsymbol{B}_{22} + \boldsymbol{B}_{21}(\boldsymbol{I}_1 - \boldsymbol{B}_{11})^{-1} \boldsymbol{B}_{12}$ that is a symmetric matrix. Then (5.29) becomes

$$K_{sc}(\boldsymbol{Y}) = \{\boldsymbol{\beta}_2 : \tilde{\boldsymbol{u}}_2^T (\boldsymbol{I}_2 - \tilde{\boldsymbol{\Phi}})^2 \tilde{\boldsymbol{u}}_2 \le \rho_s^2(\alpha), \ \tilde{\boldsymbol{u}}_2 = \boldsymbol{u}_2(\tilde{\boldsymbol{\beta}})\}.$$

This expression is consistent with the equation (3.5) of Hamilton (1986), as expected.

In (5.29), if $\boldsymbol{\beta}_2 = \boldsymbol{\beta}$ (i.e. $\boldsymbol{\beta}_1 = 0$), then $\boldsymbol{\Phi} = \boldsymbol{B}$ is a symmetric matrix and we get an alternative region of $\boldsymbol{\beta}$ dissimilar to (5.11), that is

$$K'(\boldsymbol{Y}) = \{\boldsymbol{\beta} : \tilde{\boldsymbol{u}}^T (\boldsymbol{I}_p - \boldsymbol{B})^2 \tilde{\boldsymbol{u}}, \ \tilde{\boldsymbol{u}} = \boldsymbol{u}(\tilde{\boldsymbol{\beta}})\}.$$

In Theorems 5.3 and 5.4, our derivations are all based on the replacement of $\boldsymbol{\beta}_1$ by $\tilde{\boldsymbol{\beta}}_1(\boldsymbol{\beta}_2)$. Alternatively, we may have some other choices to treat the nuisance parameter $\boldsymbol{\beta}_1$. For instance, $\boldsymbol{\beta}_1$ may be replaced by $\hat{\boldsymbol{\beta}}_1$ (Hamilton 1986). Now let us see the details. Let $\bar{\boldsymbol{\beta}} = (\hat{\boldsymbol{\beta}}_1^T, \boldsymbol{\beta}_2^T)^T$ in (5.5), then

$$\bar{\boldsymbol{u}} \triangleq \boldsymbol{u}(\bar{\boldsymbol{\beta}}) = \boldsymbol{Q}^T \boldsymbol{V}^{-1}\{\boldsymbol{\mu}(\hat{\boldsymbol{\beta}}_1, \boldsymbol{\beta}_2) - \boldsymbol{\mu}(\hat{\boldsymbol{\beta}}_1, \hat{\boldsymbol{\beta}}_2)\}.$$

The linear approximation of this equation is $\bar{\boldsymbol{u}} = \boldsymbol{R}(\bar{\boldsymbol{\beta}} - \hat{\boldsymbol{\beta}})$ whose partition is $\bar{\boldsymbol{u}}_1 = \boldsymbol{R}_{12}\Delta\boldsymbol{\beta}_2$ and $\bar{\boldsymbol{u}}_2 = \boldsymbol{R}_{22}\Delta\boldsymbol{\beta}_2$ which result in $\bar{\boldsymbol{u}}_1 = \boldsymbol{R}_{12}\boldsymbol{R}_{22}^{-1}\bar{\boldsymbol{u}}_2$, $\bar{\boldsymbol{u}}_2 = \boldsymbol{u}_2$. Since $\boldsymbol{R} = \boldsymbol{L}^{-1}$, $\boldsymbol{R}_{12} = -\boldsymbol{L}_{11}^{-1}\boldsymbol{L}_{12}\boldsymbol{L}_{22}^{-1}$, we have

$$\bar{\boldsymbol{u}}_1 = -\boldsymbol{L}_{11}^{-1}\boldsymbol{L}_{12}\bar{\boldsymbol{u}}_2, \quad \bar{\boldsymbol{u}}_2 = \boldsymbol{u}_2.$$

Then we can use this relation to replace (5.24) and to get alternative regions for $\boldsymbol{\beta}_2$ parallel to those given in Theorems 5.3 and 5.4. For instance, the counterpart of (5.29) is

$$K'_{sc}(\boldsymbol{Y}) = \{\boldsymbol{\beta}_2 : \ \bar{\boldsymbol{u}}_2^T \boldsymbol{\Psi}^T \boldsymbol{\Psi} \bar{\boldsymbol{u}}_2 \leq \rho_s^2(\alpha), \ \bar{\boldsymbol{u}}_2 = \boldsymbol{u}_2(\bar{\boldsymbol{\beta}})\},$$

where $\boldsymbol{\Psi} = \boldsymbol{I}_2 - \boldsymbol{B}_{22} + \boldsymbol{B}_{21}\boldsymbol{L}_{11}^{-1}\boldsymbol{L}_{12}$ (see also Hamilton 1986).

The method and results introduced in this chapter can be extended to more general class of models (see Chapter 7 for details).

Chapter 6

Diagnostics and Influence Analysis

Statistical models that are applied to analyze the essential features for a given data set are nearly always approximate descriptions of more complicated processes. The data may show some departures from the fitted values to some extent, therefore it is usually necessary to check the adequacy of model fit. The discrepancy appears as either a few data values showing the isolated departures from the rest or the data as a whole showing some systematic departure from the fitted values. Model diagnostics and influence analysis are useful tools to detect both isolated and systematic discrepancies of model fit in regression analysis. This chapter is devoted to a study of regression diagnostics and influence analysis as a tool to check the adequacy of model fit for exponential family nonlinear models. The basic diagnostic statistics as well as some specific problems are discussed. Section 6.1 is an introduction of regression diagnostics related to exponential family nonlinear models. Section 6.2 investigates the equivalency of case deletion model (CDM) and mean-shift outlier model (MSOM). In Section 6.3, we summarize several diagnostic statistics used in practice. In the rest of the sections, we study some specific problems of diagnostics and influence analysis, such as local influence analysis, generalized leverage and the test for varying dispersion.

6.1 Introduction

Linear regression diagnostics and influence analysis have been well developed in the past two decades. For a comprehensive introduction, the reader can refer to the literature (e.g. Cook and Weisberg 1982; Atkinson 1985; Chatterjee and Hadi 1986). However, there is not much published work on regression diagnostics for the models outside of linear regression (e.g. Pregibon 1981; Cook 1986; McCullagh and Nelder 1989; Andersen 1992; Davison and Tsai 1992). Only a few papers have some materials related to the diagnostics for exponential family nonlinear models (Gay and Welsch 1988; Davison and Tsai 1992; Wei and Shi 1994). In this section, we shall introduce the basis of model diagnostics which can be applied to exponential family nonlinear models.

The exponential family nonlinear model given by (2.1) and (2.2) may be briefly denoted by

$$g(\mu_i) = f(\boldsymbol{x}_i;\boldsymbol{\beta}) \stackrel{\Delta}{=} f_i(\boldsymbol{\beta}), \qquad i = 1, \cdots, n, \tag{6.1}$$

here $\mathrm{E}(y_i) = \mu_i$ and $y_i \sim ED(\mu_i, \sigma^2)$ is implicit. In this chapter, we shall most often use this compact form for diagnostics purposes. A fundamental approach of influence diagnostics is based on the comparison of parameter estimates $\tilde{\boldsymbol{\beta}}$ and $\tilde{\sigma}^2$ with parameter estimates $\tilde{\boldsymbol{\beta}}_{(i)}$ and $\tilde{\sigma}^2_{(i)}$ that correspond to the so-called case deletion model (CDM):

$$g(\mu_j) = f(\boldsymbol{x}_j;\boldsymbol{\beta}), \qquad j = 1, \cdots, n, \quad j \neq i. \tag{6.2}$$

This is just model (6.1) with the i-th case deleted. In what follows, if a quantity A corresponds to a full model, then we use $A_{(i)}$ to denote the quantity corresponding to the model with the i-th case deleted.

In the following, we consider regression diagnostics and influence analysis for the maximum likelihood estimators $\hat{\boldsymbol{\beta}}$ and $\hat{\sigma}^2$. To compute $\hat{\boldsymbol{\beta}}_{(i)}$ and $\hat{\sigma}^2_{(i)}$, we may use the method introduced in Section 2.3 to rerun the program with the i-th observation deleted for each $i = 1, \cdots, n$. This is usually costly but correct. For diagnostics purposes, the first order approximations $\hat{\boldsymbol{\beta}}^I_{(i)}$, $\hat{\sigma}^I_{(i)}$ of $\hat{\boldsymbol{\beta}}_{(i)}$, $\hat{\sigma}_{(i)}$ can be used in practice (see the note at the end of this section).

Lemma 6.1

The first order approximation $\hat{\boldsymbol{\beta}}_{(i)}$ of $\boldsymbol{\beta}$ in CDM (6.2) can be expressed as

$$\hat{\boldsymbol{\beta}}^I_{(i)} = \hat{\boldsymbol{\beta}} - \left\{ \frac{(\boldsymbol{D}^T \boldsymbol{V}^{-1} \boldsymbol{D})^{-1} V_i^{-\frac{1}{2}} \boldsymbol{d}_i r p_i}{1 - h_{ii}} \right\}_{\hat{\boldsymbol{\beta}}}, \tag{6.3}$$

where

$$rp_i = \{V_i^{-\frac{1}{2}}(y_i - \mu_i)\}_{\hat{\boldsymbol{\beta}}}$$

is the Pearson residual introduced in (2.23), h_{ii} is the i-th diagonal element of $\boldsymbol{H} = \boldsymbol{V}^{-\frac{1}{2}} \boldsymbol{D}(\boldsymbol{D}^T \boldsymbol{V}^{-1} \boldsymbol{D})^{-1} \boldsymbol{D}^T \boldsymbol{V}^{-\frac{1}{2}}$ and $\boldsymbol{d}_i^T$ is the i-th row of $\boldsymbol{D} = \partial \boldsymbol{\mu} / \partial \boldsymbol{\beta}^T$.

Proof. Let the log-likelihoods for $\boldsymbol{\beta}$ be

$$l(\boldsymbol{\beta}) \triangleq \sum_{j=1}^{n} l_j(\boldsymbol{\beta}) = \sum_{j=1}^{n} \log p(y_j; \mu_j(\boldsymbol{\beta}), \sigma),$$

$$l_{(i)}(\boldsymbol{\beta}) = \sum_{j \neq i} l_j(\boldsymbol{\beta}),$$

then we have $\dot{\boldsymbol{l}}_{(i)}(\hat{\boldsymbol{\beta}}_{(i)}) = \mathbf{0}$, whose first order approximation at $\hat{\boldsymbol{\beta}}$ gives

$$\dot{\boldsymbol{l}}_{(i)}(\hat{\boldsymbol{\beta}}) + \ddot{\boldsymbol{l}}_{(i)}(\hat{\boldsymbol{\beta}})(\hat{\boldsymbol{\beta}}_{(i)} - \hat{\boldsymbol{\beta}}) \approx \mathbf{0},$$

$$\hat{\boldsymbol{\beta}}_{(i)} \approx \hat{\boldsymbol{\beta}} + \{-\ddot{\boldsymbol{l}}_{(i)}(\hat{\boldsymbol{\beta}})\}^{-1}\dot{\boldsymbol{l}}_{(i)}(\hat{\boldsymbol{\beta}}).$$

By using equations (2.8), (2.9), (2.11) and $-\ddot{\boldsymbol{l}}_{(i)}(\hat{\boldsymbol{\beta}}) \approx \boldsymbol{J}_{(i)}(\hat{\boldsymbol{\beta}})$, we get

$$\begin{aligned} \hat{\boldsymbol{\beta}}_{(i)}^{I} &= \hat{\boldsymbol{\beta}} + \{\boldsymbol{J}_{(i)}^{-1}(\hat{\boldsymbol{\beta}})\}\dot{\boldsymbol{l}}_{(i)}(\hat{\boldsymbol{\beta}}) \\ &= \hat{\boldsymbol{\beta}} + \{(\boldsymbol{D}_{(i)}^T \boldsymbol{V}_{(i)}^{-1} \boldsymbol{D}_{(i)})^{-1} \boldsymbol{D}_{(i)}^T \boldsymbol{V}_{(i)}^{-1} \boldsymbol{e}_{(i)}\}_{\hat{\boldsymbol{\beta}}} \\ &= \hat{\boldsymbol{\beta}} + \{(\boldsymbol{D}^T \boldsymbol{V}^{-1} \boldsymbol{D} - V_i^{-1} \boldsymbol{d}_i \boldsymbol{d}_i^T)^{-1} (\boldsymbol{D}^T \boldsymbol{V}^{-1} \boldsymbol{e} - V_i^{-1} \boldsymbol{d}_i e_i)\}_{\hat{\boldsymbol{\beta}}}. \end{aligned}$$

Using the formula of inverse matrix (e.g. Rao 1973, p.33) and the fact $(\boldsymbol{D}^T\boldsymbol{V}^{-1}\boldsymbol{e})_{\hat{\boldsymbol{\beta}}} = \mathbf{0}$ (see (2.10)), we can get (6.3) by a few calculations. ∎

After obtaining $\hat{\boldsymbol{\beta}}_{(i)}^{I}$, $\hat{\sigma}_{(i)}^{I}$ can be computed from (2.12) and (2.13).

In the following, we introduce a fundamental diagnostic method (see Section 6.3 for further discussions).

To find the influential points, we may compute a certain "distance" between either $(\hat{\boldsymbol{\beta}}, \hat{\sigma})$ and $(\hat{\boldsymbol{\beta}}_{(i)}, \hat{\sigma}_{(i)})$ or $\hat{\boldsymbol{\beta}}$ and $\hat{\boldsymbol{\beta}}_{(i)}$. The latter is often used in practice because if a case is influential to $\hat{\boldsymbol{\beta}}$, then this case must be influential to $(\hat{\boldsymbol{\beta}}, \hat{\sigma})$. Here we introduce two kinds of distance which are used very often in influence diagnostics and can also be used for exponential family nonlinear models.

(a) Generalized Cook distance

This is a norm of $\hat{\boldsymbol{\beta}} - \hat{\boldsymbol{\beta}}_{(i)}$ with respect to certain weight matrix $\boldsymbol{M} > 0$ and defined as

$$C_i \triangleq \|\hat{\boldsymbol{\beta}} - \hat{\boldsymbol{\beta}}_{(i)}\|_M^2 = (\hat{\boldsymbol{\beta}} - \hat{\boldsymbol{\beta}}_{(i)})^T \boldsymbol{M} (\hat{\boldsymbol{\beta}} - \hat{\boldsymbol{\beta}}_{(i)}).$$

It is very natural to choose $\boldsymbol{M} = \boldsymbol{J}(\boldsymbol{\beta}) = \phi \boldsymbol{D}^T \boldsymbol{V}^{-1} \boldsymbol{D}$, the Fisher information matrix of $\boldsymbol{Y}$ for $\boldsymbol{\beta}$, which results in

$$C_i = \frac{(\hat{\boldsymbol{\beta}} - \hat{\boldsymbol{\beta}}_{(i)})^T (\boldsymbol{D}^T \boldsymbol{V}^{-1} \boldsymbol{D})_{\hat{\boldsymbol{\beta}}} (\hat{\boldsymbol{\beta}} - \hat{\boldsymbol{\beta}}_{(i)})}{\hat{\sigma}^2}, \tag{6.4}$$

that may be called the Fisher information distance. By Lemma 6.1 we can get the first order approximation of C_i as

$$C_i^I = \frac{h_{ii}}{1-h_{ii}} r_i^2, \qquad r_i = \frac{rp_i}{\hat{\sigma}\sqrt{1-h_{ii}}}, \tag{6.5}$$

where r_i is called the studentized residual or studentized standardized residual (McCullagh and Nelder 1989, p.396).

We shall see in later sections that h_{ii} and r_i play important roles in influence diagnostics. Formula (6.4) is much like the Cook distance in linear regression case (Cook and Weisberg 1982, p.117).

(b) Likelihood distance

That is defined as (Cook and Weisberg 1982, p.183)

$$LD_i(\boldsymbol{\beta}) = 2\{l(\hat{\boldsymbol{\beta}}) - l(\hat{\boldsymbol{\beta}}_{(i)})\} \tag{6.6}$$

for $\boldsymbol{\beta}$ and defined as

$$\begin{aligned} LD_i(\boldsymbol{\beta}_1|\boldsymbol{\beta}_2) &= 2\{l_p(\hat{\boldsymbol{\beta}}_1) - l_p(\hat{\boldsymbol{\beta}}_{1(i)})\} \\ &= 2\{l(\hat{\boldsymbol{\beta}}_1, \hat{\boldsymbol{\beta}}_2) - l(\hat{\boldsymbol{\beta}}_{1(i)}, \tilde{\boldsymbol{\beta}}_2(\hat{\boldsymbol{\beta}}_{1(i)}))\} \end{aligned} \tag{6.7}$$

for a parameter subset $\boldsymbol{\beta}_1$, where $\hat{\boldsymbol{\beta}}_{(i)} = (\hat{\boldsymbol{\beta}}_{1(i)}^T, \hat{\boldsymbol{\beta}}_{2(i)}^T)^T$ and $\tilde{\boldsymbol{\beta}}_2(\boldsymbol{\beta}_1)$ is the maximum likelihood estimator of $\boldsymbol{\beta}_2$ for the fixed $\boldsymbol{\beta}_1$. We can compute LD_i directly based on $\hat{\boldsymbol{\beta}}$, $\hat{\boldsymbol{\beta}}_{(i)}$ and the log-likelihood $l(\boldsymbol{\beta})$. The first order approximation of LD_i is also used very often. It is easily seen that (6.6) can be approximately expressed as

$$LD_i \approx 2[\dot{\boldsymbol{l}}^T(\hat{\boldsymbol{\beta}})(\hat{\boldsymbol{\beta}} - \hat{\boldsymbol{\beta}}_{(i)}) + \frac{1}{2}(\hat{\boldsymbol{\beta}} - \hat{\boldsymbol{\beta}}_{(i)})^T\{-\ddot{\boldsymbol{l}}(\hat{\boldsymbol{\beta}})\}(\hat{\boldsymbol{\beta}} - \hat{\boldsymbol{\beta}}_{(i)})].$$

Since $\dot{\boldsymbol{l}}(\hat{\boldsymbol{\beta}}) = \boldsymbol{0}$, $\{-\ddot{\boldsymbol{l}}(\hat{\boldsymbol{\beta}})\} \approx \boldsymbol{J}(\hat{\boldsymbol{\beta}})$, we have

$$LD_i^I(\boldsymbol{\beta}) = C_i \approx C_i^I. \tag{6.8}$$

Section 6.3 summarizes several useful diagnostic statistics (see Table 6.3) which are based on CDM and can be used for exponential family nonlinear models.

Before concluding this section, we notice the following important fact. A "good" first order approximation can often be used for diagnostics in practical situations. In fact, it is easily seen that for an appropriate diagnostic statistic, say likelihood distance, if LD_k is much greater than LD_j for any $j \neq k$, then the first order approximations of LD_k and LD_j are still expected to keep a similar relationship in most of the cases, i.e. LD_k^I is still greater enough than LD_j^I $(j \neq k)$. This is one of the reasons that the first order approximations are widely used in the literature for diagnostics purpose. Of course, the usefulness of the first order approximation is limited in some complicated cases.

6.2 Analysis of Diagnostic Models

6.2.1 Diagnostic models

The diagnostic model is the basis for constructing effective diagnostic statistics. The CDM given in (6.2) is the most important one in practice because it is very straightforward and easy to compute. Another commonly used diagnostic model is the so-called mean-shift outlier model (MSOM; Cook and Weisberg 1982, p.20). For exponential family nonlinear models, MSOM can be represented as

$$\begin{cases} g(\mu_j) = f_j(\boldsymbol{\beta}), \quad j = 1, \cdots, n; \ j \neq i, \\ g(\mu_i) = f_i(\boldsymbol{\beta}) + \gamma; \end{cases} \tag{6.9}$$

or using vector notation

$$\boldsymbol{g}(\boldsymbol{\mu}) = \boldsymbol{f}(\boldsymbol{\beta}) + \gamma \boldsymbol{c}_i,$$

where $\boldsymbol{g}(\boldsymbol{\mu}) = (g(\mu_i))$, $\boldsymbol{f}(\boldsymbol{\beta}) = (f_i(\boldsymbol{\beta}))$ are n-vectors and $\boldsymbol{c}_i$ is an n-vector having 1 at the i-th position and zeros elsewhere, γ is an extra parameter to indicate the presence of outlier. It is easily seen that the nonzero value of γ implies that the i-th case may be an outlier because the case (x_i, y_i) no longer comes from the original model (6.1). This model is usually easier to formulate than CDM. To detect outliers using MSOM (6.9), one may either estimate the parameter γ or make a testing of hypothesis $H_0 : \gamma = 0$ (we shall discuss this later). The maximum likelihood estimators of $\boldsymbol{\beta}$, γ and σ^2 in (6.9) are denoted by $\hat{\boldsymbol{\beta}}_{mi}, \hat{\gamma}_{mi}$ and $\hat{\sigma}^2_{mi}$, respectively.

The third commonly used diagnostic model is the case-weights model (CWM). Under this model, the estimates $\hat{\boldsymbol{\beta}}_\omega$ and $\hat{\sigma}^2_\omega$ maximize

$$l_\omega(\boldsymbol{\beta}, \sigma) = \sum_{j=1}^{n} \omega_j l_j(\boldsymbol{\beta}, \sigma),$$

where $l_j(\boldsymbol{\beta}, \sigma) = \log p(y_j; \boldsymbol{\beta}, \sigma)$ and ω_j is the weight used to indicate the effect of each observation on the importance of fit (Belsley, Kuh and Welsch 1980; Pregibon 1981). A special case of CWM is $\omega_i = \omega$ and $\omega_j = 1$ $(j \neq i)$, which can be expressed as

$$l_\omega(\boldsymbol{\beta}, \sigma) = \sum_{j \neq i} l_j(\boldsymbol{\beta}, \sigma) + \omega l_i(\boldsymbol{\beta}, \sigma). \tag{6.10}$$

The estimators corresponding to (6.10) are denoted by $\hat{\boldsymbol{\beta}}_{\omega i}$ and $\hat{\sigma}^2_{\omega i}$. It is easily seen that $\hat{\boldsymbol{\beta}}_{\omega i} = \hat{\boldsymbol{\beta}}$, $\hat{\sigma}^2_{\omega i} = \hat{\sigma}^2$ when $\omega = 1$ and $\hat{\boldsymbol{\beta}}_{\omega i} \to \hat{\boldsymbol{\beta}}_{(i)}$, $\hat{\sigma}^2_{\omega i} \to \hat{\sigma}^2_{(i)}$ when $\omega \to 0$. The CWM may be used to some specific situations (e.g.

Belsley, Kuh and Welsch 1980; Pregibon 1981; Cook 1986; see also Section 6.4 of this chapter).

In this chapter, we shall use these three models to construct diagnostic statistics. However, as Beckman and Cook (1983) pointed out, CDM and MSOM are central frequent models for regression diagnostics, so we shall pay more attention to these two models.

It is well known that in linear regression models, CDM and MSOM are equivalent in the following sense: the least squares estimators of parameters are equal for CDM and MSOM (Cook and Weisberg 1982; Wei, Lu and Shi 1991). However, the equivalence of CDM and MSOM outside of linear regression models has caused some concern to several authors but has not been fully studied in the literature. Storer and Crowley (1985) conjectured that $\hat{\beta}_{mi} = \hat{\beta}_{(i)}$ may hold for nonlinear regression and broader class of models. Williams (1987) used the result $\hat{\beta}_{mi} = \hat{\beta}_{(i)}$ for GLM, but he did not give the proof. Ross (1987a) and Dzieciolowski and Ross (1990) used the result $\hat{\beta}_{mi} = \hat{\beta}_{(i)}$ implicitly for nonlinear regression models and they used MSOM instead of CDM to study some diagnostic problems related to the curvature. Wei and Shi (1994) have solved this problem for many common models, in particular, for exponential family nonlinear models. In the rest of this section, we shall introduce the developments of this problem and give some further extensions as well.

6.2.2 Regression coefficients

We shall show that $\hat{\beta}_{mi} = \hat{\beta}_{(i)}$ holds for quite general class of models. For exponential family nonlinear model (6.1) (i.e. (2.2)), the log-likelihood function can be denoted by (2.5) and may be represented as

$$l(\beta, \sigma) = -\frac{1}{2}\sigma^{-2}D(y, \mu(\beta)) - \frac{1}{2}s(y, \sigma^{-2}) - \frac{1}{2}\sigma^{-2}D'(y, y), \quad g(\mu) = f(\beta).$$

It is easily seen that the maximum likelihood estimator $\hat{\beta}$ minimizes the deviance $D(y, \mu(\beta))$ but does not involve the dispersion parameter σ and may be denoted by

$$D(\beta) \triangleq D(y, \mu(\beta)) = \sum_{j=1}^{n} d_j(y_j, \mu_j(\beta)), \quad \mu_j = g^{-1}(f_j(\beta)), \tag{6.11}$$

with

$$d_j(y_j, \mu_j(\beta)) = -2\{y_j\theta_j - b(\theta_j) - c(y_j)\} + 2\{y_j\theta_j - b(\theta_j) - c(y_j)\}_{\mu_j = y_j}.$$

From the above equations we have

Theorem 6.1

For exponential family nonlinear model (6.1), if Assumptions A and B (see Sections 2.1 and 2.2) hold, then we always have $\hat{\boldsymbol{\beta}}_{mi} = \hat{\boldsymbol{\beta}}_{(i)}$ (Wei and Shi 1994).

Proof. By (6.11), the maximum likelihood estimator $\hat{\boldsymbol{\beta}}_{(i)}$ minimizes

$$D_{(i)}(\boldsymbol{\beta}) \triangleq \sum_{j\neq i} d_j(y_j, \mu_j(\boldsymbol{\beta})), \quad \mu_j = g^{-1}(f_j(\boldsymbol{\beta})). \tag{6.12}$$

So $\hat{\boldsymbol{\beta}}_{(i)}$ satisfies the following equations for any $a = 1, \cdots, p$

$$\frac{\partial D_{(i)}(\boldsymbol{\beta})}{\partial \beta_a} = \sum_{j\neq i} \frac{\partial d_j(y_j, \mu_j(\boldsymbol{\beta}))}{\partial \beta_a} = 0. \tag{6.13}$$

The maximum likelihood estimators $\hat{\boldsymbol{\beta}}_{mi}$ and $\hat{\gamma}_{mi}$ minimize

$$D_{mi}(\boldsymbol{\beta}, \gamma) \triangleq \sum_{j\neq i} \{d_j(y_j, \mu_j(\boldsymbol{\beta}))\} + d_i(y_i, \mu_i(\boldsymbol{\beta}, \gamma)) \tag{6.14}$$

with

$$\mu_i(\boldsymbol{\beta}, \gamma) = g^{-1}(f_i(\boldsymbol{\beta}) + \gamma). \tag{6.15}$$

So $\hat{\boldsymbol{\beta}}_{mi}$ and $\hat{\gamma}_{mi}$ satisfy the following equations:

$$\begin{aligned} \frac{\partial D_{mi}(\boldsymbol{\beta}, \gamma)}{\partial \beta_a} &= \frac{\partial D_{(i)}(\boldsymbol{\beta})}{\partial \beta_a} + \frac{\partial d_i(y_i, \mu_i(\boldsymbol{\beta}, \gamma))}{\partial \beta_a} = 0, \\ \frac{\partial D_{mi}(\boldsymbol{\beta}, \gamma)}{\partial \gamma} &= \frac{\partial d_i(y_i, \mu_i(\boldsymbol{\beta}, \gamma))}{\partial \gamma} = 0. \end{aligned}$$

It follows from (6.15) and (1.26) that

$$\begin{aligned} \frac{\partial d_i(y_i, \mu_i(\boldsymbol{\beta}, \gamma))}{\partial \beta_a} &= \frac{\partial d_i(y_i, \mu_i(\boldsymbol{\beta}, \gamma))}{\partial \theta_i} \frac{\partial \theta_i}{\partial \mu_i} \frac{\partial g^{-1}(f_i(\boldsymbol{\beta}) + \gamma)}{\partial f_i} \frac{\partial f_i}{\partial \beta_a} \\ &= (-2)\{y_i - \mu_i(\boldsymbol{\beta}, \gamma)\} V_i^{-1} \dot{g}^{-1}(f_i(\boldsymbol{\beta}) + \gamma) \frac{\partial f_i}{\partial \beta_a}, \\ \frac{\partial d_i(y_i, \mu_i(\boldsymbol{\beta}, \gamma))}{\partial \gamma} &= (-2)\{y_i - \mu_i(\boldsymbol{\beta}, \gamma)\} V_i^{-1} \dot{g}^{-1}(f_i(\boldsymbol{\beta}) + \gamma) = 0. \end{aligned}$$

Notice that $\dot{g}(\cdot) \neq 0$ since $g(\cdot)$ is a monotonic function. So $\hat{\boldsymbol{\beta}}_{mi}$, $\hat{\gamma}_{mi}$ must satisfy

$$\frac{\partial D_{mi}(\boldsymbol{\beta}, \gamma)}{\partial \beta_a} = \frac{\partial D_{(i)}(\boldsymbol{\beta}, \gamma)}{\partial \beta_a} = 0, \quad y_i - \mu_i(\boldsymbol{\beta}, \gamma) = 0. \tag{6.16}$$

By Assumption B, the maximum likelihood estimators of $\boldsymbol{\beta}$ exist and are unique for CDM and MSOM. So we have $\hat{\boldsymbol{\beta}}_{mi} = \hat{\boldsymbol{\beta}}_{(i)}$ because $\hat{\boldsymbol{\beta}}_{mi}$ and $\hat{\boldsymbol{\beta}}_{(i)}$ satisfy the same equation (see (6.13) and (6.16)). ∎

By this theorem, we have several corollaries.

Corollary 1

For normal nonlinear regression models and GLM, if Assumptions A and B are satisfied, we always have $\hat{\boldsymbol{\beta}}_{mi} = \hat{\boldsymbol{\beta}}_{(i)}$. ∎

The statement of Corollary 1 is of concern to many authors (e.g. Williams 1987; Dzieciolowski and Ross 1990).

Corollary 2

If the MSOM (6.9) is replaced by a more general form

$$\begin{cases} g(\mu_j) = f_j(\boldsymbol{\beta}), \quad j = 1, \cdots, n; \quad j \neq i, \\ \\ g(\mu_i) = f(x_i; \boldsymbol{\beta}, \boldsymbol{\gamma}) \triangleq f_i(\boldsymbol{\beta}, \boldsymbol{\gamma}), \end{cases} \tag{6.17}$$

where $\boldsymbol{\gamma}$ is a vector parameter and $\partial f_i / \partial \boldsymbol{\gamma} \neq 0$ everywhere in some region, then we have $\hat{\boldsymbol{\beta}}_{mi} = \hat{\boldsymbol{\beta}}_{(i)}$ (note that if $\partial f_i / \partial \boldsymbol{\gamma} = 0$ is in some region, then $f_i(\boldsymbol{\beta}, \boldsymbol{\gamma})$ does not depend on $\boldsymbol{\gamma}$).

Proof. Let $\mu_i = g^{-1}(f_i(\boldsymbol{\beta}, \boldsymbol{\gamma})) = \mu_i(\boldsymbol{\beta}, \boldsymbol{\gamma})$, then (6.16) still holds for model (6.17). So we have $\hat{\boldsymbol{\beta}}_{mi} = \hat{\boldsymbol{\beta}}_{(i)}$ as shown in Theorem 6.1. ∎

Corollary 3

For the CWM (6.10), we have $\hat{\boldsymbol{\beta}}_{mi} = \hat{\boldsymbol{\beta}}_{\omega i}$ as $\omega \to 0$. ∎

Theorem 6.1 and its corollaries can be extended to the multiple case, that is k observations are deleted, or shifted, or weighted. These models can be respectively represented as

$$\text{CDM}: \quad g(\mu_j) = f_j(\boldsymbol{\beta}), \quad j \bar{\in} I,$$

$$\text{MSOM}: \quad \begin{cases} g(\mu_j) = f_j(\boldsymbol{\beta}), & j \bar{\in} I \\ \\ g(\mu_i) = f_i(\boldsymbol{\beta}) + \gamma_i, & i \in I \end{cases}$$

$$\text{CWM}: \quad l_\omega(\boldsymbol{\beta}, \sigma) = \sum_{j \bar{\in} I} l_j(\boldsymbol{\beta}, \sigma) + \sum_{i \in I} \omega_i l_i(\boldsymbol{\beta}, \sigma),$$

where $I = \{i_1, \cdots, i_k\}$ is a subset of indices from $1, \cdots, n$. The estimates for the above models are respectively denoted by $\hat{\boldsymbol{\beta}}_{(I)}$, $\hat{\boldsymbol{\beta}}_{mI}$, and $\hat{\boldsymbol{\beta}}_{\omega I}$. By similar derivation to that of Theorem 6.1, we have the following corollary.

Corollary 4

For multiple CDM, MSOM and CWM, we always have $\hat{\beta}_{(I)} = \hat{\beta}_{mI} = \hat{\beta}_{\omega I}$ (as $\omega_i \to 0,\ i \in I$). ∎

The following corollary is the further extension of Theorem 6.1.

Corollary 5

Suppose that $\hat{\beta}$ and $\hat{\sigma}$ minimize the following objective function

$$Q(\beta, \sigma) = \sum_{i=1}^{n} \rho_i(f_i(\beta), \sigma),$$

where $\rho_i(\cdot, \cdot)$ and $f_i(\cdot)$ are known functions. If $\rho_i(f_i(\beta), \sigma) = a(\sigma)\xi_i(f_i(\beta)) + b_i(\sigma)$ and $\xi_i(\cdot)$ is a known function, then we have $\hat{\beta}_{mi} = \hat{\beta}_{(i)}$, where $\hat{\beta}_{mi}$ and $\hat{\beta}_{(i)}$ are the estimates of β respectively corresponding to MSOM and CDM as follows:

$$\text{MSOM}: \quad Q_{mi}(\beta, \sigma) = \sum_{j \neq i} \rho_j(f_j(\beta), \sigma) + \rho_i(f_i(\beta, \gamma), \sigma),$$

$$\text{CDM}: \quad Q_{(i)}(\beta, \sigma) = \sum_{j \neq i} \rho_j(f_j(\beta), \sigma).$$ ∎

The proof is quite similar to that of Theorem 6.1 and we omit it here (see Wei and Shi 1994 for more details). It is easily seen from this corollary that $\hat{\beta}_{mi} = \hat{\beta}_{(i)}$ holds for the M-estimator in nonlinear regression models (Seber and Wild, 1989).

In the following subsection, we shall discuss similar problem for dispersion parameter and deviance.

6.2.3 Deviance and dispersion parameter

It is easily shown that we usually do not have $\hat{\sigma}^2_{mi} = \hat{\sigma}^2_{(i)}$ for the maximum likelihood estimator, even in the normal linear regression case. In fact, for normal linear regression, we have $\hat{\sigma}^2_{mi} = n^{-1}RSS_{(i)}$ and $\hat{\sigma}^2_{(i)} = (n-1)^{-1}RSS_{(i)}$, where $RSS_{(i)}$ is the residual sum of squares associated with the CDM. So we have $\hat{\sigma}^2_{(i)} = (n-1)^{-1}n\hat{\sigma}^2_{mi} \approx \hat{\sigma}^2_{mi}$. Similarly, for exponential family nonlinear models, we cannot get $\hat{\sigma}^2_{mi} = \hat{\sigma}^2_{(i)}$, but have $\hat{\sigma}^2_{mi} \approx \hat{\sigma}^2_{(i)}$ because the deviances that have close connection with the maximum likelihood estimators of σ^2 are always equal for CDM and MSOM.

Theorem 6.2

Under the conditions stated in Theorem 6.1, we always have $D_{(i)}(\hat{\boldsymbol{\beta}}_{(i)}) = D_{mi}(\hat{\boldsymbol{\beta}}_{(i)}, \hat{\gamma}_{mi})$ where $D_{(i)}(\boldsymbol{\beta})$ and $D_{mi}(\boldsymbol{\beta}, \gamma)$ are respectively the deviances for CDM and MSOM given in (6.12) and (6.14).

Proof. It follows from (6.14) that

$$D_{mi}(\hat{\boldsymbol{\beta}}_{mi}, \hat{\gamma}_{mi}) = D_{(i)}(\hat{\boldsymbol{\beta}}_{mi}) + d_i(y_i, \mu_i(\hat{\boldsymbol{\beta}}_{mi}, \hat{\gamma}_{mi})).$$

From Theorem 6.1 and (6.16), we have $\hat{\boldsymbol{\beta}}_{mi} = \hat{\boldsymbol{\beta}}_{(i)}$ and $\mu_i(\hat{\boldsymbol{\beta}}_{mi}, \hat{\gamma}_{mi}) = y_i$. Substituting these results into the right-hand side of $D_{mi}(\hat{\boldsymbol{\beta}}_{mi}, \hat{\gamma}_{mi})$ gives

$$D_{mi}(\hat{\boldsymbol{\beta}}_{mi}, \hat{\gamma}_{mi}) = D_{(i)}(\hat{\boldsymbol{\beta}}_{(i)}) + d_i(y_i, y_i) = D_{(i)}(\hat{\boldsymbol{\beta}}_{(i)}).$$

Here we use the result $d_i(y_i, y_i) = 0$, which is easily shown from (1.26). ∎

Note that for normal linear and nonlinear regression models, the deviances are just the residual sum of squares.

Corollary

The deviances for CDM and MSOM are equal in GLM and the residual sum of squares of CDM and MSOM are equal in normal nonlinear regression models. ∎

As pointed out in Section 2.2 (e.g. (2.12) and (2.13)), the maximum likelihood estimator of dispersion parameter $\phi = \sigma^{-2}$ has a close connection with the deviance. By Theorem 6.2, we may compare the estimators $\hat{\phi}_{mi} = \hat{\sigma}_{mi}^{-2}$ in MSOM with $\hat{\phi}_{(i)} = \hat{\sigma}_{(i)}^{-2}$ in CDM.

Theorem 6.3

Under the conditions stated in Theorem 6.1, if (1.6) holds, that is $s(y_i, \phi) = s(\phi) + t(y_i)$ for all i, then we have

$$\dot{s}(\hat{\phi}_{(i)}) = (n-1)^{-1} n \dot{s}(\hat{\phi}_{mi}) + (n-1)^{-1} k, \tag{6.18}$$

where $k = d_i'(y_i, y_i)$ is defined in Lemma 1.5. Furthermore, for normal and inverse Gaussian nonlinear models, we have

$$\hat{\sigma}_{(i)}^2 = (n-1)^{-1} n \hat{\sigma}_{mi}^2 \approx \hat{\sigma}_{mi}^2. \tag{6.19}$$

Proof. For full model (6.1), (2.13) gives

$$-\dot{s}(\hat{\phi}) = n^{-1} D(\mathbf{Y}, \hat{\boldsymbol{\mu}}) + k.$$

So for MSOM, $\hat{\phi}_{mi}$, $\hat{\boldsymbol{\beta}}_{mi}$ and $\hat{\gamma}_{mi}$ satisfy

$$-\dot{s}(\hat{\phi}_{mi}) = n^{-1} D_{mi}(\hat{\boldsymbol{\beta}}_{mi}, \hat{\gamma}_{mi}) + k,$$

where $D_{mi}(\boldsymbol{\beta}, \gamma)$ is given in (6.14). Similarly, $\hat{\phi}_{(i)}$ and $\hat{\boldsymbol{\beta}}_{(i)}$ of CDM satisfy

$$-\dot{s}(\hat{\phi}_{(i)}) = (n-1)^{-1} D_{(i)}(\hat{\boldsymbol{\beta}}_{(i)}) + k.$$

From Theorem 6.2, we have $D_{mi}(\hat{\boldsymbol{\beta}}_{mi}, \hat{\gamma}_{mi}) = D_{(i)}(\hat{\boldsymbol{\beta}}_{(i)})$, so combining these equations gives

$$-\dot{s}(\hat{\phi}_{(i)}) = (n-1)^{-1}\{-n\dot{s}(\hat{\phi}_{mi}) - nk\} + k$$

which leads to (6.18).

For normal and inverse Gaussian models, we have $s(\phi) = -\log\phi$, $\dot{s}(\phi) = -\phi^{-1} = -\sigma^2$ and $k = 0$ (see Example 2.2). Then (6.19) can be easily obtained from (6.18). ∎

Now we show a numerical example to illustrate the results given in Theorems 6.1 and 6.2.

Example 6.1 (Fruit fly data No. 1).

These data were originally collected by Powsner (1935) and reanalyzed by McCullagh and Nelder (1989, p.306). The data came from an experiment to determine accurately the effect of temperature on the duration of the developmental stage of the fruit fly (***Drosophila melanogaster***). In these data, x_i is the experimental temperature which was kept constant over the period of the experiment, y_i is the average duration of the embryonic period measured from the time at which the eggs were laid and ω_i is the number of eggs of each batch. Following McCullagh and Nelder (1989), we use the rational function of the temperature, gamma error, log link and weighted by batch size to fit the data, that is $y_i \sim GA(\mu_i, \sigma^2\omega_i^{-1})$ and

$$\log\mu_i = \beta_1 + \beta_2 x_i + \beta_3(x_i - \beta_4)^{-1}, \qquad i = 1, \cdots, 23.$$

Here we compute estimators of $\beta_1, \beta_2, \beta_3, \beta_4$, $\phi = \sigma^{-2}$ and the deviances for CDM and MSOM respectively based on this model. The results are summarized in Tables 6.1 and 6.2. Table 6.1 gives the results of estimates and deviances based on the CDM, while Table 6.2 gives the corresponding results based on the MSOM. The results shown in the tables are consistent with Theorems 6.1 to 6.3, as expected.

In Sections 6.4 and 6.5, we will give further discussions of influence analysis for these data.

Table 6.1 Parameter estimates and deviance for case deletion model with i-th case deleted

i	β_1	β_2	β_3	β_4	*Deviance*	ϕ
1	3.29622	-0.26254	-209.509	58.2570	0.31783	69.3858
2	3.18387	-0.26301	-216.869	58.7597	0.26624	82.7986
3	3.16461	-0.26577	-220.020	58.7872	0.31797	69.3562
4	3.34038	-0.26237	-206.701	58.0657	0.29954	73.6118
5	3.21023	-0.26504	-216.783	58.6176	0.25831	85.3342
6	3.29694	-0.26269	-209.585	58.2531	0.31764	69.4277
7	3.20139	-0.26482	-217.096	58.6494	0.31422	70.1811
8	3.20060	-0.26476	-217.092	58.6528	0.31529	69.9423
9	3.19819	-0.26564	-217.492	58.5967	0.29234	75.4208
10	3.20054	-0.26550	-217.315	58.5979	0.30600	72.0613
11	3.14849	-0.26645	-221.355	58.8248	0.31344	70.3551
12	3.18322	-0.26540	-218.522	58.7033	0.31703	69.5595
13	3.20783	-0.26447	-216.449	58.6279	0.31194	70.6934
14	3.21446	-0.26409	-215.841	58.6158	0.30559	72.1577
15	3.19690	-0.26413	-217.064	58.7056	0.29683	74.2820
16	3.20766	-0.26530	-216.835	58.5872	0.28238	78.0745
17	3.07875	-0.26741	-226.641	59.1371	0.31724	69.5150
18	3.20154	-0.26488	-217.099	58.6460	0.31006	71.1195
19	3.31631	-0.26224	-208.061	58.1743	0.31650	69.6769
20	3.21211	-0.26458	-216.190	58.5982	0.31823	69.2989
21	3.02118	-0.26936	-231.293	59.2896	0.29038	75.9289
22	3.20733	-0.26604	-216.845	58.4868	0.29420	74.9459
23	3.17867	-0.26319	-218.240	58.9156	0.25840	85.3054

6.2.4 Score statistic of outlier

We can get a score statistic to detect the outlier based on the MSOM. In fact, for model (6.9), one can make a testing of hypothesis:

$$H_0 : \gamma = 0; \qquad H_1 : \gamma \neq 0.$$

If H_0 is rejected, then the i-th case may be a possible outlier because this case may not come from the original model (6.1). We now derive a score statistic for this test.

Theorem 6.4

For MSOM (6.9), the score statistic for the testing of hypothesis $H_0 : \gamma = 0$ is

$$SC_i = \frac{rp_i^2}{\hat{\sigma}^2(1 - h_{ii})} = r_i^2, \tag{6.20}$$

Table 6.2 Parameter estimates and deviance for mean-shift outlier model with i-th case shifted

i	β_1	β_2	β_3	β_4	*Deviance*	ϕ
1	3.29622	-0.26254	-209.509	58.2570	0.31783	72.5322
2	3.18392	-0.26301	-216.866	58.7595	0.26624	86.5547
3	3.16461	-0.26577	-220.020	58.7872	0.31797	72.5012
4	3.34038	-0.26237	-206.701	58.0657	0.29954	76.9502
5	3.21015	-0.26504	-216.789	58.6179	0.25831	89.2053
6	3.29694	-0.26269	-209.585	58.2531	0.31764	72.5760
7	3.20138	-0.26482	-217.096	58.6496	0.31422	73.3638
8	3.20059	-0.26476	-217.092	58.6530	0.31529	73.1141
9	3.19838	-0.26563	-217.477	58.5959	0.29234	78.8412
10	3.20068	-0.26550	-217.304	58.5973	0.30600	75.3291
11	3.14849	-0.26645	-221.355	58.8248	0.31344	73.5456
12	3.18322	-0.26540	-218.522	58.7033	0.31703	72.7138
13	3.20783	-0.26447	-216.449	58.6279	0.31194	73.8992
14	3.21446	-0.26409	-215.841	58.6158	0.30559	75.4301
15	3.19690	-0.26413	-217.064	58.7055	0.29683	77.6508
16	3.20750	-0.26530	-216.846	58.5878	0.28238	81.6156
17	3.07875	-0.26741	-226.641	59.1371	0.31724	72.6673
18	3.20154	-0.26488	-217.099	58.6461	0.31006	74.3448
19	3.31631	-0.26224	-208.061	58.1743	0.31650	72.8365
20	3.21211	-0.26458	-216.190	58.5982	0.31823	72.4413
21	3.02118	-0.26936	-231.293	59.2896	0.29038	79.3727
22	3.20680	-0.26605	-216.887	58.4890	0.29420	78.3442
23	3.17924	-0.26318	-218.196	58.9132	0.25841	89.1710

where h_{ii}, rp_i and r_i are given in (6.3) and (6.5).

Proof. Let the Fisher information matrix of $\boldsymbol{Y}$ for $\boldsymbol{\beta}$, ϕ, γ be $\boldsymbol{J}_{(\beta,\phi,\gamma)}$, then the score statistic for H_0 is

$$SC_i = \{(\frac{\partial l}{\partial \gamma})^T J^{\gamma\gamma}(\frac{\partial l}{\partial \gamma})\}_{(\hat{\beta},\hat{\sigma}^2)}, \tag{6.21}$$

where $J^{\gamma\gamma}$ is the lower right corner of $\boldsymbol{J}^{-1}_{(\beta,\phi,\gamma)}$. It is easily seen that

$$\frac{\partial l}{\partial \gamma} = (\frac{\partial l}{\partial \boldsymbol{\mu}})^T(\frac{\partial \boldsymbol{\mu}}{\partial \gamma}) = (\phi \boldsymbol{V}^{-1}\boldsymbol{e})^T(\frac{\partial \boldsymbol{\mu}}{\partial \gamma}).$$

It follows from (6.9) that

$$\frac{\partial \mu_j}{\partial \gamma} = 0 \ \ (j \neq i), \qquad \frac{\partial \mu_i}{\partial \gamma} = \dot{g}^{-1}(\mu_i).$$

So we have

$$\frac{\partial \boldsymbol{\mu}}{\partial \gamma} = \dot{g}^{-1}(\mu_i)\boldsymbol{c}_i, \quad \frac{\partial l}{\partial \gamma} = \phi V_i^{-1} e_i \dot{g}^{-1}(\mu_i),$$

where $\boldsymbol{c}_i$ is a unit vector having 1 at the i-th position and zeros elsewhere. Further, it follows from (A.10) that

$$\begin{aligned}\frac{\partial^2 l}{\partial \gamma^2} &= (\frac{\partial \boldsymbol{\mu}}{\partial \gamma})^T (\frac{\partial^2 l}{\partial \boldsymbol{\mu} \partial \boldsymbol{\mu}^T})(\frac{\partial \boldsymbol{\mu}}{\partial \gamma}) + (\frac{\partial l}{\partial \boldsymbol{\mu}})^T (\frac{\partial^2 \boldsymbol{\mu}}{\partial \gamma^2}) \\ &= -\phi(V_i^{-1} - e_i V_i^{-2} \dot{V}_i)\dot{g}^{-2}(\mu_i) + \phi V_i^{-1} e_i \ddot{g}^{-1}(\mu_i)\dot{g}^{-1}(\mu_i)\end{aligned}$$

and then

$$J_{\gamma\gamma} = \mathrm{E}(-\frac{\partial^2 l}{\partial \gamma^2}) = \phi V_i^{-1} \dot{g}^{-2}(\mu_i).$$

By similar derivation, we can get

$$\boldsymbol{J}_{(\beta,\phi,\gamma)} = \begin{pmatrix} \phi \boldsymbol{D}^T \boldsymbol{V}^{-1} \boldsymbol{D} & \boldsymbol{0} & \phi V_i^{-1} \dot{g}^{-1}(\mu_i)\boldsymbol{d}_i \\ \boldsymbol{0} & J_{\phi\phi} & 0 \\ \phi V_i^{-1} \dot{g}^{-1}(\mu_i)\boldsymbol{d}_i^T & 0 & \phi V_i^{-1} \dot{g}^{-2}(\mu_i) \end{pmatrix},$$

where $J_{\phi\phi}$ is a scalar and $\boldsymbol{D}^T = (\boldsymbol{d}_1, \cdots, \boldsymbol{d}_n)$. From this we have

$$\begin{aligned}J^{\gamma\gamma} &= \phi^{-1}\{V_i^{-1}\dot{g}^{-2}(\mu_i) - V_i^{-1}\dot{g}^{-2}(\mu_i)\boldsymbol{d}_i^T(\boldsymbol{D}^T \boldsymbol{V}^{-1} \boldsymbol{D})^{-1}\boldsymbol{d}_i V_i^{-1}\}^{-1} \\ &= \phi^{-1} V_i \dot{g}^2(\mu_i)(1 - h_{ii})^{-1}.\end{aligned}$$

Substituting $J^{\gamma\gamma}$ and $\partial l/\partial \gamma$ given above into (6.21) yields

$$\begin{aligned}SC_i &= \{\phi^2 V_i^{-2} e_i^2 \dot{g}^{-2}(\mu_i)\phi^{-1} V_i \dot{g}^2(\mu_i)(1 - h_{ii})^{-1}\}_{(\hat{\beta},\hat{\sigma}^2)} \\ &= \frac{\hat{e}_i^2}{\hat{\sigma}^2 \hat{V}_i (1 - h_{ii})} = \frac{rp_i^2}{\hat{\sigma}^2(1 - h_{ii})} = r_i^2.\end{aligned}$$ ∎

This theorem shows that the score statistic SC_i is just the square of studentized residual that is an adequate diagnostic statistic often used in linear regression diagnostics.

Table 6.4 gives some numerical results of $SC_i^{\frac{1}{2}} = r_i$.

Corollary

For the general MSOM (6.17), the expression (6.20) still holds. ∎

Theorem 6.4 can also be extended to more general shifted models (see Davison and Tsai 1992). We now consider the following model:

$$\boldsymbol{g}(\boldsymbol{\mu}) = \boldsymbol{f}(\boldsymbol{\beta}, \gamma), \tag{6.22}$$

where γ is a vector parameter. We assume that there exists $\gamma = \gamma_0$ such that $f(\boldsymbol{\beta}, \gamma_0) = f(\boldsymbol{\beta})$. In this case, $g(\mu_i) = f(x_i; \boldsymbol{\beta}, \gamma), i = 1, \cdots, n$; and the shifts are not only added on one case but also on all the cases. One can consider a testing of hypothesis $H_0 : \gamma = \gamma_0$. If H_0 is rejected, then the original model (6.1) (corresponding to $g(\boldsymbol{\mu}) = f(\boldsymbol{\beta})$) is not adequate to fit the given data. To get the test statistic, we have the following theorem.

Theorem 6.5

For model (6.22), the score statistic for the test H_0 is

$$SC = \hat{\sigma}^2 \hat{\mathbf{e}}^T \hat{\mathbf{V}}^{-\frac{1}{2}} (\mathbf{H}^+ - \mathbf{H}) \hat{\mathbf{V}}^{-\frac{1}{2}} \hat{\mathbf{e}}, \tag{6.23}$$

where $\mathbf{H}$ is given in Lemma 6.1, $\mathbf{H}^+$ is the projection matrix of $(\mathbf{V}^{-\frac{1}{2}} \mathbf{D}_\beta, \mathbf{V}^{-\frac{1}{2}} \mathbf{D}_\gamma)_{\hat{\beta}}$ and $\mathbf{D}_\beta = \partial \boldsymbol{\mu} / \partial \boldsymbol{\beta}^T$, $\mathbf{D}_\gamma = \partial \boldsymbol{\mu} / \partial \gamma^T$.

Proof. It can be shown that

$$\frac{\partial l}{\partial \boldsymbol{\beta}} = \phi \mathbf{D}_\beta^T \mathbf{V}^{-1} \mathbf{e}, \quad \frac{\partial l}{\partial \gamma} = \phi \mathbf{D}_\gamma^T \mathbf{V}^{-1} \mathbf{e},$$

and

$$\mathbf{J}_{(\beta, \phi, \gamma)} = \begin{pmatrix} \phi \mathbf{D}_\beta^T \mathbf{V}^{-1} \mathbf{D}_\beta & \mathbf{0} & \phi \mathbf{D}_\beta^T \mathbf{V}^{-1} \mathbf{D}_\gamma \\ \mathbf{0} & \mathbf{J}_{\phi\phi} & \mathbf{0} \\ \phi \mathbf{D}_\gamma^T \mathbf{V}^{-1} \mathbf{D}_\beta & \mathbf{0} & \phi \mathbf{D}_\gamma^T \mathbf{V}^{-1} \mathbf{D}_\gamma \end{pmatrix}.$$

So we have

$$\begin{aligned} \mathbf{J}^{\gamma\gamma} &= \phi^{-1} \{ \mathbf{D}_\gamma^T \mathbf{V}^{-1} \mathbf{D}_\gamma - \mathbf{D}_\gamma^T \mathbf{V}^{-1} \mathbf{D}_\beta (\mathbf{D}_\beta^T \mathbf{V}^{-1} \mathbf{D}_\beta)^{-1} \mathbf{D}_\beta^T \mathbf{V}^{-1} \mathbf{D}_\gamma \}^{-1} \\ &= \phi^{-1} (\mathbf{D}_\gamma^T \mathbf{V}^{-\frac{1}{2}} \mathbf{H}' \mathbf{V}^{-\frac{1}{2}} \mathbf{D}_\gamma)^{-1}, \end{aligned}$$

where $\mathbf{H}' = \mathbf{I}_n - \mathbf{H}$. Substituting $\mathbf{J}^{\gamma\gamma}$ and $\partial l / \partial \gamma$ into the right-hand side of (6.21) yields

$$SC = \hat{\sigma}^{-2} \{ \mathbf{e}^T \mathbf{V}^{-1} \mathbf{D}_\gamma (\mathbf{D}_\gamma^T \mathbf{V}^{-\frac{1}{2}} \mathbf{H}' \mathbf{V}^{-\frac{1}{2}} \mathbf{D}_\gamma)^{-1} \mathbf{D}_\gamma^T \mathbf{V}^{-1} \mathbf{e} \}_{\hat{\beta}}.$$

Since $(\mathbf{D}_\beta^T \mathbf{V}^{-1} \mathbf{e})_{\hat{\beta}} = 0$, we have $\mathbf{H}' \hat{\mathbf{V}}^{-\frac{1}{2}} \hat{\mathbf{e}} = \hat{\mathbf{V}}^{-\frac{1}{2}} \hat{\mathbf{e}}$. So we get

$$\begin{aligned} SC &= \hat{\sigma}^{-2} \hat{\mathbf{e}}^T \hat{\mathbf{V}}^{-\frac{1}{2}} \mathbf{H}' \hat{\mathbf{V}}^{-\frac{1}{2}} \hat{\mathbf{D}}_\gamma (\hat{\mathbf{D}}_\gamma^T \hat{\mathbf{V}}^{-\frac{1}{2}} \mathbf{H}' \hat{\mathbf{V}}^{-\frac{1}{2}} \hat{\mathbf{D}}_\gamma)^{-1} \hat{\mathbf{D}}_\gamma^T \hat{\mathbf{V}}^{-\frac{1}{2}} \mathbf{H}' \hat{\mathbf{V}}^{-\frac{1}{2}} \hat{\mathbf{e}} \\ &= \hat{\sigma}^{-2} \hat{\mathbf{e}}^T \hat{\mathbf{V}}^{-\frac{1}{2}} (\mathbf{H}^+ - \mathbf{H}) \hat{\mathbf{V}}^{-\frac{1}{2}} \hat{\mathbf{e}}, \end{aligned}$$

where we use the part (b) of Theorem A.3, that is $(\boldsymbol{H}^+ - \boldsymbol{H})$ is equal to the projection matrix of $(\boldsymbol{I}_n - \boldsymbol{H})\hat{\boldsymbol{V}}^{-\frac{1}{2}}\hat{\boldsymbol{D}}_\gamma$ (note that here we use the ordinary inner product $< \boldsymbol{a}, \boldsymbol{b} >= \boldsymbol{a}^T\boldsymbol{b}$). ∎

This result can be extended to more general models (Davison and Tsai 1992). The score statistic (6.23) may be viewed as the change in residual sum of squares when the parameter γ is added to the model. In fact (6.23) can be expressed as

$$SC = \hat{\sigma}^{-2}(RSS - RSS^+),$$

where

$$\begin{aligned} RSS &= \hat{\boldsymbol{e}}^T\hat{\boldsymbol{V}}^{-\frac{1}{2}}(\boldsymbol{I}_n - \boldsymbol{H})\hat{\boldsymbol{V}}^{-\frac{1}{2}}\hat{\boldsymbol{e}}, \\ RSS^+ &= \hat{\boldsymbol{e}}^T\hat{\boldsymbol{V}}^{-\frac{1}{2}}(\boldsymbol{I}_n - \boldsymbol{H}^+)\hat{\boldsymbol{V}}^{-\frac{1}{2}}\hat{\boldsymbol{e}}. \end{aligned}$$

6.3 Influence Diagnostics Based on Case Deletion

We have introduced the generalized Cook distance C_i and the likelihood distance LD_i in Section 6.1; and introduced the score statistic SC_i in Section 6.2. This section will summarize the diagnostic statistics which are applicable to exponential family nonlinear models. Our derivations are based on the CDM and the argument of Andersen (1992).

6.3.1 Diagnostics based on weighted least squares

The basic idea described here is first implicitly introduced by Pregibon (1981) for GLM and then developed by Andersen (1992) who discussed the diagnostics in categorical data analysis and obtained several diagnoctic statistics. We now give a further description for exponential family nonlinear models.

We now denote our model (6.1) by model (A) which will be compared with a linear model (B) obtained in the form from the iteration procedure of maximum likelihood estimator of $\boldsymbol{\beta}$ for model (A). Models (A) and (B) are summarized as

$$\text{(A)} \quad g(\mu_i) = f(x_i; \boldsymbol{\beta}), \quad y_i \sim ED(\mu_i, \sigma^2);$$

$$\text{(B)} \quad \boldsymbol{V}^{-\frac{1}{2}}\boldsymbol{Z} = \boldsymbol{V}^{-\frac{1}{2}}\boldsymbol{D}\boldsymbol{\beta} + \boldsymbol{\epsilon}, \quad \boldsymbol{\epsilon} \sim (\boldsymbol{0}, \boldsymbol{I}_n),$$

where $\boldsymbol{Z}$ is given in equations (2.20) to (2.22). To derive diagnostic statistics for model (A), we emphasis a fact that the models (A) and (B) share the following two common points:

(a) The i-th case of both models (A) and (B) corresponds to the observation $(\boldsymbol{x}_i, y_i)$ (note that $\boldsymbol{V}$ is a diagonal matrix).

(b) The maximum likelihood estimator of $\boldsymbol{\beta}$ in model (A) is equal to the least squares estimator of $\boldsymbol{\beta}$ in form obtained from model (B) (see (2.21) and (2.22)).

By these two properties, we have the following statement: assessing the influence of i-th case $(\boldsymbol{x}_i, y_i)$ on the maximum likelihood estimator in model (A) is equivalent to assessing the influence of $(\boldsymbol{x}_i, y_i)$ on the least squares estimator in a linear model (B). Therefore we may find appropriate diagnostic statistics for model (A) with the aid of model (B) that can be regarded as a weighted linear regression in form. This idea was considered by Andersen (1992) who pointed out that once model (B) has been established many important results are immediate consequences (see also Pregibon 1981). We now summarize the commonly used diagnostic statistics for model (A), obtained from model (B) in the form using linear regression diagnostics (see (2.20) to (2.21)).

(a) Residuals and studentized residuals

$$\boldsymbol{rp} = (\boldsymbol{V}^{-\frac{1}{2}}\boldsymbol{Z} - \boldsymbol{V}^{-\frac{1}{2}}\boldsymbol{D}\boldsymbol{\beta})_{\hat{\beta}} = \hat{\boldsymbol{V}}^{-\frac{1}{2}}\hat{\boldsymbol{e}},$$

$$r_i = \frac{rp_i}{\hat{\sigma}\sqrt{1-h_{ii}}}.$$

(b) Hat matrix

$$\boldsymbol{H} = \{\boldsymbol{V}^{-\frac{1}{2}}\boldsymbol{D}(\boldsymbol{D}^T\boldsymbol{V}^{-1}\boldsymbol{D})^{-1}\boldsymbol{D}^T\boldsymbol{V}^{-\frac{1}{2}}\}_{\hat{\beta}}.$$

Since $(\boldsymbol{D}^T\boldsymbol{V}^{-1}\boldsymbol{e})_{\hat{\beta}} = \boldsymbol{0}$, we have

$$\boldsymbol{H}\boldsymbol{rp} = \boldsymbol{0}, \quad (\boldsymbol{I}_n - \boldsymbol{H})\boldsymbol{rp} = \boldsymbol{rp}.$$

(c) Cook distance

$$C_i = \frac{(\hat{\boldsymbol{\beta}} - \hat{\boldsymbol{\beta}}_{(i)})^T(\boldsymbol{D}^T\boldsymbol{V}^{-1}\boldsymbol{D})(\hat{\boldsymbol{\beta}} - \hat{\boldsymbol{\beta}}_{(i)})}{p\hat{\sigma}^2},$$

$$C_i^I = \frac{h_{ii}}{1-h_{ii}}\frac{r_i^2}{p}, \tag{6.24}$$

where C_i^I is the first order approximation of C_i.

(d) AP statistic

$$AP_i = \left\{ \frac{|\boldsymbol{X}^{*T}_{(i)} \boldsymbol{X}^*_{(i)}|}{|\boldsymbol{X}^{*T} \boldsymbol{X}^*|} \right\}_{\hat{\beta}},$$

where $|\cdot|$ denotes the determinant, $\boldsymbol{X}^* = (\boldsymbol{V}^{-\frac{1}{2}}\boldsymbol{D},\ \boldsymbol{V}^{-\frac{1}{2}}\boldsymbol{Z})$ and $\boldsymbol{X}^*_{(i)}$ is obtained from $\boldsymbol{X}^*$ with i-th row deleted. We can get a formula for AP_i just like the linear regression case.

Theorem 6.6

Let h^*_{ii} be the diagonal element of the projection matrix $\boldsymbol{H}^*$ obtained from $\boldsymbol{X}^*$, then we have

$$AP_i = 1 - h^*_{ii} = 1 - h_{ii} - rp_i^2/(\boldsymbol{rp}^T\boldsymbol{rp}). \tag{6.25}$$

Proof. Let the i-th row of $\boldsymbol{X}^*$ be $\boldsymbol{x}_i^{*T}$, then we have $\boldsymbol{X}^{*T}_{(i)} \boldsymbol{X}^*_{(i)} = \boldsymbol{X}^{*T}\boldsymbol{X}^* - \boldsymbol{x}_i^*\boldsymbol{x}_i^{*T}$. So by the expansion of a determinant we get

$$|\boldsymbol{X}^{*T}_{(i)} \boldsymbol{X}^*_{(i)}| = |\boldsymbol{X}^{*T}\boldsymbol{X}^*||1 - \boldsymbol{x}_i^{*T}(\boldsymbol{X}^{*T}\boldsymbol{X}^*)^{-1}\boldsymbol{x}_i^*|,$$

$$AP_i = \left\{ \frac{|\boldsymbol{X}^{*T}_{(i)} \boldsymbol{X}^*_{(i)}|}{|\boldsymbol{X}^{*T} \boldsymbol{X}^*|} \right\}_{\hat{\beta}} = 1 - h^*_{ii}.$$

To get the expression of h^*_{ii}, we use Theorem A.3 of Appendix A to the matrix $\boldsymbol{X}^* = (\boldsymbol{V}^{-\frac{1}{2}}\boldsymbol{D},\ \boldsymbol{V}^{-\frac{1}{2}}\boldsymbol{Z})$, which leads to

$$\boldsymbol{H}^* = \boldsymbol{H} + \boldsymbol{P}_{(I_n - H)V^{-\frac{1}{2}}Z}.$$

Note that $\boldsymbol{Z} = (\boldsymbol{D\beta} + \boldsymbol{e})_{\hat{\beta}}$ (see (2.20)), $\boldsymbol{Hrp} = \boldsymbol{0}$, so $(\boldsymbol{I}_n - \boldsymbol{H})\boldsymbol{V}^{-\frac{1}{2}}\boldsymbol{Z}$ can be expressed as

$$\begin{aligned} \{(\boldsymbol{I}_n - \boldsymbol{H})\boldsymbol{V}^{-\frac{1}{2}}\boldsymbol{Z}\}_{\hat{\beta}} &= \{(\boldsymbol{I}_n - \boldsymbol{H})(\boldsymbol{V}^{-\frac{1}{2}}\boldsymbol{D\beta} + \boldsymbol{V}^{-\frac{1}{2}}\boldsymbol{e})\}_{\hat{\beta}} \\ &= (\boldsymbol{I}_n - \boldsymbol{H})\boldsymbol{rp} = \boldsymbol{rp}. \end{aligned}$$

Hence we have $\boldsymbol{P}_{(I_n - H)V^{-\frac{1}{2}}Z} = \boldsymbol{P}_{rp}$ and

$$\boldsymbol{H}^* = \boldsymbol{H} + \boldsymbol{rp}(\boldsymbol{rp}^T\boldsymbol{rp})^{-1}\boldsymbol{rp}^T,$$

which results in

$$h^*_{ii} = h_{ii} + rp_i^2/(\boldsymbol{rp}^T\boldsymbol{rp})$$

and yields the desired result. ∎

6.3.2 Diagnostics based on deviance

The diagnostics related to the deviance introduced here are quite similar to those introduced by Pregibon (1981) for GML.

(a) Deviance residuals

$$rD_i = \{\text{sign}(y_i - \hat{\mu}_i)\}d_i(y_i, \mu_i(\hat{\boldsymbol{\beta}})) \tag{6.26}$$

It is easily seen that rD_i reflects the adequacy fitted to the given data by using the underlying model (Davison and Gigli 1989; McCullagh and Nelder 1989).

(b) Difference of deviance

$$\Delta_i D = D(\hat{\boldsymbol{\beta}}) - D_{(i)}(\hat{\boldsymbol{\beta}}_{(i)}),$$

where $D(\boldsymbol{\beta})$ and $D_{(i)}(\boldsymbol{\beta})$ are given in (6.11) and (6.12), respectively. $\Delta_i D$ actually reflects the difference of the maximum likelihood estimators of σ^2 under the original model and under the CDM. In fact, for normal linear and nonlinear regression, we have $D(\boldsymbol{\beta}) = RSS$, the residual sum of squares, and

$$\begin{aligned} \Delta_i D &= RSS - RSS_{(i)} = n\hat{\sigma}^2 - (n-1)\hat{\sigma}^2_{(i)} \\ &= (n-1)(\hat{\sigma}^2 - \hat{\sigma}^2_{(i)}) + \hat{\sigma}^2. \end{aligned}$$

Similarly, let us consider the case where $s(y_i, \phi) = s(\phi) + t(y_i)$ is satisfied in model (6.1). Then it follows from (2.13) that $\hat{\phi}$ and $\hat{\phi}_{(i)}$ respectively satisfy

$$-n\dot{s}(\hat{\phi}) = D(\boldsymbol{\beta}) + nk,$$

$$-(n-1)\dot{s}(\hat{\phi}_{(i)}) = D_{(i)}(\boldsymbol{\beta}_{(i)}) + (n-1)k,$$

which lead to

$$\Delta_i D = -(n-1)\{\dot{s}(\hat{\phi}) - \dot{s}(\hat{\phi}_{(i)})\} - \{\dot{s}(\hat{\phi}) + k\}.$$

That also reflects the difference between $\hat{\phi}$ and $\hat{\phi}_{(i)}$ to some extent.

Pregibon (1981, p.720) obtained a formula of $\Delta_i D$ for GLM with canonical link. We now extend his result to the general exponential family nonlinear models.

Theorem 6.7

For model (6.1), the first order approximation of $\Delta_i D$ can be expressed as

$$\Delta_i D^I = d_i(y_i, \hat{\mu}_i) + h_{ii} r_i^2. \tag{6.27}$$

Proof. We use the notation given in Lemma 6.1 and Theorem 6.1. Then $\Delta_i D$ can be expressed as

$$\begin{aligned} \Delta_i D &= d_i(y_i, \hat{\mu}_i) + D_{(i)}(\hat{\boldsymbol{\beta}}) - D_{(i)}(\hat{\boldsymbol{\beta}}_{(i)}) \\ &= d_i + \phi^{-1}\{l_{(i)}(\hat{\boldsymbol{\beta}}) - l_{(i)}(\hat{\boldsymbol{\beta}}_{(i)})\}. \end{aligned}$$

Here we use the result $LD(\boldsymbol{\mu}) = \phi D(\boldsymbol{y}, \boldsymbol{\mu})$ for the CDM (see (1.23) and those nearby). By Taylor's expansion we get

$$\Delta_i D \approx d_i + \phi^{-1}(\hat{\boldsymbol{\beta}} - \hat{\boldsymbol{\beta}}_{(i)})^T\{-\ddot{l}_{(i)}(\hat{\boldsymbol{\beta}}_{(i)})\}(\hat{\boldsymbol{\beta}} - \hat{\boldsymbol{\beta}}_{(i)}).$$

Using the first order approximation of $\hat{\boldsymbol{\beta}} - \hat{\boldsymbol{\beta}}_{(i)}$ given in (6.3) and the fact

$$\begin{aligned} -\ddot{l}_{(i)}(\boldsymbol{\beta}) \approx \boldsymbol{J}_{(i)}(\boldsymbol{\beta}) &= \phi\{\boldsymbol{D}_{(i)}^T \boldsymbol{V}_{(i)}^{-1} \boldsymbol{D}_{(i)}\} \\ &= \phi(\boldsymbol{D}^T \boldsymbol{V}^{-1} \boldsymbol{D} - V_i^{-1} \boldsymbol{d}_i \boldsymbol{d}_i^T), \end{aligned}$$

we get (6.27) by a little calculation. ∎

Note that the idea described in this section can be applied to some other models where we can get an iteration equation such as (2.16) which results in (2.21) and (2.22).

Table 6.3 summarizes the influence diagnostics introduced in Sections 6.1 to 6.3 for exponential family nonlinear models. There are eight diagnostic statistics in this table. In summary, statistics rp_i, r_i, rD_i, h_{ii}, h_{ii}^* and AP_i may directly reflect the influence of the i-th observation on the model fit; C_i and $LD_i(\boldsymbol{\beta})$ measure the difference between $\hat{\boldsymbol{\beta}}$ and $\hat{\boldsymbol{\beta}}_{(i)}$; while $\Delta_i D$ measures the difference between $\hat{\sigma}^2$ and $\hat{\sigma}^2_{(i)}$. However, for diagnostics purpose, there are not many distinctions among these influence measures in practice.

Now let us look at two examples for computing diagnostic statistics.

Example 6.2 (Grass yield data No. 1).

These data were originally published by Welch, Adams and Corman (1963) and reanalyzed by McCullagh and Nelder (1989, p.383). They studied the results of a 4^3 factorial experiment with the three major plant nutrients (in lb/acre), x_1: nitrogen (N), x_2: phosphorus (P), and x_3: potassium(K), on the yield y (in tons/acre) of coastal Bermuda grass, where the yield y is averaged over the three years 1955–1957 and three replicates. McCullagh and Nelder (1989) fitted these data by a gamma nonlinear model

$$\mu_i^{-1} = \beta_0 + \beta_1(x_{i1} + \alpha_1)^{-1} + \beta_2(x_{i2} + \alpha_2)^{-1} + \beta_3(x_{i3} + \alpha_3)^{-1}. \quad (6.28)$$

Table 6.3 Summary of diagnostic statistics

Diagnostics	Formula	Notes
Pearson residuals	$rp_i = \hat{V}_i^{-\frac{1}{2}}\hat{e}_i$	Lemma 6.1
Studentized residuals	$r_i = \frac{rp_i}{\hat{\sigma}\sqrt{1-h_{ii}}}$	(6.20)
Deviance residuals	$rD_i = \{\text{sign } (y_i - \hat{\mu}_i)\}d_i(y_i, \mu_i(\hat{\beta}))$	(6.26)
Hat matrix and its	$\boldsymbol{H} = \boldsymbol{V}^{-\frac{1}{2}}\boldsymbol{D}(\boldsymbol{D}^T\boldsymbol{V}^{-1}\boldsymbol{D})^{-1}\boldsymbol{D}^T\boldsymbol{V}^{-\frac{1}{2}}$	Lemma 6.1
diagonal elements	$h_{ii} = V_i^{-1}\boldsymbol{d}_i^T(\boldsymbol{D}^T\boldsymbol{V}^{-1}\boldsymbol{D})^{-1}\boldsymbol{d}_i$	
AP statistic	$AP_i = 1 - h_{ii}^* = 1 - h_{ii} - \frac{rp_i^2}{(\boldsymbol{rp}^T\boldsymbol{rp})}$	(6.25)
	$h_{ii}^* = h_{ii} + \frac{rp_i^2}{(\boldsymbol{rp}^T\boldsymbol{rp})}$	
Cook distance	$C_i^I = \frac{h_{ii}}{1-h_{ii}}\frac{r_i^2}{p}$	(6.4), (6.24)
Likelihood distance	$LD_i^I(\boldsymbol{\beta}) = pC_i$	(6.6), (6.8)
Difference of deviance	$\Delta_i D^I = d_i + h_{ii}r_i^2$	Theorem 6.7

We have computed the diagnostic statistics and the results are shown in Table 6.4. Since the results are quite similar, only the statistics r_i, AP_i, C_i^I, and $\Delta_i D^I$ are listed in this table. All the results show that case 15 may be a possible outlier. This is consistent with the argument of McCullagh and Nelder (1989, p.383) who guessed that the value y_{15} may be 2.49, but not 2.94. In Section 6.6, we will give more detailed discussions about these data. ∎

Example 6.3 (Leukemia data No. 2).

These data have been studied in Example 2.5. Following this example, we compute two diagnostic measures: studentized residuals r_i and likelihood distance LD_i. The results are given in Table 6.5. The results of LD_i show that case 15 is a high influential point, and case 19 is also influential. It

Table 6.4 Some diagnostic statistics for grass yield data

i	r_i 10^{-1}	AP_i	C_i^I 10^{-3}	$\Delta_i D^I$ 10^{-2}	i	r_i 10^{-1}	AP_i	C_i^I 10^{-3}	$\Delta_i D^I$ 10^{-2}
1	-0.68	0.92	0.05	0.17	33	0.95	0.87	0.16	0.35
2	-0.25	0.92	0.01	0.02	34	1.08	0.87	0.21	0.46
3	0.00	0.93	0.00	0.00	35	1.22	0.89	0.21	0.55
4	-2.01	0.86	0.50	1.63	36	1.28	0.85	0.33	0.65
5	1.73	0.88	0.36	1.08	37	1.21	0.87	0.27	0.57
6	-0.40	0.92	0.02	0.06	38	1.25	0.87	0.27	0.60
7	-1.76	0.88	0.36	1.23	39	0.70	0.92	0.06	0.18
8	1.95	0.86	0.49	1.37	40	-0.06	0.88	0.00	0.00
9	-0.22	0.93	0.01	0.02	41	-0.35	0.91	0.02	0.05
10	1.83	0.88	0.39	1.20	42	0.69	0.92	0.06	0.17
11	-1.09	0.91	0.13	0.46	43	0.36	0.96	0.01	0.05
12	0.62	0.92	0.05	0.14	44	0.70	0.91	0.06	0.18
13	-0.09	0.92	0.00	0.00	45	0.70	0.87	0.10	0.20
14	0.91	0.91	0.11	0.30	46	-0.62	0.88	0.07	0.16
15	4.21	0.67	2.16	5.99	47	-0.01	0.92	0.00	0.00
16	0.47	0.91	0.03	0.08	48	1.28	0.85	0.34	0.65
17	1.39	0.87	0.33	0.74	49	-1.00	0.83	0.27	0.44
18	-0.15	0.89	0.00	0.01	50	0.24	0.84	0.02	0.03
19	-1.75	0.86	0.45	1.26	51	-0.97	0.85	0.21	0.40
20	-0.25	0.88	0.01	0.03	52	1.03	0.81	0.33	0.46
21	1.68	0.85	0.50	1.08	53	-1.36	0.81	0.53	0.84
22	0.82	0.88	0.12	0.26	54	0.59	0.83	0.10	0.15
23	-0.23	0.91	0.01	0.02	55	0.47	0.87	0.05	0.09
24	-0.31	0.88	0.02	0.04	56	0.03	0.81	0.00	0.00
25	0.17	0.91	0.00	0.01	57	-1.06	0.85	0.26	0.48
26	-0.71	0.90	0.07	0.20	58	-0.23	0.87	0.01	0.02
27	-0.47	0.93	0.02	0.08	59	-0.75	0.91	0.07	0.22
28	-1.66	0.86	0.42	1.13	60	-0.33	0.86	0.02	0.04
29	0.32	0.88	0.02	0.04	61	-0.09	0.82	0.00	0.00
30	0.32	0.88	0.02	0.04	62	-0.94	0.80	0.28	0.40
31	0.18	0.90	0.00	0.01	63	0.73	0.85	0.12	0.22
32	-0.18	0.87	0.01	0.01	64	0.67	0.79	0.17	0.21

agrees with the fact that case 15 has the maximum value of WBC (white blood count) among AG positive patients and case 19 has the minimum value of WBC among AG negative patients. The results of r_i show that six cases, indexed by 5, 9, 15, 16, 17 and 19, may need further investigation. In Example 6.5, we will use local influence approach to analyze these data.

Table 6.5 r_i and LD_i measures for leukemia data

i	r_i	LD_i	i	r_i	LD_i
1	0.35	0.015	16	2.03	0.853
2	0.14	0.003	17	1.59	0.637
3	0.54	0.033	18	-0.61	0.056
4	0.38	0.018	19	-1.74	16.859
5	-1.67	0.366	20	-0.36	0.012
6	0.95	0.101	21	-0.50	0.031
7	0.92	0.096	22	-0.33	0.010
8	-0.92	0.098	23	-0.23	0.003
9	-1.78	0.427	24	-0.19	0.002
10	0.74	0.063	25	-0.19	0.002
11	0.89	0.090	26	-0.18	0.001
12	-0.67	0.049	27	-0.20	0.002
13	-0.64	0.046	28	-0.22	0.003
14	-0.55	0.038	29	-0.12	0.000
15	1.67	49.103	30	-0.16	0.002

6.4 Local Influence Analysis

The local influence approach was presented by Cook (1986) and developed further by several authors (Thomas and Cook 1989; Wei, Lu and Shi 1991; Escobar and Meeker 1992; Wu and Luo 1993a,b; Wu and Wan 1994; Wei and Hickernell 1996). In this section we first review the basic idea and formulas of local influence approach, and then apply those to exponential family nonlinear models by considering a random perturbation scheme.

6.4.1 Perturbed models

Let $\boldsymbol{\alpha} = (\boldsymbol{\beta}^T, \sigma^2)^T$ and $l(\boldsymbol{\alpha})$ be the log-likelihood corresponding to (6.1), to be called the postulated model in this section. Let $\boldsymbol{\omega}$ denote an n-vector defined in an open subset Ω in R^n, and employed to quantify a perturbation into the model (Pregibon 1981; Cook 1986). For instance, $\boldsymbol{\omega}$ may be a case weights or a minor shift as given in Section 6.2. Let $l(\boldsymbol{\alpha}|\boldsymbol{\omega})$ denote the log-likelihood corresponding to the perturbed model for a given $\boldsymbol{\omega}$ and $\hat{\boldsymbol{\alpha}}_\omega = (\hat{\boldsymbol{\beta}}_\omega^T, \hat{\sigma}_\omega^2)^T$ be the maximum likelihood estimator under $l(\boldsymbol{\alpha}|\boldsymbol{\omega})$. We assume that there exists an $\boldsymbol{\omega}_0$ of $\boldsymbol{\omega}$ such that $l(\boldsymbol{\alpha}|\boldsymbol{\omega}_0) = l(\boldsymbol{\alpha})$, $\hat{\boldsymbol{\alpha}}_{\omega_0} = \hat{\boldsymbol{\alpha}}$, which means that $\boldsymbol{\omega}_0$ corresponds to the postulated model (null model). The derivatives of $l(\boldsymbol{\alpha}|\boldsymbol{\omega})$ with respect to $\boldsymbol{\alpha}$ and $\boldsymbol{\omega}$ are assumed to be existent up to the third order. Note that $\hat{\alpha}_\omega$ must satisfy

$$\left.\frac{\partial l(\alpha|\omega)}{\partial \alpha}\right|_{\alpha=\hat{\alpha}_\omega} = \mathbf{0} \quad \text{for} \quad \omega \in \Omega. \tag{6.29}$$

To assess the influence of perturbation ω on the maximum likelihood estimator $\hat{\alpha} = (\hat{\beta}^T, \hat{\sigma}^2)^T$, the likelihood distance is usually preferable (Pregibon 1981; Cook and Weisberg 1982). Following Cook (1986) and Escobar and Meeker (1992), we consider

$$LD(\omega) = 2\{l(\hat{\alpha}) - l(\hat{\alpha}_\omega)\} \tag{6.30}$$

that measures the distance between $\hat{\alpha}$ and $\hat{\alpha}_\omega$ in terms of the log-likelihood difference. If $LD(\omega) > \chi^2(p+1, 1-\gamma)$, then the perturbation ω results in a $\hat{\alpha}_\omega$ that lies outside of the null perturbation likelihood region for α (at the level $1-\gamma$). So the large $LD(\omega)$ indicates the great influence of ω on the fit for the postulated model. In practice, one may use $LD(\omega) > \frac{1}{2}\chi^2(p+1, 0.5)$ as a warning signal that perturbation could be importantly influential (Escobar and Meeker 1992). For a specific perturbed model (6.10), if $\omega \to 0$, then $\hat{\alpha}_{\omega i} \to \hat{\alpha}_{(i)}$ and (6.30) becomes (6.6).

To analyze the local behavior of $LD(\omega)$, the most direct method is the Taylor series expansion. It is easily seen from (6.30) that $LD(\omega_0) = 0$ and

$$\left\{\frac{\partial LD(\omega)}{\partial \omega^T}\right\}_{\omega_0} = (-2)\left\{\frac{\partial l(\alpha)}{\partial \alpha}\frac{\partial \hat{\alpha}_\omega}{\partial \omega^T}\right\}_{(\omega_0, \hat{\alpha})} = \mathbf{0}. \tag{6.31}$$

Now let

$$\ddot{A} = \left\{\frac{\partial^2 LD(\omega)}{\partial \omega \partial \omega^T}\right\}_{\omega_0} = -2\ddot{F}, \qquad \ddot{F} = \left\{\frac{\partial^2 l(\hat{\alpha}_\omega)}{\partial \omega \partial \omega^T}\right\}_{\omega_0}.$$

Then $\ddot{A}$ and $-\ddot{F}$ are positive definite and are referred to as influence matrices. Further, we set $\omega = \omega_0 + h$ where h is a direction of perturbation in Ω and we may assume $||h||^2 = h^T h = 1$ without loss of generality (in this chapter, we always use the ordinary inner product). Then we have the following theorem.

Theorem 6.8

The second order approximation of likelihood distance $LD(\omega)$ given in (6.30) can be expressed as

$$LD^{II}(\omega) = \frac{1}{2}h^T \ddot{A} h = -h^T \ddot{F} h, \tag{6.32}$$

where

$$\ddot{F} = \Delta^T \ddot{l}^{-1}(\hat{\alpha}) \Delta = G^T \ddot{l}(\hat{\alpha}) G, \tag{6.33}$$

$$\Delta = \left\{ \frac{\partial^2 l(\alpha|\omega)}{\partial \alpha \partial \omega^T} \right\}_{(\omega_0, \hat{\alpha})}, \quad G = \left(\frac{\partial \hat{\alpha}_\omega}{\partial \omega^T} \right)_{\omega_0}. \tag{6.34}$$

Proof. Equation (6.32) is directly obtained from the Taylor series expansion because $LD(\omega_0) = 0$ and $\{\partial LD(\omega)/\partial \omega^T\}_{\omega_0} = \mathbf{0}$. We now prove (6.33). It follows from (A.9) and (A.10) that

$$\dot{F} = \left\{ \frac{\partial l(\hat{\alpha}_\omega)}{\partial \omega^T} \right\}_{(\omega_0, \hat{\alpha})} = \left\{ \frac{\partial l(\alpha)}{\partial \alpha^T} \frac{\partial \hat{\alpha}_\omega}{\partial \omega^T} \right\}_{(\omega_0, \hat{\alpha})} = \mathbf{0},$$

$$\begin{aligned} \ddot{F} &= \left\{ (\frac{\partial \hat{\alpha}_\omega}{\partial \omega^T})^T \frac{\partial^2 l(\alpha)}{\partial \alpha \partial \alpha^T} (\frac{\partial \hat{\alpha}_\omega}{\partial \omega^T}) + [\frac{\partial l(\alpha)}{\partial \alpha^T}][\frac{\partial^2 \hat{\alpha}_\omega}{\partial \omega \partial \omega^T}] \right\}_{(\omega_0, \hat{\alpha})} \\ &= G^T \ddot{l}(\hat{\alpha}) G. \end{aligned}$$

Furthermore, by taking derivative with respect to ω in (6.29), we have

$$\left\{ \frac{\partial^2 l(\alpha|\omega)}{\partial \alpha \partial \omega^T} + \frac{\partial^2 l(\alpha|\omega)}{\partial \alpha \partial \alpha^T} \frac{\partial \hat{\alpha}_\omega}{\partial \omega^T} \right\}_{(\omega_0, \hat{\alpha})} = 0$$

which leads to $\Delta = -\ddot{l}(\hat{\alpha}) G$. So we get (6.33). ∎

This theorem can be used to identify the influential cases. Now let $\omega = \omega_0 + h_i$ where h_i is a unit vector having 1 at the i-th position and zeros elsewhere, which means that only the i-th case is perturbed. Then (6.32) gives

$$LD^{II}(\omega) = \frac{1}{2} h_i^T \ddot{A} h_i = \frac{1}{2} \ddot{A}_{ii},$$

where $\ddot{A}_{ii}$ is the i-th diagonal element of $\ddot{A}$. Escobar and Meeker (1992) proposed an index plot of $\ddot{A}_{ii}$, $i = 1, \cdots, n$; which gives a general description of each case. A large value $\ddot{A}_{ii}$ indicates that the specified standard perturbation for the i-th case could be importantly influential. Figure 6.1(a) of Example 6.4 gives such a plot.

Theorem 6.8 has a geometric interpretation presented by Cook (1986) whose idea is based on the directional curvature and the maximum curvature defined by Bates and Watts (1980). Now we give a brief introduction. Given a perturbation ω, $\eta(\omega) = (\omega^T, LD(\omega))^T$ may be regarded as an n-dimensional surface π_ω in R^{n+1}, which is referred to as the influence graph by Cook (1986). Let $\eta_h(b) = \eta(\omega_0 + bh)$ where b is a real number and h is defined earlier. Then the directional normal curvature at ω_0 along h and the maximum curvature are respectively defined as

$$C_h = \|\ddot{\eta}_h^N\| / \|\dot{\eta}_h\|^2, \quad C_{\max} = \max_{\|h\|=1} C_h, \tag{6.35}$$

where $\dot{\eta}_h = \{d\eta_h(b)/db\}_{b=0}$, $\ddot{\eta}_h = \{d^2\eta_h(b)/db^2\}_{b=0}$ and $\ddot{\eta}_h^N$ is the normal component of $\ddot{\eta}_h$. These definitions are quite similar to Definition 3.2 given in Chapter 3. Both C_h and C_{max} are referred to as the influence curvatures. Then we have the following theorem.

Theorem 6.9

The normal curvature of π_ω at ω_0 defined in (6.35) can be expressed as

$$C_h = \boldsymbol{h}^T \ddot{\boldsymbol{A}} \boldsymbol{h} = -2\boldsymbol{h}^T \ddot{\boldsymbol{F}} \boldsymbol{h}. \tag{6.36}$$

Proof. Let $\omega = \omega_0 + b\boldsymbol{h}$, then $\omega = \omega_0$ as $b = 0$. So we have

$$\left\{\frac{dLD(\omega_0 + b\boldsymbol{h})}{db}\right\}_{b=0} = \left\{\frac{\partial LD(\omega)}{\partial \omega^T}\frac{d\omega}{db}\right\}_{\omega_0} = 0,$$

$$\left\{\frac{d^2LD(\omega_0 + b\boldsymbol{h})}{db^2}\right\}_{b=0} = \left\{(\frac{d\omega}{db})^T \frac{\partial^2 LD(\omega)}{\partial\omega\partial\omega^T}(\frac{d\omega}{db})\right\}_{b=0} = \boldsymbol{h}^T \ddot{\boldsymbol{A}} \boldsymbol{h}.$$

Since $\eta_h(b) = (\omega_0^T + b\boldsymbol{h}^T,\ LD(\omega_0 + b\boldsymbol{h}))^T$, we have

$$\dot{\eta}_h = (\boldsymbol{h}^T, 0)^T, \quad \ddot{\eta}_h = (0, \cdots, 0,\ \boldsymbol{h}^T \ddot{\boldsymbol{A}} \boldsymbol{h})^T.$$

Note that the tangent space of π_ω at ω_0 is spanned by $\{\partial\eta(\omega)/\partial\omega^T\}_{\omega_0} = (\boldsymbol{I}_n, \boldsymbol{0})^T$, so the normal vector is just $(0, \cdots, 0, 1)^T$. Then we have $\ddot{\eta}_h^N = \ddot{\eta}_h$. Substituting these results into (6.35) yields (6.36). ∎

This theorem can be applied to identify the influential cases. Let $\boldsymbol{h}_{max}$ be the eigenvector associated with C_{max} that is just the largest eigenvalue of $\ddot{\boldsymbol{A}}$. Then the index plot of $(\boldsymbol{h}_{max})_i, i = 1, \cdots, n$ can identify the influential cases, because it indicates how to perturb the postulated model to obtain the greatest local change in the likelihood distance (Cook 1986). This approach has been widely used in the literature.

In summary, by the local influence approach described here, the influence matrix $\ddot{\boldsymbol{A}} = -2\ddot{\boldsymbol{F}}$ is the most important quantity to identify the influential cases. For a given perturbation scheme, both the diagonal elements of $\ddot{\boldsymbol{A}}$ and the direction $\boldsymbol{h}_{max}$ can be applied as the influence diagnostics according to Theorems 6.8 and 6.9. The procedures are: (a) to compute the observed information $-\ddot{\boldsymbol{l}}(\hat{\alpha})$ for the postulated model; (b) to consider a specific perturbation and then compute $\boldsymbol{\Delta}$ or $\boldsymbol{G}$ given in (6.34) under the perturbed model; and (c) to compute $\ddot{\boldsymbol{A}} = -2\ddot{\boldsymbol{F}}$ from (6.33) and plot $(i,\ \boldsymbol{A}_{ii})$ or $(i,\ (\boldsymbol{h}_{\max})_i)$, or both, $i = 1, \cdots, n$. In Example 6.4, we show some index plots based on these procedures. Note that one can find the analytical expression of $\boldsymbol{h}_{max}$ only for a few special cases but one can always compute $\boldsymbol{h}_{max}$ numerically.

Now let us apply these results to exponential family nonlinear models.

(a) For the postulated model (6.1), the observed information $-\ddot{l}(\hat{\alpha})$ has been given in Section 2.2 and Lemma 5.1 (see (5.6)). So we have

$$-\ddot{l}(\hat{\alpha}) = \begin{pmatrix} \sigma^{-2}\boldsymbol{R}^T(\boldsymbol{I}_p - \boldsymbol{B}_\theta)\boldsymbol{R} & \boldsymbol{0} \\ \boldsymbol{0} & -\ddot{l}_{\phi\phi} \end{pmatrix}_{(\hat{\beta},\hat{\sigma}^2)}, \tag{6.37}$$

where $-\ddot{l}_{\beta\beta}(\hat{\alpha})$ is given in (5.6), $-\ddot{l}_{\beta\phi}(\hat{\alpha}) = (\boldsymbol{D}^T\boldsymbol{V}^{-1}\boldsymbol{e})_{\hat{\beta}} = \boldsymbol{0}$ and $\ddot{l}_{\phi\phi}$ is a scalar.

(b) Compute the influence matrix $\ddot{\boldsymbol{A}}$ for a given perturbed model. As an example, we consider perturbing the response vector $\boldsymbol{Y}$ by adding a vector $\boldsymbol{\omega} \in R^n$ to $\boldsymbol{Y}$. Then the log-likelihood corresponding to the perturbed model is

$$l(\boldsymbol{\alpha}|\boldsymbol{\omega}) = \sum_{i=1}^{n}[\phi\{(y_i+\omega_i)\theta_i - b(\theta_i) + c(y_i+\omega_i)\} - \frac{1}{2}s(\phi) - \frac{1}{2}t(y_i+\omega_i)] \tag{6.38}$$

with $\boldsymbol{\alpha} = (\boldsymbol{\beta}^T, \sigma^2)^T$, $g(\mu_i) = f(\boldsymbol{x}_i;\boldsymbol{\beta})$ and $\boldsymbol{\omega}_0 = \boldsymbol{0}$. Here we assume $s(\phi, y_i) = s(\phi) + t(y_i)$ for simplicity. In this case, $-\ddot{l}_{\phi\phi} = \frac{1}{2}n\ddot{s}(\phi)$. It follows from (6.38) that

$$\dot{l}_\beta = \phi\boldsymbol{D}^T\boldsymbol{V}^{-1}(\boldsymbol{Y} + \boldsymbol{\omega} - \boldsymbol{\mu}(\boldsymbol{\beta})), \quad \ddot{l}_{\beta\omega} = \phi\boldsymbol{D}^T\boldsymbol{V}^{-1} = \phi\dot{\boldsymbol{\theta}}^T,$$

$$\left\{\frac{\partial^2 l}{\partial\phi\partial\omega_i}\right\}_{(\omega_0,\hat{\alpha})} = \hat{\theta}_i - \dot{c}(y_i) = \hat{\theta}_i - \tilde{\theta}_i.$$

Here we use the fact $\dot{c}(y_i) = \dot{b}^{-1}(y_i)$ (see Lemma 1.2) and set $\tilde{\boldsymbol{\theta}} = (\tilde{\theta}_i) = (\dot{b}^{-1}(y_i))$. So we have $\ddot{\boldsymbol{l}}_{\phi\omega} = \hat{\boldsymbol{\theta}}^T - \tilde{\boldsymbol{\theta}}^T$ and then

$$\boldsymbol{\Delta}^T = (\phi\dot{\boldsymbol{\theta}},\ \boldsymbol{\theta} - \tilde{\boldsymbol{\theta}})_{(\hat{\theta},\hat{\sigma}^2)}.$$

(c) Substituting $-\ddot{l}(\hat{\alpha})$ and $\boldsymbol{\Delta}$ into (6.33), we get

$$\ddot{\boldsymbol{A}} = 2\hat{\phi}\hat{\boldsymbol{V}}^{-1}\boldsymbol{Q}(\boldsymbol{I}_p - \boldsymbol{B}_\theta)^{-1}\boldsymbol{Q}^T\hat{\boldsymbol{V}}^{-1} + 4n^{-1}\ddot{s}^{-1}(\hat{\phi})(\boldsymbol{\theta} - \tilde{\boldsymbol{\theta}})(\boldsymbol{\theta} - \tilde{\boldsymbol{\theta}})^T. \tag{6.39}$$

Then we can compute the values of A_{ii} and $(\boldsymbol{h}_{max})_i$, $i = 1, \cdots, n$ and make index plots to find influential points. Now let us see an example.

Example 6.4 (Fruit fly data No. 2)

The data have been discussed in Example 6.1. Here we consider perturbing the response vector $\boldsymbol{Y}$ by adding a vector $\boldsymbol{\omega}$ to $\boldsymbol{Y}$, as discussed in (6.38). The values of $\ddot{\boldsymbol{A}}$ are computed by (6.39). So the index plots of $\ddot{A}_{ii}$ and $(\boldsymbol{h}_{max})_i$ (here it is multiplied by 100) can be obtained. Figure 6.1(a) and (b) give such plots, both of which show that case 23 has the greatest influence. This is reasonable because case 23 corresponds to the highest temperature and this case is also an outlying point in the plot of McCullagh and Nelder (1989, p.308).

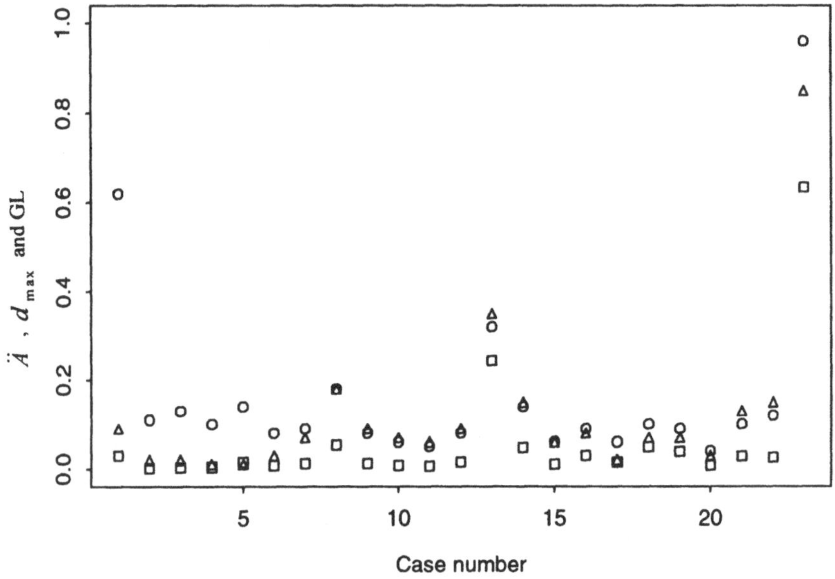

Figure 6.1 INDEX PLOTS OF DIAGNOSTICS FOR FRUIT FLY DATA
(a) $\ddot{A}_{ii}$, SHOWN BY RECTANGLES
(b) ABSOLUTE ELEMENTS OF 100 $\boldsymbol{d}_{max}$, SHOWN BY TRIANGLES
(c) $(GL(\hat{\beta}))_{ii}$, SHOWN BY CIRCLES.

6.4.2 Random perturbation scheme

A random perturbation model was presented by Wu and Wan (1994) for extracting the information from the interaction between cases in a study of local influence in nonlinear regression. This perturbation scheme can also be applied to exponential family nonlinear models. We give a brief introduction.

The postulated model (6.1) may be represented as

$$\boldsymbol{Y} = \boldsymbol{\mu}(\boldsymbol{\beta}) + \sigma \boldsymbol{V}^{\frac{1}{2}}(\boldsymbol{\beta})\boldsymbol{\epsilon}, \quad \boldsymbol{\epsilon} \sim (\boldsymbol{0}, \boldsymbol{I}_n), \tag{6.40}$$

where $\boldsymbol{\epsilon}$ is the standard random error and $\boldsymbol{Y} \sim ED(\boldsymbol{\mu}(\boldsymbol{\beta}), \sigma^2)$ is the observed vector. Consider perturbing $\boldsymbol{\epsilon}$ by adding another random vector $\boldsymbol{\rho}$ in it to indicate the interaction between cases. Then the perturbed model with a random "noise" ρ can be expressed as

$$Y = \mu(\beta) + \sigma V^{\frac{1}{2}}(\beta)(\epsilon + \rho). \tag{6.41}$$

Here the random vector ρ may be regarded as a noise into the model. For further discussion, we assume that

$$\mathrm{E}(\rho) = 0, \ Var(\rho) = a^2 I_n, \ \mathrm{E}(\rho\epsilon^T) = \Omega,$$

where a is a constant and $\Omega = (\omega_{kj}) = (\Omega_1, \cdots, \Omega_n)$ with $\omega_{kj} = E(\rho_k \epsilon_j)$ that may reflect the interaction of cases k and j via a noise from the k-th case, and $\Omega_k = \mathrm{E}(\rho_k \epsilon)$ reflects the interaction between cases via a noise from the k-th case, where ρ_k is the k-th component of ρ. This model may be called RP model for short. Obviously, since the distribution of ρ is unknown, the distribution of Y is no longer necessary to belong to the exponential family, so we usually cannot get the maximum likelihood estimators of β and σ for RP model (6.41). As an alternative, the quasi-likelihood estimator of β is an adequate choice (McCullagh and Nelder 1989; see also Chapter 7). It is easily seen that the variances of $\epsilon + \rho$ and Y are respectively

$$\begin{aligned} \mathrm{Var}(\epsilon + \rho) &\triangleq \Sigma_{RP} = I_n + \Omega + \Omega^T + a^2 I_n, \\ Var(Y) &\triangleq \sigma^2 V_{RP} = \sigma^2 V^{\frac{1}{2}} \Sigma_{RP} V^{\frac{1}{2}}. \end{aligned} \tag{6.42}$$

Then the QLE $\hat{\beta}_{RP}$ of β is defined as a solution to the following quasi-score equation

$$QSE(\beta, \Sigma_{RP}) = D^T(\beta) V_{RP}{}^{-1}(\beta)(Y - \mu(\beta)) = 0. \tag{6.43}$$

Here σ^2 is treated as a nuisance parameter, estimated by

$$\hat{\sigma}^2_{RP} = n^{-1} \hat{e}^T_{RP} V_{RP}{}^{-1}(\hat{\beta}_{RP}) \hat{e}_{RP}$$

with $\hat{e}_{RP} = Y - \mu(\hat{\beta}_{RP})$. Note that equation (6.43) is very similar to equation (2.10) which the maximum likelihood estimator $\hat{\beta}$ satisfies. So $\hat{\beta}_{RP}$ can be computed by using the iteration procedure as given in (2.16) for $\hat{\beta}$ except that every quantity must have a subscript "RP" added in (2.16). However, it is not necessary to do that here, because we only need $\hat{\beta}$ that corresponds to the null model in which the noise ρ disappears. Obviously, if the random perturbation ρ vanishes, then model (6.41) goes back to the null model (6.40) and we have $\Sigma_{RP} = I_n$, $V_{RP} = V$, $\hat{\beta}_{RP} = \hat{\beta}$, and $\hat{e}_{RP} = \hat{e}$, respectively.

Following (6.30), the likelihood distance under perturbed model (6.41) is denoted by

$$LD_{RP}(\Omega) = 2\{l(\hat{\beta}) - l(\hat{\beta}_{RP})\}$$

which is a function of $\boldsymbol{\Omega} = (\omega_{kj})$ and a. To get valuable diagnostic statistics, we prefer to consider the noise ρ_k $(k = 1, \cdots, n)$ separately. It means that $LD_{RP}(\boldsymbol{\Omega})$ is regarded as a function of the perturbation $\boldsymbol{\omega}^{(k)} = \boldsymbol{\Omega}_k = (\omega_{k1}, \cdots, \omega_{kn})^T$ at a time, $k = 1, \cdots, n$. In other words, we separately consider the partial influence graph

$$\eta(\boldsymbol{\omega}^{(k)}) = \begin{pmatrix} \boldsymbol{\omega}^{(k)} \\ LD_{RP}(\boldsymbol{\Omega}) \end{pmatrix}, \quad k = 1, \cdots, n \tag{6.44}$$

with $\boldsymbol{\omega}^{(k)} = \boldsymbol{\Omega}_k$ and $\boldsymbol{\omega}_0{}^{(k)} = \mathbf{0}$. In fact, $\eta(\boldsymbol{\omega}^{(k)})$ corresponds to the model in which only the k-th case is perturbed by a random noise ρ_k. Following the notation introduced in the previous subsection, the influence matrices and the influence curvature corresponding to (6.44) are denoted by $\ddot{\boldsymbol{A}}^{k} = -2\ddot{\boldsymbol{F}}^{(k)}$, $C_h^{(k)}$, $C_{max}^{(k)}$ and $\boldsymbol{h}_{max}^{(k)}$, respectively. From Theorem 6.8 and the quasi-score equation (6.43), it is not too hard to get these quantities. We have the following theorem.

Theorem 6.10

For partial influence graph (6.44), $\ddot{\boldsymbol{A}}^{(k)}$ can be expressed as

$$\ddot{\boldsymbol{A}}^{(k)} = 2\hat{\phi}^2\{\tilde{h}_{kk}(\boldsymbol{rp})(\boldsymbol{rp})^T + rp_k\tilde{\boldsymbol{h}}_k\boldsymbol{rp}^T + rp_k\boldsymbol{rp}\tilde{\boldsymbol{h}}_k^T + rp_k^2\tilde{\boldsymbol{H}}\}, \tag{6.45}$$

where

$$\tilde{\boldsymbol{H}} = (\tilde{\boldsymbol{h}}_1, \cdots, \tilde{\boldsymbol{h}}_n) = (\tilde{h}_{kj}) = \{\boldsymbol{V}^{-\frac{1}{2}}\boldsymbol{D}(-\ddot{\boldsymbol{l}}_{\beta\beta}{}^{-1})\boldsymbol{D}^T\boldsymbol{V}^{-\frac{1}{2}}\}_{\hat{\beta}}.$$

Proof. It follows from (6.33) and (6.34) that $\frac{1}{2}\ddot{\boldsymbol{A}}^{(k)} = \frac{1}{2}(\ddot{A}_{ij}^{(k)})$ can be expressed as

$$\frac{1}{2}\ddot{A}_{ij}^{(k)} = \left\{(\frac{\partial\hat{\boldsymbol{\beta}}_{RP}}{\partial\omega_{ki}})^T(-\ddot{\boldsymbol{l}}_{\beta\beta})(\frac{\partial\hat{\boldsymbol{\beta}}_{RP}}{\partial\omega_{kj}})\right\}_{(\omega_0^{(k)},\hat{\beta})}, \tag{6.46}$$

where $\partial\hat{\boldsymbol{\beta}}_{RP}/\partial\omega_{ki}$ can be obtained by taking derivatives for quasi-score equation (6.43) in $\omega^{(k)}$. To do so, we note that

$$\left\{\frac{\partial\boldsymbol{\Sigma}_{RP}^{-1}}{\partial\omega_{ki}}\right\}_{\omega_0^{(k)}} = \left\{-\boldsymbol{\Sigma}_{RP}^{-1}\frac{\partial\boldsymbol{\Sigma}_{RP}}{\partial\omega_{ki}}\boldsymbol{\Sigma}_{RP}^{-1}\right\}_{\omega_0^{(k)}} = -(\boldsymbol{\delta}_k\boldsymbol{\delta}_i^T + \boldsymbol{\delta}_i\boldsymbol{\delta}_k^T),$$

which is obtained from (6.43), and $\boldsymbol{\delta}_i$ is a unit vector having 1 at the i-th position and zeros elsewhere, as before. Then it follows from (6.43) that

$$\left\{\frac{\partial}{\partial\omega_{ki}}QSE(\boldsymbol{\beta}_{RP}, \boldsymbol{\Sigma}_{RP})\right\}_{(\omega_0^{(k)},\hat{\beta})} = 0,$$

which gives

$$\left\{\frac{\partial}{\partial \boldsymbol{\beta}^T} QSE(\boldsymbol{\beta}, \boldsymbol{\Sigma}_{RP}) \frac{\partial \hat{\boldsymbol{\beta}}_{RP}}{\partial \omega_{ki}} + \boldsymbol{D}^T \boldsymbol{V}^{-\frac{1}{2}} \frac{\partial \boldsymbol{\Sigma}_{RP}^{-1}}{\partial \omega_{ki}} \boldsymbol{V}^{-\frac{1}{2}} \boldsymbol{e}\right\}_{(\omega_0^{(k)}, \hat{\beta})} = \boldsymbol{0},$$

and then

$$\left\{\phi^{-1} \ddot{\boldsymbol{l}}_{\beta\beta} \frac{\partial \hat{\boldsymbol{\beta}}_{RP}}{\partial \omega_{ki}} - \boldsymbol{D}^T \boldsymbol{V}^{-\frac{1}{2}} (\boldsymbol{\delta}_k \boldsymbol{\delta}_i^T + \boldsymbol{\delta}_i \boldsymbol{\delta}_k^T) \boldsymbol{rp}\right\}_{(\omega_0^{(k)}, \hat{\beta})} = \boldsymbol{0}.$$

So we get

$$\left(\frac{\partial \hat{\boldsymbol{\beta}}_{RP}}{\partial \omega_{ki}}\right)_{(\omega_0^{(k)}, \hat{\beta})} = \hat{\phi} \hat{\ddot{\boldsymbol{l}}}_{\beta\beta}^{-1} \hat{\boldsymbol{D}}^T \hat{\boldsymbol{V}}^{-\frac{1}{2}} (\boldsymbol{\delta}_k \boldsymbol{\delta}_i^T + \boldsymbol{\delta}_i \boldsymbol{\delta}_k^T) \boldsymbol{rp}.$$

Substituting this result into (6.46) yields

$$\begin{aligned}\frac{1}{2} \ddot{A}_{ij}^{(k)} &= \hat{\phi}^2 \boldsymbol{rp}^T (\boldsymbol{\delta}_i \boldsymbol{\delta}_k^T + \boldsymbol{\delta}_k \boldsymbol{\delta}_i^T) \tilde{\boldsymbol{H}} (\boldsymbol{\delta}_k \boldsymbol{\delta}_j^T + \boldsymbol{\delta}_j \boldsymbol{\delta}_k^T) \boldsymbol{rp} \\ &= \hat{\phi}^2 (rp_i rp_j \tilde{h}_{kk} + rp_i rp_k \tilde{h}_{kj} + rp_k^2 \tilde{h}_{ij} + rp_k rp_j \tilde{h}_{ik}),\end{aligned}$$

which results in (6.45). ∎

The influence matrix $\ddot{\boldsymbol{A}}^{(k)}$ can be applied to local influence analysis as shown after Theorems 6.8 and 6.9. In practice, the observed information $-\ddot{\boldsymbol{l}}_{\beta\beta}$ usually can be approximated by Fisher information matrix $\boldsymbol{J}_\beta(\boldsymbol{Y}) = \phi \boldsymbol{D}^T \boldsymbol{V}^{-1} \boldsymbol{D}$ in (6.45). So we have $\tilde{\boldsymbol{H}} \approx \hat{\phi}^{-1} \boldsymbol{H}$ and then

$$\ddot{\boldsymbol{A}}^{(k)} \approx 2\hat{\sigma}^{-2} \{h_{kk} (\boldsymbol{rp})(\boldsymbol{rp})^T + rp_k \boldsymbol{h}_k \boldsymbol{rp}^T + rp_k \boldsymbol{rp} \boldsymbol{h}_k^T + rp_k^2 \boldsymbol{H}\}. \quad (6.47)$$

Our discussion in Theorem 6.11 will be based on this approximation. From this, we can find the maximum curvature $C_{max}^{(k)}$ and the corresponding direction $\boldsymbol{h}_{max}^{(k)}$.

Theorem 6.11

For partial influence graph (6.44), $C_{max}^{(k)} = C^{(k)}$ and $\boldsymbol{h}_{max}^{(k)} = \boldsymbol{h}^{(k)}$ can be approximately expressed as

$$C^{(k)} \approx 2\hat{\phi}\{(rp_k)^2 + h_{kk} (\boldsymbol{rp})^T (\boldsymbol{rp})\}, \quad (6.48)$$

$$\boldsymbol{h}^{(k)} \approx (2\hat{\phi})^{\frac{1}{2}} (\boldsymbol{h}_k rp_k + h_{kk} \boldsymbol{rp}) \{h_{kk} C^{(k)}\}^{-\frac{1}{2}}. \quad (6.49)$$

Proof. From (6.47), we can find the largest eigenvalue of $\ddot{\boldsymbol{A}}^{(k)}$. In fact, by a little calculation, (6.47) can be represented as

$$\ddot{\boldsymbol{A}}^{(k)} \approx 2\hat{\phi}\{h_{kk}^{-1} \boldsymbol{u}_k \boldsymbol{u}_k^T + (rp_k)^2 (\boldsymbol{H} - h_{kk}^{-1} \boldsymbol{h}_k \boldsymbol{h}_k^T)\} = \ddot{\boldsymbol{A}}_a^{(k)},$$

where $u_k = (rp_k)h_k + h_{kk}(rp)$. From this decomposition we can see that (a) $H_k^* = (H - h_{kk}^{-1}h_k h_k^T)$ is a projection matrix and $H_k^* u_k = 0$ because $H(rp) = 0$ (see Subsection 6.3.1); (b) u_k is an eigenvector of $\ddot{A}_a^{(k)}$ with eigenvalue

$$\lambda_1 = 2\hat{\phi}h_{kk}^{-1}u_k^T u_k = 2\hat{\phi}\{(rp_k)^2 + h_{kk}(rp)^T(rp)\};$$

and (c) another eigenvalue of $\ddot{A}_a^{(k)}$ is $\lambda_2 = 2\hat{\phi}(rp_k)^2$ corresponding to eigenvectors which are located in the space spanned by H_k^*. Since $\lambda_1 > \lambda_2$, we get (6.49) from Theorem 6.9, while (6.49) can be obtained from $h^{(k)} = u_k(u_k^T u_k)^{-\frac{1}{2}}$. ∎

Theorem 6.11 can be used to measure interaction between cases as described by Wu and Wan (1994). By the local influence approach of Cook (1986), the vector $h^{(k)}$ reflects the interaction between cases when the k-th case is perturbed by a random noise ρ_k. The size of the j-th element $h_j^{(k)}$ represents the share contributed to the influence curvature $C^{(k)}$ by case j when perturbing case k. To get influence measures, following Wu and Wan (1994), we introduce an interaction matrix $Z = (Z_{kj})$ with elements

$$Z_{kj} = C^{(k)}(h_j^{(k)})^2, \quad k, j = 1, \cdots, n.$$

Here the element Z_{kj} measures the influence of the j-th case when there is a noise from case k. The matrix Z is very helpful to detect the interaction and the joint influence among cases. To use the matrix Z more effectively, we now define vectors ACZ and ARZ as measures of interaction between cases with elements

$$ACZ_j = n^{-1}\sum_{k=1}^{n} Z_{kj}, \quad ARZ_k = n^{-1}\sum_{j=1}^{n} Z_{kj}. \tag{6.50}$$

Note that the vectors ACZ and ARZ have different meanings. ACZ_j measures the average contribution of case j to the influence of other cases, while ARZ_k measures the average contribution of other cases to the influence of case k. We have formulas to compute these measures.

Theorem 6.12

If the dispersion parameter ϕ is estimated by $\hat{\phi}^{-1} = \hat{\sigma}^2 = n^{-1}(rp)^T(rp)$, then we have

$$\begin{aligned} ACZ_j &= 2(rp^T rp)^{-1}\{p(rp_j)^2 + \sum_{k=1}^{n} h_{kj}^2(rp_k)^2 h_{kk}^{-1}\}, \\ ARZ_k &= 2\{h_{kk} + (rp_k)^2(rp^T rp)^{-1}\}. \end{aligned} \tag{6.51}$$

Proof. It follows from (6.49) and (6.49) that

$$\begin{aligned} ARZ_k &= 2(nh_{kk}\hat{\sigma}^2)^{-1}\sum_{j=1}^{n}(h_{kj}rp_k + h_{kk}rp_j)^2 \\ &= 2(nh_{kk}\hat{\sigma}^2)^{-1}\sum_{j=1}^{n}(h_{kj}^2 rp_k^2 + 2h_{kj}h_{kk}rp_k rp_j + h_{kk}^2 rp_j^2). \end{aligned}$$

It follows from $\boldsymbol{H}^2 = \boldsymbol{H}$ and $\boldsymbol{H}(\boldsymbol{rp}) = \boldsymbol{0}$ that

$$h_{kk} = \sum_{j=1}^{n} h_{kj}^2, \quad \sum_{j=1}^{n} h_{kj}rp_j = 0.$$

Substituting these results into ARZ_k above gives the second equation of (6.51). Similarly, we can get the first equation. ∎

To detect the joint influence of cases a and b, we may compare the values of Z_{aa}, Z_{ab}, Z_{ba} and Z_{bb} with other values of columns a and b. To do so, we introduce the following measures:

$$\begin{aligned} Z_{(a,b),a} &= (Z_{aa} + Z_{ab})/\textstyle\sum_{j=1}^{n} Z_{aj}, \\ Z_{(a,b),b} &= (Z_{ba} + Z_{bb})/\textstyle\sum_{j=1}^{n} Z_{bj}. \end{aligned} \tag{6.52}$$

If the values of both $Z_{(a,b),a}$, and $Z_{(a,b),b}$ are large enough, say they are greater than 0.5, and the values of the other elements Z_{aj}, Z_{bj} , $j \neq a, b$, are all small, then cases a and b may be jointly influential. A similar argument can also be used for detecting the subset influence of three or more cases (see Example 6.5 for further discussions).

Example 6.5 (Leukemia data No. 3).

These data have been studied in Examples 2.5 and 6.3. Here we only take care for the influence of cases; no attempt is made to provide a complete analysis of the data.

To find influential subsets, we apply the interaction matrix $\boldsymbol{Z}$ and formulas (6.51) and (6.52) to the data. Figure 6.2 plots ARZ versus ACZ, which picks out six cases, indexed by 5, 9, 15, 16, 17 and 19, for further investigation. The cases outside this group only have a very small contribution to the interaction matrix $\boldsymbol{Z}$, so we will just consider these six cases. This is consistent with the result of studentized residuals given in Table 6.5 of Example 6.3. As also shown in Example 6.3, case 15 is a high influential point, which corresponds to the maximum value of WBC among AG positive patients, so we will just consider the rest, that is, cases 5, 9, 16, 17 and 19.

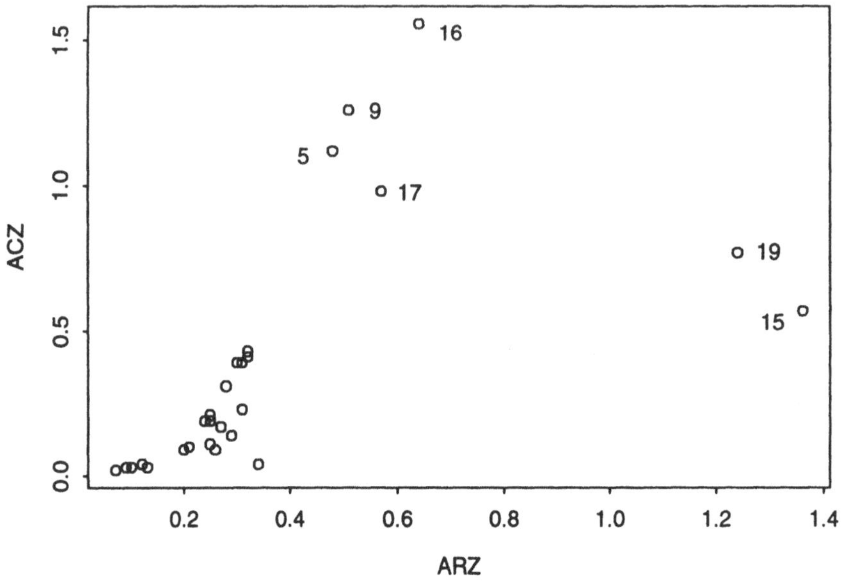

Figure 6.2 *ARZ-ACZ* PLOT FROM FITTING A NONLINEAR LOGISTIC MODEL TO THE LEUKEMIA DATA

Table 6.6 lists the rows of matrix $\boldsymbol{Z}$ corresponding to the cases 5, 9, 16, 17 and 19. A simple calculation based on (6.52) shows that

$$Z_{(16,17),16} = 0.572, \qquad Z_{(16,17),17} = 0.563,$$

which forcefully suggests a high joint influence of cases 16 and 17. In fact, cases 16 and 17 are only two AG negative survivors (see Table 5.4.1 of Cook and Weisberg 1982, p.193). If these two patients were absent, one could conclude that the probability of survival for AG negative patients is zero or near zero. The result of likelihood distance also justifies the above statement, which gives $LD_{16,17} = 121.62$. It is large enough to indicate the high joint influence of cases 16 and 17. It is interesting to note that the individual likelihood distances of cases 16 and 17 are very small, which are just $LD_{16} = 0.853$ and $LD_{17} = 0.637$, respectively (see Table 6.5).

Now let us see the influence of cases 5, 9 and 19. Table 6.7 gives the values of $Z_{(a,b),a}$, $Z_{(a,b),b}$ and corresponding likelihood distances related to these three cases. From these results, we cannot say that any pair is significantly influential. However, the case-triplet $\{5, 9, 19\}$ is obviously a high influential subset, because $LD_{5,9,19} = 76.28$ is large enough and the values $Z_{a,5} + Z_{a,9} + Z_{a,19}$ $(a = 5, 9, 19)$ are also large enough, contributing to each of ARZ_a (see Table 6.6). This is consistent with the fact that the cases 5, 9 and 19 correspond to three patients having small values of WBC but surviving less than 52 weeks.

Table 6.6 Selected elements Z_{aj} of interaction matrix Z

j	Z_{5j}	Z_{9j}	Z_{16j}	Z_{17j}	Z_{19j}
1	0.123	0.176	0.017	0.109	0.009
2	0.017	0.004	0.002	0.044	0.078
3	0.016	0.212	0.078	0.152	0.078
4	0.130	0.185	0.026	0.116	0.004
5	1.613°	1.716°	0.397	0.523	1.850°
6	0.053	0.058	0.230	0.221	0.672
7	0.058	0.068	0.221	0.212	0.624
8	0.656	0.656	0.116	0.250	0.336
9	1.716°	1.823°	0.476	0.608	2.190°
10	0.123	0.152	0.152	0.185	0.325
11	0.073	0.084	0.203	0.212	0.563
12	0.260	0.240	0.096	0.185	0.152
13	0.221	0.203	0.096	0.176	0.137
14	0.090	0.084	0.102	0.152	0.109
15	0.281	0.314	0.026	0.090	0.740
16	0.476	0.476	2.856*	2.496*	1.664
17	0.221	0.194	2.434*	2.250*	0.518
18	0.096	0.096	0.372	0.123	0.624
19	0.372°	0.504°	0.084	0.116	3.764°
20	0.012	0.006	0.230	0.040	0.123
21	0.053	0.048	0.336	0.084	0.348
22	0.008	0.004	0.212	0.032	0.096
23	0.000	0.002	0.102	0.012	0.036
24	0.002	0.006	0.068	0.006	0.026
25	0.003	0.006	0.063	0.006	0.026
26	0.004	0.008	0.058	0.005	0.023
27	0.002	0.005	0.073	0.006	0.026
28	0.000	0.003	0.090	0.010	0.032
29	0.012	0.017	0.014	0.009	0.014
30	0.029	0.040	0.020	0.002	0.029

* : Large values of Z_{aj} in case-pair $\{16, 17\}$

◇ : Large values of Z_{aj} in case-triplet $\{5, 9, 19\}$

Table 6.7 Selected values of $Z_{(a,b),a}$, $Z_{(a,b),b}$ and LD measures

$Z_{(5,9),5}$	$Z_{(5,9),9}$	$Z_{(5,19),5}$	$Z_{(5,19),19}$	$Z_{(9,19),9}$	$Z_{(9,19),19}$
0.495	0.479	0.295	0.369	0.315	0.391

$LD_{5,9} = 2.34$ $\quad LD_{5,19} = 14.63$ $\quad LD_{9,19} = 19.25$

$LD_{5,9,19} = 76.28$

6.5 Generalized Leverage

Leverage plays an important role in linear regression diagnostics (Cook and Weisberg 1982; Chatterjee and Hadi 1986). It is measured by the diagonal elements h_{ii} of the hat matrix $\boldsymbol{H}$ and it can be used for assessing the importance of individual observations. In Section 6.3, we have introduced a hat matrix $\boldsymbol{H} = \boldsymbol{V}^{-\frac{1}{2}}\boldsymbol{D}(\boldsymbol{D}^T\boldsymbol{V}^{-1}\boldsymbol{D})^{-1}\boldsymbol{D}^T\boldsymbol{V}^{-\frac{1}{2}}$ for exponential family nonlinear models. However, this quantity does not fit very well with the characteristic of leverage. In fact, in linear regression, say $y_i = \boldsymbol{x}_i^T\boldsymbol{\beta} + \epsilon_i$, $i = 1, \cdots, n$, the leverage has several nice properties, such as $\text{Var}(y_i) = \sigma^2 h_{ii}$, $\text{Var}(\hat{e}_i) = \sigma^2(1 - h_{ii})$, $\rho(y_i, \hat{y}_i) = \sqrt{h_{ii}}$, $h_{ij} = \partial\hat{y}_i/\partial y_j$ and so on, where $\hat{y}_i = \boldsymbol{x}_i^T\hat{\boldsymbol{\beta}}$, $\hat{e}_i = y_i - \boldsymbol{x}_i^T\hat{\boldsymbol{\beta}}$ and $\rho(y_i, \hat{y}_i)$ is the correlation coefficient of y_i and $\hat{y}_i$. It is easily seen that these properties no longer hold for $\boldsymbol{H} = \boldsymbol{V}^{-\frac{1}{2}}\boldsymbol{D}(\boldsymbol{D}^T\boldsymbol{V}^{-1}\boldsymbol{D})^{-1}\boldsymbol{D}^T\boldsymbol{V}^{-\frac{1}{2}} = (h_{ij})$ in exponential family nonlinear models. So we need to find, if possible, a more adequate definition of leverage for exponential family nonlinear models. This section, in a straightforward way, introduces a general definition for a general class of models. Our definition can be applied to exponential family nonlinear models and (a) is consistent with the ordinary leverage of linear regression models; and (b) can be used as a measure for assessing the importance of individual observations. There is also a close connection between this leverage and the local influence of a mean-shift perturbation.

6.5.1 Definition and computation

As pointed out by many authors (e.g. Hoaglin and Welsch 1978; Cook and Weisberg 1982; Emerson, Hoaglin and Kempthorne 1984; Yoshioze 1991; St. Laurent and Cook 1992), the property $h_{ii} = \partial\hat{y}_i/\partial y_i$ most directly reflects the influence of the observation y_i on the fit, so Definition 6.1 will be based on this point of view.

Definition 6.1

Let $\boldsymbol{Y} = (y_1, \cdots, y_n)^T$ be an n-vector of observable response with the probability density function $p(\boldsymbol{y}; \boldsymbol{\alpha})$ and $\text{E}(\boldsymbol{Y}) = \boldsymbol{\mu} = \boldsymbol{\mu}(\boldsymbol{\alpha})$, where $\boldsymbol{\alpha}$ is an unknown parameter. An estimator of $\boldsymbol{\alpha}$ is denoted by $\tilde{\boldsymbol{\alpha}} = \tilde{\boldsymbol{\alpha}}(\boldsymbol{Y})$ and $\tilde{\boldsymbol{Y}} = \boldsymbol{\mu}(\tilde{\boldsymbol{\alpha}})$ may be regarded as the predicted response vector. Then we define

$$GL(\tilde{\boldsymbol{\alpha}}) = \partial\tilde{\boldsymbol{Y}}/\partial\boldsymbol{Y}^T = (\partial\tilde{y}_i/\partial y_j). \tag{6.53}$$

as the generalized leverage of $\tilde{\boldsymbol{\alpha}}$. ∎

By this definition, one can see that the generalized leverage GL_{ij} of an estimator at (i, j) is the instantaneous rate of change in the i-th predicted value with respect to the j-th response value. This characteristic is exactly consistent with that of ordinary leverage in linear regression models. So the generalized leverage still measures the influence of observations on the fit of the model under the estimator $\tilde{\boldsymbol{\alpha}}$. Obviously, definition (6.53) can be used for quite general estimators and models, including exponential family nonlinear models. The observation with large value of $GL_{ii} = \partial\tilde{y}_i/\partial y_i$ will also be called the leverage point. Furthermore, it is easily seen from $\tilde{\boldsymbol{Y}} = \boldsymbol{\mu}(\tilde{\boldsymbol{\alpha}})$ and $\tilde{\boldsymbol{\alpha}} = \tilde{\boldsymbol{\alpha}}(\boldsymbol{Y})$ that

$$GL(\tilde{\boldsymbol{\alpha}}) = \left\{(\frac{\partial\boldsymbol{\mu}}{\partial\boldsymbol{\alpha}^T})\frac{\partial\tilde{\boldsymbol{\alpha}}(\boldsymbol{Y})}{\partial\boldsymbol{Y}^T}\right\}_{\alpha=\tilde{\alpha}},$$

which is invariant under reparameterizations, that is if $\boldsymbol{\alpha} = \boldsymbol{\alpha}(\boldsymbol{\gamma})$ is a one-to-one mapping and set $\tilde{\boldsymbol{\gamma}} = \boldsymbol{\gamma}(\tilde{\boldsymbol{\alpha}})$, where $\boldsymbol{\gamma}(\boldsymbol{\alpha})$ is the inverse of $\boldsymbol{\alpha}(\boldsymbol{\gamma})$, then

$$GL(\tilde{\boldsymbol{\alpha}}) = \{(\partial\boldsymbol{\mu}/\partial\boldsymbol{\gamma}^T)(\partial\tilde{\boldsymbol{\gamma}}(\boldsymbol{Y})/\partial\boldsymbol{Y}^T)\}_{\tilde{\gamma}} = GL(\tilde{\boldsymbol{\gamma}}).$$

In the literature, Yoshizoe (1991) defined the leverage for the least squares estimator in nonlinear regression models by using influence function while St. Laurent and Cook (1992) defined what they called the tangent plane leverage and the Jacobian leverage for nonlinear regression models by using directional derivatives. However, definition (6.35) might be more straightforward and simpler. Moreover, it is also easily applicable to estimators and models much more generally than the least squares estimator in linear and nonlinear regression models. For the general maximum likelihood estimators, we have the following basic lemma.

Lemma 6.2

Let $l(\boldsymbol{\alpha};\boldsymbol{y})$ be the log-likelihood of $\boldsymbol{Y}$ and $\hat{\boldsymbol{\alpha}} = \hat{\boldsymbol{\alpha}}(\boldsymbol{Y})$ be the unique maximum likelihood estimator of $\boldsymbol{\alpha}$. If $l(\boldsymbol{\alpha};\boldsymbol{y})$ has second order continuous derivatives with respect to $\boldsymbol{\alpha}$ and $\boldsymbol{y}$, then we have

$$GL(\hat{\boldsymbol{\alpha}}) = \{(\boldsymbol{D}_\alpha)(-\ddot{\boldsymbol{l}}_{\alpha\alpha}^{-1})(\ddot{\boldsymbol{l}}_{\alpha Y})\}_{\hat{\alpha}}, \tag{6.54}$$

where

$$\boldsymbol{D}_\alpha = \partial\boldsymbol{\mu}/\partial\boldsymbol{\alpha}^T,\ \ddot{\boldsymbol{l}}_{\alpha\alpha} = \partial^2 l(\boldsymbol{\alpha};\boldsymbol{Y})/\partial\boldsymbol{\alpha}\partial\boldsymbol{\alpha}^T \text{ and } \ddot{\boldsymbol{l}}_{\alpha Y} = \partial^2 l(\boldsymbol{\alpha};\boldsymbol{Y})/\partial\boldsymbol{\alpha}\partial\boldsymbol{Y}^T.$$

Proof. It follows from (6.53) that

$$GL(\hat{\boldsymbol{\alpha}}) = \left\{\frac{\partial\boldsymbol{\mu}}{\partial\boldsymbol{\alpha}^T}\frac{\partial\hat{\boldsymbol{\alpha}}(\boldsymbol{Y})}{\partial\boldsymbol{Y}^T}\right\}_{\alpha=\hat{\alpha}}. \tag{6.55}$$

Since $\hat{\boldsymbol{\alpha}}(\boldsymbol{Y})$ is the maximum likelihood estimator of $\boldsymbol{\alpha}$, we have

$$\left.\frac{\partial l(\boldsymbol{\alpha};\boldsymbol{Y})}{\partial\boldsymbol{\alpha}}\right|_{\alpha=\hat{\alpha}(Y)} = \boldsymbol{0}$$

for all $\boldsymbol{Y}$. Differentiating this equation with respect to $\boldsymbol{Y}$ and evaluating at $\boldsymbol{\alpha} = \hat{\boldsymbol{\alpha}}(\boldsymbol{Y})$ yield

$$\left\{ \frac{\partial^2 l(\boldsymbol{\alpha}; \boldsymbol{Y})}{\partial \boldsymbol{\alpha} \partial \boldsymbol{Y}^T} + \frac{\partial^2 l(\boldsymbol{\alpha}; \boldsymbol{Y})}{\partial \boldsymbol{\alpha} \partial \boldsymbol{\alpha}^T} \frac{\partial \hat{\boldsymbol{\alpha}}(\boldsymbol{Y})}{\partial \boldsymbol{Y}^T} \right\}_{\hat{\alpha}} = \mathbf{0}.$$

So we have

$$\frac{\partial \hat{\boldsymbol{\alpha}}(\boldsymbol{Y})}{\partial \boldsymbol{Y}^T} = \{(-\ddot{\boldsymbol{l}}_{\alpha\alpha}^{-1})(\ddot{\boldsymbol{l}}_{\alpha Y})\}_{\hat{\alpha}}.$$

Substituting this results into (6.55), we obtain (6.54). ∎

This lemma reveals the essential connection between generalized leverage and observed information matrix, which usually leads to the curvature in nonlinear models (see (6.56)). The formula (6.54) can be applied to quite a general class of models. In the following, we shall use this lemma to derive a useful expression of leverage for exponential family nonlinear models.

Theorem 6.13

For exponential family nonlinear models (6.1), the leverage of the maximum likelihood estimator $\hat{\boldsymbol{\alpha}} = (\hat{\beta}^T, \hat{\sigma}^2)^T$ can be expressed as

$$GL(\hat{\boldsymbol{\alpha}}) = \{\boldsymbol{Q}(\boldsymbol{I}_p - \boldsymbol{B}_\theta)^{-1} \boldsymbol{Q}^T \boldsymbol{V}^{-1}\}_{\hat{\beta}}. \tag{6.56}$$

Proof. Lemma 6.3 can be applied to (6.1). In this case, $\boldsymbol{\mu} = \dot{\boldsymbol{b}}(\boldsymbol{\theta}(\boldsymbol{\beta}))$ results in

$$\boldsymbol{D}_\alpha = \frac{\partial \boldsymbol{\mu}}{\partial \boldsymbol{\alpha}^T} = (\frac{\partial \boldsymbol{\mu}}{\partial \boldsymbol{\beta}^T}, \frac{\partial \boldsymbol{\mu}}{\partial \sigma^2}) = (\boldsymbol{D}, \mathbf{0}). \tag{6.57}$$

$-\ddot{\boldsymbol{l}}_{\alpha\alpha}$ has been given in (6.37) and it follows from (2.8) that $\ddot{\boldsymbol{l}}_{\beta Y} = \phi \boldsymbol{D}^T \boldsymbol{V}^{-1}$. Substituting these results into (6.54) yields

$$\begin{aligned} GL(\hat{\boldsymbol{\alpha}}) &= (\boldsymbol{D}, \mathbf{0}) \begin{pmatrix} \phi \boldsymbol{R}^T (\boldsymbol{I}_p - \boldsymbol{B}_\theta) \boldsymbol{R} & \mathbf{0} \\ \mathbf{0} & -\ddot{\boldsymbol{l}}_{\phi\phi} \end{pmatrix}^{-1} \begin{pmatrix} \phi \boldsymbol{D}^T \boldsymbol{V}^{-1} \\ \ddot{\boldsymbol{l}}_{\phi Y} \end{pmatrix} \\ &= \{\boldsymbol{D}\boldsymbol{L}(\boldsymbol{I}_p - \boldsymbol{B}_\theta)^{-1} \boldsymbol{L}^T \boldsymbol{D}^T \boldsymbol{V}^{-1}\}_{\hat{\beta}}. \end{aligned}$$

Since $\boldsymbol{D} = \boldsymbol{Q}\boldsymbol{R}$, we get (6.56). ∎

This theorem shows that the generalized leverage is closely connected with the nonlinearity of the model because $\boldsymbol{B}_\theta = [\hat{\boldsymbol{e}}^T \boldsymbol{N}_\theta][\boldsymbol{A}_\theta{}^I]$ is the effective residual curvature matrix (see (5.7)). From this theorem, we have several useful results.

(1) Since $\boldsymbol{\mu} = \dot{\boldsymbol{b}}(\boldsymbol{\theta}(\boldsymbol{\beta}))$ does not depend on σ^2 which is orthogonal to $\boldsymbol{\beta}$ (i.e. $\ddot{\boldsymbol{l}}_{\beta\sigma} = \mathbf{0}$ at $\hat{\boldsymbol{\beta}}$), from (6.57) we have

$$GL(\hat{\boldsymbol{\alpha}}) = GL(\hat{\boldsymbol{\beta}}) = \left\{ \frac{\partial \boldsymbol{\mu}}{\partial \boldsymbol{\beta}^T} (-\ddot{\boldsymbol{l}}_{\beta\beta})^{-1} \ddot{\boldsymbol{l}}_{\beta Y} \right\}_{\beta = \hat{\beta}},$$

where $GL(\hat{\beta})$ is the leverage of $\hat{\beta}$ with known σ^2. This means that the leverage of $\hat{\alpha} = (\hat{\beta}^T, \hat{\sigma}^2)^T$ is independent of the dispersion parameter σ^2. ∎

(2) In (6.54), if we replace the observed information $-\ddot{l}_{\beta\beta}$ by the Fisher information $\boldsymbol{J}_\beta(\boldsymbol{Y}) = \phi \boldsymbol{D}^T \boldsymbol{V}^{-1} \boldsymbol{D}$, then $GL(\hat{\alpha}) \approx \boldsymbol{D}(\boldsymbol{D}^T \boldsymbol{V}^{-1} \boldsymbol{D})^{-1} \boldsymbol{D}^T \boldsymbol{V}^{-1}$ which has the same diagonal element as $\boldsymbol{H} = \boldsymbol{V}^{-\frac{1}{2}} \boldsymbol{D}(\boldsymbol{D}^T \boldsymbol{V}^{-1} \boldsymbol{D})^{-1} \boldsymbol{D}^T \boldsymbol{V}^{-\frac{1}{2}}$ and $\boldsymbol{H}$ is called the tangent plane leverage by St. Laurent and Cook (1992). It is interesting to note that the difference between observed information and Fisher information induces the difference between generalized leverage and tangent plane leverage (St. Laurent and Cook 1992). ∎

(3) For GML with canonical link, we have $\boldsymbol{\theta} = \boldsymbol{X}\boldsymbol{\beta}$, $\boldsymbol{A}_\theta^I = \boldsymbol{0}$ and $\boldsymbol{B}_\theta = \boldsymbol{0}$. Thus $GL(\hat{\alpha}) = \boldsymbol{X}(\boldsymbol{X}^T \boldsymbol{V}^{-1} \boldsymbol{X})^{-1} \boldsymbol{X}^T \boldsymbol{V}^{-1}$ and

$$\boldsymbol{H} = \boldsymbol{V}^{-\frac{1}{2}} \boldsymbol{X}(\boldsymbol{X}^T \boldsymbol{V}^{-1} \boldsymbol{X})^{-1} \boldsymbol{X}^T \boldsymbol{V}^{-\frac{1}{2}}$$

have the same diagonal elements, as expected. ∎

(4) For normal nonlinear regression models, since $\boldsymbol{V} = \boldsymbol{I}_n$, $\boldsymbol{B}_\theta = \boldsymbol{B}$, we have $\boldsymbol{H} = \boldsymbol{D}(\boldsymbol{D}^T \boldsymbol{D})^{-1} \boldsymbol{D}^T$ and $GL(\hat{\alpha}) = \boldsymbol{Q}^T (\boldsymbol{I}_p - \boldsymbol{B})^{-1} \boldsymbol{Q}$. These results are exactly consistent with the tangent plane leverage and Jacobian leverage introduced by St. Laurent and Cook (1992), as expected. Note that the leverage defined by Yoshizoe (1991) actually leads to the tangent plane leverage. Moreover, for normal linear regression models, we have $GL(\hat{\alpha}) = \boldsymbol{H} = \boldsymbol{X}(\boldsymbol{X}^T \boldsymbol{X})^{-1} \boldsymbol{X}^T$ that coincides with the well-known hat matrix of linear models. ∎

We have introduced a general definition of leverage to a quite general class of models and applied it to exponential family nonlinear models. Obviously, this definition can also be applied to some other estimators and models. In Wei, Hu and Fung (1997), they discuss the leverage of M-estimator and Bayesian estimator in linear and nonlinear regression models based on (6.53); and the leverage connected with censored data and correlated data is also investigated there. Definition (6.53) still keeps the basic characteristic as in linear regression models, because $GL_{ii} = \partial \tilde{y}_i / \partial y_i$ still reflects the influence of individual observations on the fit. But note that there must be some differences between generalized leverage and the ordinary leverage in linear regression. Obviously, $\text{Var}(\hat{y}_i) = \sigma^2 h_{ii}$, $\text{Var}(e_i) = \sigma^2(1 - h_{ii})$ and $\rho(y_i, \hat{y}_i) = \sqrt{h_{ii}}$ no longer hold for generalized leverage in general regression models. Furthermore, we emphasize the following two points: (a) the generalized leverage is no longer independent of the response $\boldsymbol{Y}$ (in linear regression, $\boldsymbol{H} = \boldsymbol{X}(\boldsymbol{X}^T \boldsymbol{X})^{-1} \boldsymbol{X}^T$ depends only on $\boldsymbol{X}$); and (b) the generalized leverage is usually connected with the nonlinearity (i.e. curvature) of the model.

6.5.2 Generalized leverage and local influence

As St. Laurent and Cook (1993) pointed out, there is a close connection between leverage and local influence analysis (see Section 6.4). Consider the perturbed model (6.38), that is the response vector $\boldsymbol{Y}$ is perturbed by adding a vector $\boldsymbol{\omega}$ to it as discussed after (6.38). It is easily seen from (6.38) that $\boldsymbol{\Delta} = \partial^2 l / \partial\boldsymbol{\alpha}\partial\boldsymbol{\omega}^T$ given in (6.34) is just $\ddot{\boldsymbol{l}}_{\alpha Y}$ in (6.54). So it follows from (6.33) and (6.54) that

$$\begin{aligned} -\ddot{\boldsymbol{F}} &= \boldsymbol{\Delta}^T(\boldsymbol{D}_\alpha^T\boldsymbol{D}_\alpha)^{-1}(\boldsymbol{D}_\alpha^T\boldsymbol{D}_\alpha)(-\ddot{\boldsymbol{l}}_{\alpha\alpha}^{-1})\boldsymbol{\Delta} \\ &= \boldsymbol{\Delta}^T(\boldsymbol{D}_\alpha^T\boldsymbol{D}_\alpha)^{-1}\boldsymbol{D}_\alpha^T GL(\hat{\boldsymbol{\alpha}}) \end{aligned}$$

if $\boldsymbol{D}_\alpha^T\boldsymbol{D}_\alpha$ is nonsingular, and

$$\begin{aligned} GL(\hat{\boldsymbol{\alpha}}) &= \boldsymbol{D}_\alpha(\boldsymbol{\Delta}\boldsymbol{\Delta}^T)^{-1}(\boldsymbol{\Delta}\boldsymbol{\Delta}^T)(-\ddot{\boldsymbol{l}}_{\alpha\alpha}^{-1})\boldsymbol{\Delta} \\ &= \boldsymbol{D}_\alpha(\boldsymbol{\Delta}\boldsymbol{\Delta}^T)^{-1}\boldsymbol{\Delta}(-\ddot{\boldsymbol{F}}) \end{aligned}$$

if $\boldsymbol{\Delta}\boldsymbol{\Delta}^T$ is nonsingular. We now consider a commonly encountered case in which $s(y_i, \phi) = s(\phi) + t(y_i)$ in (6.1). Then it follows from (6.39) and (6.56) that

$$\ddot{\boldsymbol{A}} = 2\hat{\phi}\hat{\boldsymbol{V}}^{-1}GL(\hat{\boldsymbol{\alpha}}) + 4n^{-1}\ddot{s}^{-1}(\hat{\phi})(\hat{\boldsymbol{\theta}} - \tilde{\boldsymbol{\theta}})(\hat{\boldsymbol{\theta}} - \tilde{\boldsymbol{\theta}})^T.$$

In particular, for normal nonlinear regression models, we have $s(\phi) = -\log\phi$, $\boldsymbol{V} = \boldsymbol{I}_n$, $\boldsymbol{\mu} = \boldsymbol{\theta}$ and $\tilde{\boldsymbol{\mu}} = \boldsymbol{Y}$. Then $\ddot{\boldsymbol{A}}$ reduces to (St. Laurent and Cook 1993)

$$\ddot{\boldsymbol{A}} = 2\hat{\sigma}^{-2}GL(\hat{\boldsymbol{\alpha}}) + 4n^{-1}\hat{\sigma}^{-4}\hat{\boldsymbol{e}}\hat{\boldsymbol{e}}^T.$$

Now we give two numerical examples to illustrate our theoretical results.

Example 6.6 (Fruit fly data No. 3).

These data have been discussed in Example 6.4. We now compute GL_{ii} by using formula (6.56). The results are given by an index plot, the plot (c) of Figure 6.1 (with circles). This plot is compared with plots (a) and (b) of Figure 6.1 obtained from local influence analysis. The results show that all the plots identify the same influential point, case 23, as expected. ∎

Example 6.7 (Leukemia data No. 4).

In Example 2.5, we fit these data by using a nonlinear logistic regression. Here we apply formula (6.56) to compute GL_{ii}. The results are given by an index plot, which is shown in Figure 6.3. There are two high leverage points: cases 15 and 19. This is quite similar to the result of likelihood distance given in Example 6.3, where we mentioned that case 15 has the maximum value of WBC among AG positive patients and case 19 has the minimum value of WBC among AG negative patients. ∎

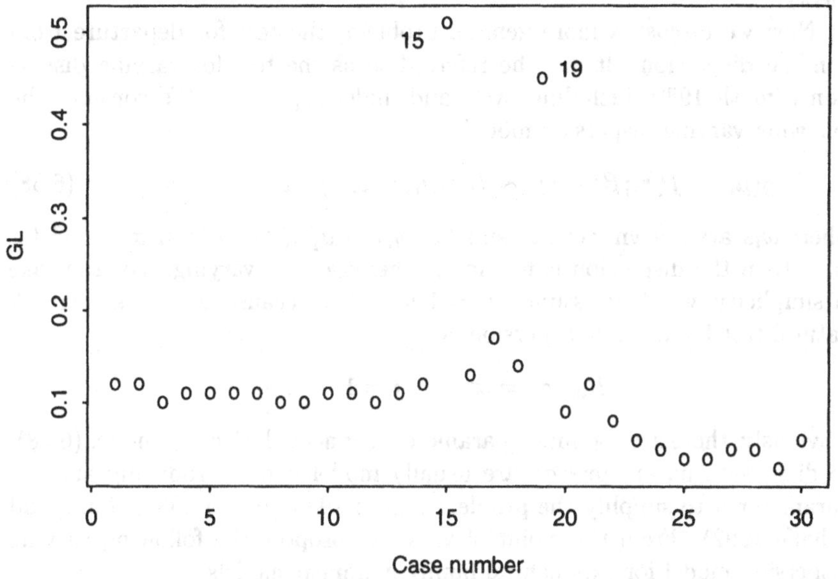

Figure 6.3 INDEX PLOT OF GL_{ii} FROM FITTING A NONLINEAR LOGISTIC MODEL TO THE LEUKEMIA DATA

6.6 Diagnostics for Varying Dispersion

Testing to detect nonconstant variance (heteroscedasticity) is a common problem in regression diagnostics (Cook and Weisberg 1982; 1983). For GML and exponential family nonlinear models, because the nominal variance $\mathrm{Var}(y_i) = \sigma^2 V(\mu_i), i = 1, \cdots, n$ are always nonconstant (except normal distribution), it is not necessary to detect nonconstant variance. However, the variance problem still exists with these models. Overdispersion has been a common problem of concern in recent years. The problem goes back to Fisher (1950) and has been investigated by many authors in the past decade (e.g. Cox 1983; Efron 1986; 1991; 1992; Dean and Lawless 1989; McCullagh and Nelder 1989; Zelterman and Chen 1989; Breslow 1990; Gelfand and Dalal 1990; Dean 1992; Ganio and Schaffe 1992; Smith and Heitjan 1993; Wei *et al.* 1997). The term overdispersion means that the actual variance of a measured response exceeds its nominal variance (for example, the nominal binomial variance is $mp_i(1-p_i)$ if $y_i \sim b(m, p_i)$). Similarly, as a counterpart, underdispersion means that the actual variance of response is less than its nominal variance. As Cox (1983) pointed out, very common practical complication is the presence of overdispersion (or more rarely underdispersion), which will lead to a failure of variance-mean relation and a possible loss of efficiency (see also McCullagh and Nelder 1989). So the test for over- or underdispersion is an important problem in both practice and theory.

Now we discuss a more general problem, the test for departure from nominal dispersion. It may be referred to as the test for varying dispersion (Smyth 1989) including over- and underdispersion. We consider the following varying dispersion model

$$g(\mu_i) = f(\boldsymbol{x}_i; \boldsymbol{\beta}), \quad y_i \sim ED(\mu_i, \sigma_i^2 \omega_i^{-1}), \quad i = 1, \cdots, n, \tag{6.58}$$

where ω_is are known weights and $\mathrm{Var}(y_i) = \sigma_i^2 \omega_i^{-1} V(\mu_i)$. If $\sigma_i^2 = \sigma^2$ for all i, then the dispersion is nominal, otherwise it is varying. For the sake of simplicity, we shall assume $\omega_i = 1$ for all i, because ω_is are known. A natural test for varying dispersion is

$$H_0 : \sigma_i^2 = \sigma^2, \qquad i = 1, \cdots, n.$$

Obviously, there are too many parameters connected with the model (6.58), so dispersion parameters σ_i^2 are usually modeled by introducing another parameter γ to simplify the problem (Efron 1986; Smyth 1989; Ganio and Schaffe 1992). From this point of view, we propose the following varying dispersion model for exponential family nonlinear models:

$$\begin{cases} y_i \sim ED(\mu_i, \phi_i^{-1}), & \phi_i = \phi m_i, \\ g(\mu_i) = f(\boldsymbol{x}_i; \boldsymbol{\beta}), & m_i = m(\boldsymbol{z}_i; \gamma); \end{cases} \tag{6.59}$$

where $\boldsymbol{z}_i$s are covariates, γ is a q-vector parameter and $m(\cdot;\cdot)$ is a known function. In this model, $\phi = \sigma^{-2}$ may be regarded as the nominal dispersion and m_is are added to reflect the variation of dispersion. It is assumed that there is a unique value γ_0 of γ such that $m(\boldsymbol{z}_i; \gamma_0) = 1$ for all i. If $\gamma = \gamma_0$, then $\mathrm{Var}(y_i) = \sigma^2 V(\mu_i)$ for all i and the dispersion is nominal. Hence the test for varying dispersion is equivalent to a test of hypothesis

$$H_0 : \gamma = \gamma_0; \quad H_1 : \gamma \neq \gamma_0.$$

It is easily seen that if y_i is normally distributed as $y_i \sim N(\boldsymbol{x}_i^T \boldsymbol{\beta},\ \sigma^2 m_i^{-1})$ in model (6.59), then H_0 is reduced to a test of heteroscedasticity in linear regression, which has been studied by many authors (e.g. Cook and Weisberg 1983; Simonoff and Tsai 1994; and the references therein). In the rest of this section, we shall derive several test statistics for H_0 based on the varying dispersion model (6.59).

6.6.1 Likelihood ratio and score statistics

For simplicity, we assume $s(y_i, \phi) = s(\phi) + t(y_i)$ holds throughout this section. It follows from (1.6) and (1.27) that the log-likelihood of model

(6.59) can be expressed as

$$\begin{aligned} l(\gamma,\phi,\beta) &= \sum_{i=1}^{n}\left[\phi m_i\{y_i\theta_i - b(\theta_i) - c(y_i)\} - \frac{1}{2}s(\phi m_i) - \frac{1}{2}t(y_i)\right] \\ &= -\frac{1}{2}\sum_{i=1}^{n}\{\phi m_i d_i' + s(\phi m_i) + t(y_i)\}, \qquad (6.60) \end{aligned}$$

where $d_i' = -2\{y_i\theta_i - b(\theta_i) - c(y_i)\}$ (see (1.26) and nearby). In this section, we set $\boldsymbol{\alpha} = (\gamma^T, \phi, \boldsymbol{\beta}^T)^T$ and the maximum likelihood estimator of $\boldsymbol{\alpha}$ for model (6.59) are denoted by $\hat{\boldsymbol{\alpha}} = (\hat{\gamma}^T, \hat{\phi}, \hat{\boldsymbol{\beta}}^T)^T$. If H_0 holds, the maximum likelihood estimator of $\boldsymbol{\alpha}$ are denoted by $\hat{\boldsymbol{\alpha}}_0 = (\gamma_0^T, \hat{\phi}_0, \hat{\boldsymbol{\beta}}_0^T)^T$. Further, we set $\tilde{\boldsymbol{\alpha}} = \tilde{\boldsymbol{\alpha}}(\gamma) = (\gamma^T, \tilde{\phi}(\gamma), \tilde{\boldsymbol{\beta}}^T(\gamma))^T$ where $\tilde{\phi}(\gamma)$, $\tilde{\boldsymbol{\beta}}^T(\gamma)$ are the maximum likelihood estimators of ϕ and $\boldsymbol{\beta}$ for any given γ.

Theorem 6.14

For varying dispersion models (6.59) and (6.60), the likelihood ratio statistic for H_0 is

$$LR = n\{s(\hat{\phi}_0) - \hat{\phi}_0\dot{s}(\hat{\phi}_0)\} - \{\mathbf{1}^T\boldsymbol{s}_m - \phi\mathbf{1}^T\boldsymbol{M}\dot{\boldsymbol{s}}_m\}_{\hat{\alpha}}, \qquad (6.61)$$

where $\boldsymbol{M} = \text{diag}(m_1, \cdots, m_n)$, $\boldsymbol{s}_m = (s(\phi m_1), \cdots, s(\phi m_n))^T$, $\dot{\boldsymbol{s}}_m = (\dot{s}(\phi m_1), \cdots, \dot{s}(\phi m_n))^T$ and $\mathbf{1} = (1, \cdots, 1)^T$, respectively.

Proof. The log-likelihood (6.60) can be expressed as

$$l(\boldsymbol{\alpha}) = -\frac{1}{2}\{\phi\mathbf{1}^T\boldsymbol{M}\boldsymbol{d}' + \mathbf{1}^T\boldsymbol{s}_m + \mathbf{1}^T\boldsymbol{t}(\boldsymbol{y})\},$$

where $\boldsymbol{t}(\boldsymbol{y}) = (t(y_1), \cdots, t(y_n))^T$ and $\boldsymbol{d}' = (d_1', \cdots, d_n')^T$. Then we have

$$\begin{aligned} LR &= 2\{l(\hat{\boldsymbol{\alpha}}) - l(\hat{\boldsymbol{\alpha}}_0)\} \\ &= \{\phi\mathbf{1}^T\boldsymbol{M}\boldsymbol{d}' + \mathbf{1}^T\boldsymbol{s}_m\}_{\hat{\alpha}_0} - \{\phi\mathbf{1}^T\boldsymbol{M}\boldsymbol{d}' + \mathbf{1}^T\boldsymbol{s}_m\}_{\hat{\alpha}}. \end{aligned}$$

On the other hand, we have

$$\frac{\partial l}{\partial \phi} = -\frac{1}{2}(\mathbf{1}^T\boldsymbol{M}\boldsymbol{d}' + \mathbf{1}^T\boldsymbol{M}\dot{\boldsymbol{s}}_m).$$

So the maximum likelihood estimators $\hat{\phi}$ and $\hat{\phi}_0$ respectively meet

$$(\mathbf{1}^T\boldsymbol{M}\boldsymbol{d}')_{\hat{\alpha}} = -(\mathbf{1}^T\boldsymbol{M}\dot{\boldsymbol{s}}_m)_{\hat{\alpha}}; \quad (\mathbf{1}^T\boldsymbol{M}\boldsymbol{d}')_{\hat{\alpha}_0} = -(\mathbf{1}^T\dot{\boldsymbol{s}}_m)_{\hat{\alpha}_0}. \qquad (6.62)$$

Using these results, LR reduces to

$$LR = \{\mathbf{1}^T\boldsymbol{s}_m - \phi\mathbf{1}^T\dot{\boldsymbol{s}}_m\}_{\hat{\alpha}_0} - \{\mathbf{1}^T\boldsymbol{s}_m - \phi\mathbf{1}^T\boldsymbol{M}\dot{\boldsymbol{s}}_m\}_{\hat{\alpha}}. \qquad (6.63)$$

Note that if H_0 holds, we have $\gamma = \gamma_0$ and $m_i = 1$ for all i, which results in $\boldsymbol{M} = \boldsymbol{I}_n$, $\boldsymbol{s}_m = s(\phi)\boldsymbol{1}$ and $\dot{\boldsymbol{s}}_m = \dot{s}(\phi)\boldsymbol{1}$. Then (6.61) is obtained from (6.63) and the theorem is proved. ∎

Corollary

For normal and inverse Gaussian model, (6.61) reduces to

$$LR = n\log(\hat{\phi}\hat{m}_g/\hat{\phi}_0), \qquad m_g = (\prod_{i=1}^n m_i)^{\frac{1}{n}}. \tag{6.64}$$

Proof. Substituting $s(\phi) = -\log\phi$, $\dot{s}(\phi) = -\phi^{-1}$ into (6.61) gives

$$\begin{aligned} LR &= n(-\log\hat{\phi}_0 + 1) - \sum_{i=1}^n \{-\log(\hat{\phi}\hat{m}_i) + 1\} \\ &= n\log(\hat{\phi}/\hat{\phi}_0) + \sum_{i=1}^n \log(\hat{m}_i), \end{aligned}$$

which leads to (6.64). ∎

In model (6.59), γ is the parameter of interest and $(\phi, \boldsymbol{\beta}^T)^T$ are nuisance parameters. So we may get the score statistic for H_0. As usual, the score statistic for H_0 is (see (6.21))

$$SC = \left\{\left(\frac{\partial l}{\partial\gamma}\right)^T \boldsymbol{J}^{\gamma\gamma}\left(\frac{\partial l}{\partial\gamma}\right)\right\}_{\hat{\alpha}_0}, \tag{6.65}$$

where $\boldsymbol{J}^{\gamma\gamma}$ is the upper left corner of $\boldsymbol{J}^{-1}$ and $\boldsymbol{J} = \boldsymbol{J}(\gamma, \phi, \boldsymbol{\beta})$ is the Fisher information matrix of $\boldsymbol{Y}$ for $\boldsymbol{\alpha}$. Then we have the following theorem.

Theorem 6.15

For varying dispersion models (6.59) and (6.60), the score statistic for the test H_0 can be expressed as

$$SC = \frac{1}{2}\ddot{s}^{-1}(\hat{\phi}_0)(\boldsymbol{d}^T\boldsymbol{P}_c\boldsymbol{d})_{\hat{\alpha}_0}, \tag{6.66}$$

where $\boldsymbol{d} = (d_1, \cdots, d_n)^T$, $d_i = d_i(y_i, \mu_i) = d_i' + k$, $\boldsymbol{P}_c$ is the projection matrix of $\boldsymbol{\Psi}_c = (\boldsymbol{I}_n - \boldsymbol{1}\boldsymbol{1}^T/n)\boldsymbol{\Psi}$ and $\boldsymbol{\Psi} = (\Psi_{ia})$ with $\Psi_{ia} = \partial m_i/\partial\gamma_a$, $i = 1, \cdots, n$, $a = 1, \cdots, q$.

Proof. To get SC from (6.65), we need to find $(\partial l/\partial \gamma)$ and $\boldsymbol{J}(\gamma, \phi, \boldsymbol{\beta})$. It follows from (6.60) that

$$\begin{aligned}
\frac{\partial l}{\partial \gamma_a} &= -\frac{1}{2}\sum_{i=1}^{n}\{\phi d_i' + \phi \dot{s}(\phi m_i)\}\frac{\partial m_i}{\partial \gamma_a}, \quad \gamma = 1, \cdots, q, \\
\frac{\partial l}{\partial \boldsymbol{\gamma}} &= -\frac{1}{2}\phi \boldsymbol{\Psi}^T(\boldsymbol{d}' + \dot{\boldsymbol{s}}_m), \\
\left(\frac{\partial l}{\partial \boldsymbol{\gamma}}\right)_{\hat{\alpha}_0} &= -\frac{1}{2}\hat{\phi}_0\{\boldsymbol{\Psi}^T(\boldsymbol{d}' + \dot{s}(\phi)\mathbf{1})\}_{\hat{\alpha}_0}.
\end{aligned}$$

It follows from (6.62) that

$$(\mathbf{1}^T\boldsymbol{d}')_{\hat{\alpha}_0} = -n\dot{s}(\hat{\phi}_0)$$

which results in

$$\left(\frac{\partial l}{\partial \boldsymbol{\gamma}}\right)_{\hat{\alpha}_0} = -\frac{1}{2}\hat{\phi}_0\{\boldsymbol{\Psi}^T(\boldsymbol{I}_n - \mathbf{1}\mathbf{1}^T/n)\boldsymbol{d}'\}_{\hat{\alpha}_0}.$$

Then we have

$$\left(\frac{\partial l}{\partial \boldsymbol{\gamma}}\right)_{\hat{\alpha}_0} = -\frac{1}{2}\hat{\phi}_0(\boldsymbol{\Psi}_c^T\boldsymbol{d}')_{\hat{\alpha}_0} = -\frac{1}{2}\hat{\phi}_0(\boldsymbol{\Psi}_c^T\boldsymbol{d})_{\hat{\alpha}_0}, \tag{6.67}$$

here we use the following result: from (1.28) we have $\boldsymbol{d}' = \boldsymbol{d} + k\mathbf{1}$, so

$$\boldsymbol{\Psi}_c^T\boldsymbol{d}' = \boldsymbol{\Psi}_c^T\boldsymbol{d} + \boldsymbol{\Psi}^T(\boldsymbol{I}_n - \mathbf{1}\mathbf{1}^T/n)(k\mathbf{1}) = \boldsymbol{\Psi}_c^T\boldsymbol{d}.$$

Now by direct calculation from (6.60), we can get the Fisher information matrix of $\boldsymbol{Y}$ for $\boldsymbol{\alpha} = (\boldsymbol{\gamma}^T, \phi, \boldsymbol{\beta}^T)^T$:

$$\boldsymbol{J}(\gamma, \phi, \boldsymbol{\beta}) = \begin{pmatrix} \frac{1}{2}\phi^2\boldsymbol{\Psi}^T\ddot{\boldsymbol{S}}_m\boldsymbol{\Psi} & \frac{1}{2}\phi\boldsymbol{\Psi}^T\ddot{\boldsymbol{S}}_m\boldsymbol{M}\mathbf{1} & \mathbf{0} \\ \frac{1}{2}\phi\mathbf{1}^T\boldsymbol{M}\ddot{\boldsymbol{S}}_m\boldsymbol{\Psi} & \frac{1}{2}\mathbf{1}^T\boldsymbol{M}\ddot{\boldsymbol{S}}_m\boldsymbol{M}\mathbf{1} & \mathbf{0} \\ \mathbf{0} & \mathbf{0} & \phi\dot{\boldsymbol{\theta}}^T\boldsymbol{V}\boldsymbol{M}\dot{\boldsymbol{\theta}} \end{pmatrix} \tag{6.68}$$

where $\ddot{\boldsymbol{S}}_m = \text{diag}(\ddot{s}(\phi m_1), \cdots, \ddot{s}(\phi m_n))$ and $\dot{\boldsymbol{\theta}} = \boldsymbol{D}_\theta = \partial\boldsymbol{\theta}/\partial\boldsymbol{\beta}^T$. When H_0 holds, we have $(\ddot{\boldsymbol{S}}_m)_{\hat{\alpha}_0} = \ddot{s}(\hat{\phi}_0)\boldsymbol{I}_n$, $(\boldsymbol{M})_{\hat{\alpha}_0} = \boldsymbol{I}_n$. Then it follows from (6.68) that

$$\begin{aligned}
(\boldsymbol{J}^{\gamma\gamma})_{\hat{\alpha}_0} &= \left\{\frac{1}{2}\phi^2\ddot{s}(\phi)\boldsymbol{\Psi}^T\boldsymbol{\Psi} - \frac{1}{2n}\phi^2\ddot{s}(\phi)\boldsymbol{\Psi}^T\mathbf{1}\mathbf{1}^T\boldsymbol{\Psi}\right\}_{\hat{\alpha}_0}^{-1} \\
&= 2\hat{\phi}_0^{-2}\ddot{s}^{-1}(\hat{\phi}_0)(\boldsymbol{\Psi}_c^T\boldsymbol{\Psi}_c)^{-1}.
\end{aligned} \tag{6.69}$$

Substituting this result and (6.67) into (6.65) yields

$$SC = \frac{1}{2}\ddot{s}^{-1}(\hat{\phi}_0)\{\boldsymbol{d}^T\boldsymbol{\Psi}_c(\boldsymbol{\Psi}_c^T\boldsymbol{\Psi}_c)^{-1}\boldsymbol{\Psi}_c^T\boldsymbol{d}\}_{\hat{\alpha}_0},$$

which results in (6.66). ∎

Equation (6.66) is a useful formula to detect varying dispersion. To compute it, one only needs the statistics under null hypothesis. Furthermore, the quantity $\boldsymbol{d}^T\boldsymbol{P}_c\boldsymbol{d} = \boldsymbol{d}^T\boldsymbol{\Psi}_c(\boldsymbol{\Psi}_c^T\boldsymbol{\Psi}_c)^{-1}\boldsymbol{\Psi}_c^T\boldsymbol{d}$ is just the sum of squares for the regression of $\boldsymbol{d}$ on $\boldsymbol{\Psi}_c$ in the constructed model "$\boldsymbol{d} = \beta_0\mathbf{1} + \boldsymbol{\Psi}_c\boldsymbol{\beta} + \boldsymbol{\epsilon}$" and thus it can be easily obtained using the standard regression program.

From this theorem we have several useful results.

(1) Equation (6.66) can be used for commonly encountered nonlinear models. For normal model, we have $\ddot{s}(\phi) = \phi^{-2} = \sigma^4$ and $d_i = e_i^2 = (y_i - \mu_i)^2$, then (6.66) becomes

$$SC = \frac{1}{2}(\boldsymbol{u}^T\boldsymbol{P}_c\boldsymbol{u})_{\hat{\alpha}_0} \tag{6.70}$$

with $\boldsymbol{u} = (u_i)$, $u_i = e_i^2/\sigma^2$. This result coincides with equation (8) of Cook and Weisberg (1983) for the test of heteroscedasticity in normal linear regression, as expected. But (6.70) is also valid for normal nonlinear regression case. For inverse Gaussian nonlinear model $y_i \sim IG(\mu_i, \sigma^2 m_i^{-1})$, we have $\ddot{s}(\phi) = \phi^{-2} = \sigma^4$, $d_i = (y_i - \mu_i)^2/(y_i\mu_i^2)$ and $\hat{\sigma}_0^2 = n^{-1}(\sum_i d_i)_{\hat{\alpha}_0}$. If we set $\boldsymbol{u} = (u_i) = (d_i/\sigma^2)$, then (6.70) still holds with u_i defined above. For gamma nonlinear model $y_i \sim GA(\mu_i, \sigma^2 m_i^{-1})$, we have $s(\phi) = -2\{\phi\log\phi - \log\Gamma(\phi)\}$, $d_i = 2\{(y_i - \mu_i)/\mu_i - \log(y_i/\mu_i)\}$, and SC can be computed by formula (6.66). ∎

(2) We may rewrite (6.66) as a standardized form like (6.70). Since $\mathrm{E}(d_i) = -\dot{s}(\phi) - k$, $\mathrm{Var}(d_i) = 2\ddot{s}(\phi)$ (see Lemma 2.2) and $\boldsymbol{P}_c\mathbf{1} = \mathbf{0}$, if we set $\boldsymbol{d}_s = (d_{si})$ with $d_{si} = \{d_i - \mathrm{E}(d_i)\}/\mathrm{Var}^{\frac{1}{2}}(d_i)$, then we have $SC = (\boldsymbol{d}_s^T\boldsymbol{P}_c\boldsymbol{d}_s)_{\hat{\alpha}_0}$, that is the norm of the standardized deviance $\boldsymbol{d}_s$ evaluated at $\hat{\boldsymbol{\alpha}}_0$. ∎

(3) Expression (6.66) can be extended to more general models in which the condition $s(y_i, \phi) = s(\phi) + t(y_i)$ may not hold. By some calculations we get

$$SC = \frac{1}{2}\{(\boldsymbol{d}' + \boldsymbol{s}')^T\tilde{\boldsymbol{S}}^{-\frac{1}{2}}\tilde{\boldsymbol{P}}\tilde{\boldsymbol{S}}^{-\frac{1}{2}}(\boldsymbol{d}' + \boldsymbol{s}')\}_{\hat{\alpha}_0},$$

where $\tilde{\boldsymbol{S}} = \mathrm{diag}(\tilde{S}_i)$, $\tilde{S}_i = \mathrm{E}\{\partial^2 s(\phi, y_i)/\partial\phi^2\}$, $\boldsymbol{s}' = (s_i')$, $s_i' = \partial s(\phi, y_i)/\partial\phi$, $\tilde{\boldsymbol{P}} = \tilde{\boldsymbol{\Psi}}(\tilde{\boldsymbol{\Psi}}^T\tilde{\boldsymbol{\Psi}})^{-1}\tilde{\boldsymbol{\Psi}}^T$, $\tilde{\boldsymbol{\Psi}} = (\boldsymbol{I}_n - \tilde{\boldsymbol{P}}_1)\tilde{\boldsymbol{S}}^{\frac{1}{2}}\boldsymbol{\Psi}$ and $\tilde{\boldsymbol{P}}_1 = \tilde{\boldsymbol{S}}^{\frac{1}{2}}\mathbf{1}(\mathbf{1}^T\tilde{\boldsymbol{S}}\mathbf{1})^{-1}\mathbf{1}^T\tilde{\boldsymbol{S}}^{\frac{1}{2}}$. It is not too hard to show that the expression (6.66) is the special case of this formula. ∎

6.6.2 Adjusted likelihood ratio and score statistics

As pointed out by Cox and Reid (1987), the adjusted profile likelihood is needed if many nuisance parameters are being fitted. In this subsection, we first introduce the basic results of Cox and Reid (1987; 1993) and then apply them to the varying dispersion model (6.59).

It is easily seen from (6.68) that the parameter γ is orthogonal to $\boldsymbol{\beta}$ but not orthogonal to ϕ, i.e. $\boldsymbol{J}_{\gamma\beta} = \mathbf{0}$ but $\boldsymbol{J}_{\gamma\phi} \neq \mathbf{0}$ (here $\boldsymbol{J}$ is naturally partitioned according to γ, ϕ and $\boldsymbol{\beta}$). To get the adjusted profile likelihood, we need a one-to-one transformation from $(\phi, \boldsymbol{\beta}^T)^T$ to a vector parameter $\boldsymbol{\lambda}$ such that γ is orthogonal to $\boldsymbol{\lambda}$ (Cox and Reid 1987).

Now let $\boldsymbol{\alpha} = (\boldsymbol{\alpha}_1^T, \boldsymbol{\alpha}_2^T)^T$ with $\boldsymbol{\alpha}_1 = \gamma$, $\boldsymbol{\alpha}_2 = (\phi, \boldsymbol{\beta}^T)^T = \boldsymbol{\alpha}_2(\boldsymbol{\alpha}_1, \boldsymbol{\lambda})$ and $\boldsymbol{\alpha}_{10} = \gamma_0$, $\hat{\boldsymbol{\alpha}}_{20} = (\hat{\phi}_0, \hat{\boldsymbol{\beta}}_0^T)^T$. Then the log-likelihood of $\boldsymbol{Y}$ can be expressed as

$$l(\boldsymbol{\alpha}_1, \boldsymbol{\alpha}_2) = l(\boldsymbol{\alpha}_1, \boldsymbol{\alpha}_2(\boldsymbol{\alpha}_1, \boldsymbol{\lambda})) = l^*(\boldsymbol{\alpha}_1, \boldsymbol{\lambda}). \tag{6.71}$$

The maximum likelihood estimators of $\boldsymbol{\alpha}_2$ and $\boldsymbol{\lambda}$ for any given $\boldsymbol{\alpha}_1$ are denoted by $\tilde{\boldsymbol{\alpha}}_2(\boldsymbol{\alpha}_1)$ and $\tilde{\boldsymbol{\lambda}}(\boldsymbol{\alpha}_1)$, respectively. Then $\tilde{\boldsymbol{\alpha}}_2(\hat{\boldsymbol{\alpha}}_1) = \hat{\boldsymbol{\alpha}}_2$, $\tilde{\boldsymbol{\alpha}}_2(\boldsymbol{\alpha}_{10}) = \hat{\boldsymbol{\alpha}}_{20}$ and $\tilde{\boldsymbol{\lambda}}(\hat{\boldsymbol{\alpha}}_1) = \hat{\boldsymbol{\lambda}}$. We set $\tilde{\boldsymbol{\lambda}}(\hat{\boldsymbol{\alpha}}_{10}) = \hat{\boldsymbol{\lambda}}_0$ and $\tilde{\boldsymbol{\alpha}}(\boldsymbol{\alpha}_1) = (\boldsymbol{\alpha}_1^T, \tilde{\boldsymbol{\alpha}}_2^T(\boldsymbol{\alpha}_1))^T$, then $\tilde{\boldsymbol{\alpha}}(\hat{\boldsymbol{\alpha}}_1) = (\hat{\boldsymbol{\alpha}}_1^T, \hat{\boldsymbol{\alpha}}_2^T)^T$, $\tilde{\boldsymbol{\alpha}}(\boldsymbol{\alpha}_{10}) = (\boldsymbol{\alpha}_{10}^T, \hat{\boldsymbol{\alpha}}_{20}^T)^T$ (here $\boldsymbol{\alpha}_{10} = \gamma_0$). The Fisher information matrix corresponding to $l^*(\boldsymbol{\alpha}_1, \boldsymbol{\lambda})$ is denoted by $\boldsymbol{J}^*(\boldsymbol{\alpha}_1, \boldsymbol{\lambda})$ whose partitions are $\boldsymbol{J}_{11}^*$, $\boldsymbol{J}_{1\lambda}^*$ and $\boldsymbol{J}_{\lambda\lambda}^*$, respectively. Similarly, $\boldsymbol{J} = \boldsymbol{J}(\boldsymbol{\alpha}_1, \boldsymbol{\alpha}_2)$ is partitioned as $\boldsymbol{J}_{11}$, $\boldsymbol{J}_{12}$ and $\boldsymbol{J}_{22}$, respectively. Then we have the following lemma.

Lemma 6.3

Suppose that $\boldsymbol{\alpha}_2 = \boldsymbol{\alpha}_2(\boldsymbol{\alpha}_1, \boldsymbol{\lambda})$ transforms $\boldsymbol{\alpha} = (\boldsymbol{\alpha}_1^T, \boldsymbol{\alpha}_2^T)^T$ to $(\boldsymbol{\alpha}_1^T, \boldsymbol{\lambda}^T)^T$ such that $\boldsymbol{\alpha}_1$ is orthogonal to $\boldsymbol{\lambda}$, then $\boldsymbol{\alpha}_2 = \boldsymbol{\alpha}_2(\boldsymbol{\alpha}_1, \boldsymbol{\lambda})$ must satisfy

$$\frac{\partial \boldsymbol{\alpha}_2}{\partial \boldsymbol{\alpha}_1^T} = -\boldsymbol{J}_{22}^{-1}\boldsymbol{J}_{21}. \tag{6.72}$$

Further, $\boldsymbol{J}^*(\boldsymbol{\alpha}_1, \boldsymbol{\lambda})$ and $\ddot{\boldsymbol{l}}_{\lambda\lambda}^*$ are given by

$$\boldsymbol{J}_{11}^* = \boldsymbol{J}_{11} - \boldsymbol{J}_{12}\boldsymbol{J}_{22}^{-1}\boldsymbol{J}_{21} = (\boldsymbol{J}^{11})^{-1}, \qquad \boldsymbol{J}_{1\lambda}^* = \mathbf{0}, \tag{6.73}$$

$$\boldsymbol{J}_{\lambda\lambda}^* = (\frac{\partial \boldsymbol{\alpha}_2}{\partial \boldsymbol{\lambda}^T})^T \boldsymbol{J}_{22} (\frac{\partial \boldsymbol{\alpha}_2}{\partial \boldsymbol{\lambda}^T}), \tag{6.74}$$

and

$$\ddot{\boldsymbol{l}}_{\lambda\lambda}^*(\boldsymbol{\alpha}_1, \tilde{\boldsymbol{\lambda}}(\boldsymbol{\alpha}_1)) = \left\{(\frac{\partial \boldsymbol{\alpha}_2}{\partial \boldsymbol{\lambda}^T})^T \ddot{\boldsymbol{l}}_{22} (\frac{\partial \boldsymbol{\alpha}_2}{\partial \boldsymbol{\lambda}^T})\right\}_{(\alpha_1, \tilde{\lambda}(\alpha_1))}. \tag{6.75}$$

Proof. It follows from (6.71), (A.9) and (A.10) that

$$\frac{\partial l^*}{\partial \boldsymbol{\alpha}_1} = \frac{\partial l}{\partial \boldsymbol{\alpha}_1} + (\frac{\partial \boldsymbol{\alpha}_2}{\partial \boldsymbol{\alpha}_1^T})^T \frac{\partial l}{\partial \boldsymbol{\alpha}_2},$$

$$\begin{aligned}
\frac{\partial^2 l^*}{\partial \boldsymbol{\alpha}_1 \partial \boldsymbol{\lambda}^T} &= \frac{\partial^2 l}{\partial \boldsymbol{\alpha}_1 \partial \boldsymbol{\alpha}_2^T} \frac{\partial \boldsymbol{\alpha}_2}{\partial \boldsymbol{\lambda}^T} + (\frac{\partial \boldsymbol{\alpha}_2}{\partial \boldsymbol{\alpha}_1^T})^T \frac{\partial^2 l}{\partial \boldsymbol{\alpha}_2 \partial \boldsymbol{\alpha}_2^T} (\frac{\partial \boldsymbol{\alpha}_2}{\partial \boldsymbol{\lambda}^T}) + \\
& \left[\frac{\partial l}{\partial \boldsymbol{\alpha}_2^T}\right] \left[\frac{\partial^2 \boldsymbol{\alpha}_2}{\partial \boldsymbol{\alpha}_1 \partial \boldsymbol{\lambda}^T}\right],
\end{aligned}$$

which results in

$$\boldsymbol{J}_{1\lambda}^* = \boldsymbol{J}_{12} \frac{\partial \boldsymbol{\alpha}_2}{\partial \boldsymbol{\lambda}^T} + (\frac{\partial \boldsymbol{\alpha}_2}{\partial \boldsymbol{\alpha}_1^T})^T \boldsymbol{J}_{22} (\frac{\partial \boldsymbol{\alpha}_2}{\partial \boldsymbol{\lambda}^T}).$$

From $\boldsymbol{J}_{1\lambda}^* = 0$ and $\partial \boldsymbol{\alpha}_2 / \partial \boldsymbol{\lambda}^T \neq 0$, we get

$$\boldsymbol{J}_{12} + (\frac{\partial \boldsymbol{\alpha}_2}{\partial \boldsymbol{\alpha}_1^T})^T \boldsymbol{J}_{22} = \mathbf{0},$$

that leads to (6.72). On the other hand, we have

$$\begin{aligned}
\frac{\partial^2 l^*}{\partial \boldsymbol{\alpha}_1 \partial \boldsymbol{\alpha}_1^T} &= \frac{\partial^2 l}{\partial \boldsymbol{\alpha}_1 \partial \boldsymbol{\alpha}_1^T} + \frac{\partial^2 l}{\partial \boldsymbol{\alpha}_1 \partial \boldsymbol{\alpha}_2^T} (\frac{\partial \boldsymbol{\alpha}_2}{\partial \boldsymbol{\alpha}_1^T}) + (\frac{\partial \boldsymbol{\alpha}_2}{\partial \boldsymbol{\alpha}_1^T})^T \frac{\partial^2 l}{\partial \boldsymbol{\alpha}_2 \partial \boldsymbol{\alpha}_1^T} + \\
& (\frac{\partial \boldsymbol{\alpha}_2}{\partial \boldsymbol{\alpha}_1^T})^T \frac{\partial^2 l}{\partial \boldsymbol{\alpha}_2 \partial \boldsymbol{\alpha}_2^T} (\frac{\partial \boldsymbol{\alpha}_2}{\partial \boldsymbol{\alpha}_1^T}) + \left[\frac{\partial l}{\partial \boldsymbol{\alpha}_2^T}\right] \left[\frac{\partial^2 \boldsymbol{\alpha}_2}{\partial \boldsymbol{\alpha}_1 \partial \boldsymbol{\alpha}_1^T}\right].
\end{aligned}$$

Taking expectation for this equation and using (6.72), we get

$$\begin{aligned}
\boldsymbol{J}_{11}^* &= \boldsymbol{J}_{11} + \boldsymbol{J}_{12}(-\boldsymbol{J}_{22}^{-1}\boldsymbol{J}_{21}) + (-\boldsymbol{J}_{22}^{-1}\boldsymbol{J}_{21})^T \boldsymbol{J}_{21} + \\
& (-\boldsymbol{J}_{22}^{-1}\boldsymbol{J}_{21})^T \boldsymbol{J}_{22} (-\boldsymbol{J}_{22}^{-1}\boldsymbol{J}_{21}) \\
&= \boldsymbol{J}_{11} - \boldsymbol{J}_{12}\boldsymbol{J}_{22}^{-1}\boldsymbol{J}_{21},
\end{aligned}$$

which results in (6.73). Equations (6.74) and (6.75) can be obtained by similar derivations. ∎

From Lemma 6.3, we can get the adjusted profile likelihood for $\boldsymbol{\alpha}_1$ as follows (Cox and Reid 1987; 1993; see also (6.75)):

$$l_A(\boldsymbol{\alpha}_1) = l_p(\boldsymbol{\alpha}_1) - \frac{1}{2} \log\left[\det\{-\ddot{\boldsymbol{l}}_{\lambda\lambda}^*(\boldsymbol{\alpha}_1, \tilde{\boldsymbol{\lambda}}(\boldsymbol{\alpha}_1))\}\right], \tag{6.76}$$

$$\begin{aligned}
\log\left[\det\{-\ddot{\boldsymbol{l}}_{\lambda\lambda}^*(\boldsymbol{\alpha}_1, \tilde{\boldsymbol{\lambda}}(\boldsymbol{\alpha}_1))\}\right] &= \log\left[\det\{-\ddot{\boldsymbol{l}}_{22}(\boldsymbol{\alpha}_1, \tilde{\boldsymbol{\alpha}}_2(\boldsymbol{\alpha}_1))\}\right] + \\
& 2\log\left[\det\{\partial \boldsymbol{\alpha}_2/\partial \boldsymbol{\lambda}^T\}\right]_{(\alpha_1, \tilde{\lambda}(\alpha_1))}
\end{aligned} \tag{6.77}$$

where $l_p(\boldsymbol{\alpha}_1) = l(\boldsymbol{\alpha}_1, \tilde{\boldsymbol{\alpha}}_2(\boldsymbol{\alpha}_1)) = l(\tilde{\boldsymbol{\alpha}})$. The adjusted term here actually comes from Barndorff-Nielsen (1983). Once the adjusted profile likelihood $l_A(\boldsymbol{\alpha}_1)$ is obtained, many important results are the immediate consequence. The adjusted likelihood ratio statistic LR_A can be obtained from (6.76) while the adjusted score statistic SC_A is the first order approximation of LR_A. Now let us see the details.

For the test $H_0 : \boldsymbol{\alpha}_1 = \boldsymbol{\alpha}_{10}$ (i.e. $\boldsymbol{\gamma} = \boldsymbol{\gamma}_0$), we have

$$LR_A = 2\{l_A(\hat{\boldsymbol{\alpha}}_1) - l_A(\boldsymbol{\alpha}_{10})\} = LR - \Delta LR,$$

$$\Delta LR = \log\left[\det\{-\ddot{\boldsymbol{l}}^*_{\lambda\lambda}(\hat{\boldsymbol{\alpha}}_1, \hat{\boldsymbol{\lambda}})\}/\det\{-\ddot{\boldsymbol{l}}^*_{\lambda\lambda}(\boldsymbol{\alpha}_{10}, \hat{\boldsymbol{\lambda}}_0\}\right], \tag{6.78}$$

and

$$SC_A = SC - \Delta SC,$$

where ΔSC is the first order approximation of ΔLR and SC is given in (6.65). These results can be applied to model (6.59).

As Cox and Reid (1987; 1993) pointed out, to use formula (6.76) directly, one needs to solve for $\boldsymbol{\lambda}$ the partial differential equation (6.72). It is usually a major handicap to the use of (6.76). However we find out that the adjusted score statistic can always be obtained indirectly without solving for equation (6.72), which has not been seen in the literature. We summarize this result as the following lemma.

Lemma 6.4

The adjusted score statistic based on (6.76) can be expressed as $SC_A = SC - \Delta SC$ with

$$\begin{aligned}\Delta SC = & \sum_{k=1}^{q}\left[\mathrm{tr}\{\boldsymbol{J}_{22}^{-1}\frac{\partial}{\partial\gamma_k}\boldsymbol{J}_{22}(\boldsymbol{\gamma}, \tilde{\boldsymbol{\alpha}}_2(\boldsymbol{\gamma}))\} - \right. \\ & \left. 2\mathrm{tr}\{\frac{\partial}{\partial\boldsymbol{\alpha_2}^T}(\boldsymbol{J}_{22}^{-1}\boldsymbol{J}_{21}\boldsymbol{\delta}_k)\}\right]_{\hat{\boldsymbol{\alpha}}_0}(\hat{\gamma}_k - \gamma_{0k}),\end{aligned} \tag{6.79}$$

where $\boldsymbol{\delta}_k$ is a unit vector having 1 at the k-th position and zeros elsewhere, $\boldsymbol{\alpha}_{10} = \boldsymbol{\gamma}_0 = (\gamma_{01}, \cdots, \gamma_{0q})^T$ and $\hat{\boldsymbol{\alpha}}_1 = \hat{\boldsymbol{\gamma}} = (\hat{\gamma}_1, \cdots, \hat{\gamma}_q)^T$, respectively.

Proof. It follows from (6.77) and (6.78) that the first order Taylor expansion of ΔLR can be expressed as $\Delta SC = \Delta SC_1 + 2\Delta SC_2$, where

$$\begin{aligned}\Delta SC_1 &= \sum_{k=1}^{q}\left[\frac{\partial}{\partial\gamma_k}\log\left[\det\{-\ddot{\boldsymbol{l}}_{22}(\boldsymbol{\gamma}, \tilde{\boldsymbol{\alpha}}_2(\boldsymbol{\gamma}))\}\right]\right]_{\hat{\boldsymbol{\alpha}}_0}(\hat{\gamma}_k - \gamma_{0k}), \\ \Delta SC_2 &= \sum_{k=1}^{q}\left[\frac{\partial}{\partial\gamma_k}\log\{\det(\partial\boldsymbol{\alpha}_2/\partial\boldsymbol{\lambda}^T)\}\right]_{\hat{\boldsymbol{\alpha}}_0}(\hat{\gamma}_k - \gamma_{0k}).\end{aligned}$$

Using the common approximation $-\ddot{\boldsymbol{l}}_{22}(\boldsymbol{\gamma}, \tilde{\boldsymbol{\alpha}}_2(\boldsymbol{\gamma})) \approx \boldsymbol{J}_{22}(\boldsymbol{\gamma}, \tilde{\boldsymbol{\alpha}}_2(\boldsymbol{\gamma}))$, we get

$$\Delta SC_1 = \sum_{k=1}^{q}\mathrm{tr}\left[\boldsymbol{J}_{22}^{-1}\frac{\partial}{\partial\gamma_k}\boldsymbol{J}_{22}(\boldsymbol{\gamma}, \tilde{\boldsymbol{\alpha}}_2(\boldsymbol{\gamma}))\right]_{\hat{\boldsymbol{\alpha}}_0}(\hat{\gamma}_k - \gamma_{0k}).$$

Here we use the formula (A.12)

$$\frac{\partial}{\partial t}\log\{\det(\boldsymbol{A})\} = \mathrm{tr}\{\boldsymbol{A}^{-1}\frac{\partial}{\partial t}(\boldsymbol{A})\}.$$

Similarly, using this formula and (6.72) with $\boldsymbol{\alpha}_1 = \boldsymbol{\gamma}$ yields

$$\begin{aligned}
\Delta SC_2 &= \sum_{k=1}^{q}\mathrm{tr}\left\{(\frac{\partial\boldsymbol{\alpha}_2}{\partial\boldsymbol{\lambda}^T})^{-1}\frac{\partial}{\partial\gamma_k}(\frac{\partial\boldsymbol{\alpha}_2}{\partial\boldsymbol{\lambda}^T})\right\}_{\hat{\alpha}_0}(\hat{\gamma}_k-\gamma_{0k})\\
&= \sum_{k=1}^{q}\mathrm{tr}\left\{(\frac{\partial\boldsymbol{\alpha}_2}{\partial\boldsymbol{\lambda}^T})^{-1}\frac{\partial}{\partial\boldsymbol{\lambda}^T}(\frac{\partial\boldsymbol{\alpha}_2}{\partial\boldsymbol{\gamma}^T}\boldsymbol{\delta}_k)\right\}_{\hat{\alpha}_0}(\hat{\gamma}_k-\gamma_{0k})\\
&= \sum_{k=1}^{q}\mathrm{tr}\left\{(\frac{\partial\boldsymbol{\alpha}_2}{\partial\boldsymbol{\lambda}^T})^{-1}\frac{\partial}{\partial\boldsymbol{\alpha}_2^T}(\frac{\partial\boldsymbol{\alpha}_2}{\partial\boldsymbol{\gamma}^T}\boldsymbol{\delta}_k)(\frac{\partial\boldsymbol{\alpha}_2}{\partial\boldsymbol{\lambda}^T})\right\}_{\hat{\alpha}_0}(\hat{\gamma}_k-\gamma_{0k})\\
&= -\sum_{k=1}^{q}\mathrm{tr}\left\{\frac{\partial}{\partial\boldsymbol{\alpha}_2^T}(\boldsymbol{J}_{22}^{-1}\boldsymbol{J}_{21}\boldsymbol{\delta}_k)\right\}_{\hat{\alpha}_0}(\hat{\gamma}_k-\gamma_{0k}).
\end{aligned}$$

Combining ΔSC_1 and ΔSC_2, we get (6.79). ∎

Note that Lemma 6.4 and formula (6.79) can be used for quite general parametric models. For model (6.59), we have the following theorem.

Theorem 6.16

For varying dispersion models (6.59) and (6.60), the adjusted score statistic for the test H_0 can be expressed as

$$SC_A = \frac{1}{2}\ddot{s}^{-1}(\hat{\phi}_0)\{(\boldsymbol{d}+2\phi^{-1}\boldsymbol{h})^T\boldsymbol{P}_c\boldsymbol{d}\}_{\hat{\alpha}_0} \tag{6.80}$$

where $\boldsymbol{h} = (h_{11}, \cdots, h_{nn})^T$, h_{ii} is the i-th diagonal element of $\boldsymbol{H}$ (see Table 6.3) and $\boldsymbol{P}_c$ is defined in Theorem 6.15.

Proof. We shall respectively calculate the first term ΔSC_1 and the second term $2\Delta SC_2$ of (6.79) given in Lemma 6.6. To get ΔSC_1, we need $\boldsymbol{J}_{22}$ and $\partial\tilde{\boldsymbol{\alpha}}_2/\partial\boldsymbol{\gamma}^T$. It follows from (6.68) and $\boldsymbol{\alpha}_2 = (\phi, \boldsymbol{\beta}^T)^T$ that

$$\begin{aligned}
\boldsymbol{J}_{22} &= \mathrm{diag}(\frac{1}{2}\boldsymbol{1}^T\boldsymbol{M}\ddot{\boldsymbol{S}}_m\boldsymbol{M}\boldsymbol{1},\ \phi\dot{\boldsymbol{\theta}}^T\boldsymbol{M}\boldsymbol{V}\dot{\boldsymbol{\theta}}), && (6.81)\\
(\boldsymbol{J}_{22}^{-1})_{\hat{\alpha}_0} &= \mathrm{diag}(2n^{-1}\ddot{s}^{-1}(\phi),\ (\phi\dot{\boldsymbol{\theta}}^T\boldsymbol{V}\dot{\boldsymbol{\theta}})^{-1})_{\hat{\alpha}_0}, && (6.82)
\end{aligned}$$

$$\boldsymbol{J}_{22}^{-1}\boldsymbol{J}_{21} = \begin{pmatrix}(\boldsymbol{1}^T\boldsymbol{M}\ddot{\boldsymbol{S}}_m\boldsymbol{M}\boldsymbol{1})^{-1}\phi\boldsymbol{1}^T\boldsymbol{M}\ddot{\boldsymbol{S}}_m\boldsymbol{\Psi}\\ \boldsymbol{0}\end{pmatrix}. \tag{6.83}$$

To get $\partial\tilde{\boldsymbol{\alpha}}_2(\gamma)/\partial\gamma^T$, we use the equation $\partial l(\boldsymbol{\alpha}_1, \boldsymbol{\alpha}_2)/\partial\boldsymbol{\alpha}_2 = \mathbf{0}$ at $\tilde{\boldsymbol{\alpha}}_2(\gamma)$ (note that here $\boldsymbol{\alpha}^T = (\boldsymbol{\alpha}_1^T, \boldsymbol{\alpha}_2^T)$ and $\boldsymbol{\alpha}_1 = \gamma$). Taking derivative in γ for this equation yields

$$\ddot{l}_{21}(\boldsymbol{\alpha}_1, \tilde{\boldsymbol{\alpha}}_2(\boldsymbol{\alpha}_1)) + \ddot{l}_{22}(\boldsymbol{\alpha}_1, \tilde{\boldsymbol{\alpha}}_2(\boldsymbol{\alpha}_1))\frac{\partial\tilde{\boldsymbol{\alpha}}_2}{\partial\boldsymbol{\alpha}_1^T} = \mathbf{0},$$

which results in

$$\frac{\partial\tilde{\boldsymbol{\alpha}}_2}{\partial\boldsymbol{\alpha}_1^T} \approx -\{\boldsymbol{J}_{22}^{-1}\boldsymbol{J}_{21}\}_{\tilde{\alpha}}, \quad (\boldsymbol{\alpha}_1 = \gamma).$$

Since $\boldsymbol{\alpha}_2 = (\phi, \boldsymbol{\beta}^T)^T$, it follows from (6.83) that

$$\left\{\frac{\partial\tilde{\phi}(\gamma)}{\partial\gamma^T}\right\}_{\hat{\alpha}_0} = -n^{-1}(\phi\mathbf{1}^T\boldsymbol{\Psi})_{\hat{\alpha}_0}, \quad \frac{\partial\boldsymbol{\beta}(\gamma)}{\partial\gamma^T} = \mathbf{0}. \tag{6.84}$$

Now we can calculate ΔSC_1 as follows. After some calculations based on (6.82) and (6.84) we get

$$\left\{\frac{\partial}{\partial\gamma_k}\boldsymbol{J}_{22}(\gamma,\ \tilde{\boldsymbol{\alpha}}_2(\gamma))\right\}_{\hat{\alpha}_0}$$

$$= \operatorname{diag}\left[\ddot{s}(\phi)\mathbf{1}^T\boldsymbol{\Psi}_k,\quad \phi\dot{\boldsymbol{\theta}}^T\{\boldsymbol{V}D(\boldsymbol{\Psi}_k) - n^{-1}\mathbf{1}\boldsymbol{\Psi}\boldsymbol{V}\}\dot{\boldsymbol{\theta}}\right]_{\hat{\alpha}_0}, \tag{6.85}$$

where $\boldsymbol{\Psi}_k$ is the k-th column of $\boldsymbol{\Psi}$ and $D(\boldsymbol{\Psi}_k) = \operatorname{diag}(\Psi_{1k}, \cdots, \Psi_{nk})$. It follows from (6.83) and (6.85) that

$$\begin{aligned}
\operatorname{tr}\left\{\boldsymbol{J}_{22}^{-1}\frac{\partial}{\partial\gamma_k}\boldsymbol{J}_{22}(\gamma, \tilde{\boldsymbol{\alpha}}_2(\gamma))\right\}_{\hat{\alpha}_0} &= \frac{2}{n}\mathbf{1}^T\boldsymbol{\Psi}_k + \boldsymbol{h}^T\boldsymbol{\Psi}_k - \frac{p}{n}\mathbf{1}^T\boldsymbol{\Psi}_k \\
&= \frac{2}{n}\mathbf{1}^T\boldsymbol{\Psi}_k + \boldsymbol{h}^T(\boldsymbol{I}_n - \mathbf{1}\mathbf{1}^T/n)\boldsymbol{\Psi}_k.
\end{aligned}$$

Here we use the fact $p = \boldsymbol{h}^T\mathbf{1} = \sum_i h_{ii}$, then we get

$$\begin{aligned}
\Delta SC_1 &= \sum_{k=1}^{q}\{\frac{2}{n}\mathbf{1}^T + \boldsymbol{h}^T(\boldsymbol{I}_n - \mathbf{1}\mathbf{1}^T/n)\}\boldsymbol{\Psi}\boldsymbol{\delta}_k\boldsymbol{\delta}_k^T(\hat{\gamma} - \gamma_0) \\
&= (\frac{2}{n}\mathbf{1}^T\boldsymbol{\Psi} + \boldsymbol{h}^T\boldsymbol{\Psi}_c)(\hat{\gamma} - \gamma_0).
\end{aligned}$$

Next, we calculate ΔSC_2. It follows from (6.83) that $\partial(-\boldsymbol{J}_{22}^{-1}\boldsymbol{J}_{21}\boldsymbol{\delta}_k)/\partial\boldsymbol{\beta} = \mathbf{0}$. Then we have

$$\Delta SC_2 = -\sum_{k=1}^{q}\left[\frac{\partial}{\partial\phi}\{(\mathbf{1}^T\boldsymbol{M}\ddot{\boldsymbol{S}}_m\boldsymbol{M}\mathbf{1})^{-1}(\phi\mathbf{1}^T\boldsymbol{M}\ddot{\boldsymbol{S}}_m\boldsymbol{\Psi}\boldsymbol{\delta}_k)\}\right]_{\hat{\alpha}_0}(\hat{\gamma}_k - \gamma_{0k}).$$

After some calculations we get

$$\Delta SC_2 = -\sum_{k=1}^{q} \frac{1}{n} \mathbf{1}^T \boldsymbol{\Psi}_k (\hat{\gamma}_k - \gamma_{0k}) = -\frac{1}{n} \mathbf{1}^T \boldsymbol{\Psi} (\hat{\gamma} - \gamma_0).$$

Combining ΔSC_1 and ΔSC_2 yields

$$\Delta SC = \Delta SC_1 + 2\Delta SC_2 = \boldsymbol{h}^T \boldsymbol{\Psi}_c (\hat{\gamma} - \gamma_0). \tag{6.86}$$

On the other hand, we have $\hat{\gamma} - \gamma_0 \approx \{\boldsymbol{J}^{\gamma\gamma}(\partial l / \partial \gamma)\}_{\hat{\alpha}_0}$ (Cox and Reid 1987). It follows from (6.67) and (6.69) that

$$\hat{\gamma} - \gamma_0 = -\{\phi^{-1} \ddot{s}^{-1}(\phi)(\boldsymbol{\Psi}_c^T \boldsymbol{\Psi}_c)^{-1} \boldsymbol{\Psi}_c^T \boldsymbol{d}\}_{\hat{\alpha}_0}.$$

Substituting this result into (6.86) yields

$$\Delta SC = -\{\phi^{-1} \ddot{s}^{-1}(\phi) \boldsymbol{h}^T \boldsymbol{P}_c \boldsymbol{d}\}_{\hat{\alpha}_0}.$$

Since $SC_A = SC - \Delta SC$, then (6.80) follows from (6.66) and the above expression. ∎

This theorem can be applied to the test for varying dispersion for normal, inverse Gaussian and gamma nonlinear models.

Corollary

For normal and inverse Gaussian nonlinear models, the adjusted score statistic (6.80) becomes

$$SC_A = \frac{1}{2}\{(\boldsymbol{u} + 2\boldsymbol{h})^T \boldsymbol{P}_c \boldsymbol{u}\}_{\hat{\alpha}_0}, \tag{6.87}$$

where $\boldsymbol{u}$ is defined in (6.70).

Proof. Since $s(\phi) = -\log \phi$ for normal and inverse Gaussian models, substituting $\ddot{s}(\phi) = \phi^{-2}$ into (6.80) yields (6.87). ∎

As a special case of the above corollary, (6.87) can be applied to the test for heteroscedasticity in linear and nonlinear regression. For linear regression, (6.87) is the same as equation (10) of Simonoff and Tsai (1994), as expected.

See Subsections 6.6.3 and 6.6.4 for numerical examples and simulation studies.

We now discuss the adjusted likelihood ratio statistic. To get this statistic, we must solve for equation (6.72). This is practicable only for some special cases. Fortunately, normal and inverse Gaussian nonlinear models are these cases.

Lemma 6.5

For varying dispersion model (6.59) with normal or inverse Gaussian distribution. The parameter γ is orthogonal to the new parameters $\boldsymbol{\lambda} = (\psi, \boldsymbol{\beta}^T)^T$ under the transformation

$$\psi = \phi m_g(\gamma), \qquad m_g(\gamma) = \prod_{i=1}^{n} \{m_i(\gamma)\}^{\frac{1}{n}}. \tag{6.88}$$

The corresponding Fisher information matrix for $(\gamma^T, \psi, \boldsymbol{\beta}^T)^T$ is

$$\begin{aligned} \boldsymbol{J}(\gamma, \psi, \boldsymbol{\beta}) = \text{diag}(&\frac{1}{2}\boldsymbol{\Psi}^T \boldsymbol{M}^{-1}(\boldsymbol{I}_n - \mathbf{1}\mathbf{1}^T/n)\boldsymbol{M}^{-1}\boldsymbol{\Psi}, \\ &\frac{1}{2}n\psi^{-2}, \quad \psi m_g^{-1}\dot{\boldsymbol{\theta}}^T \boldsymbol{V}\boldsymbol{M}\dot{\boldsymbol{\theta}}). \end{aligned} \tag{6.89}$$

Proof. We use Lemma 6.3 to find parameters $\boldsymbol{\lambda}$. For normal and inverse Gaussian models, $s(\phi) = -\log\phi$, so the Fisher information matrix (6.68) becomes

$$\boldsymbol{J}(\gamma, \phi, \boldsymbol{\beta}) = \begin{pmatrix} \frac{1}{2}\boldsymbol{\Psi}^T \boldsymbol{M}^{-2}\boldsymbol{\Psi} & \frac{1}{2}\phi^{-1}\boldsymbol{\Psi}^T \boldsymbol{M}^{-1}\mathbf{1} & \mathbf{0} \\ \frac{1}{2}\phi^{-1}\mathbf{1}^T \boldsymbol{M}^{-1}\boldsymbol{\Psi} & \frac{1}{2}n\phi^{-2} & \mathbf{0} \\ \mathbf{0} & \mathbf{0} & \phi\dot{\boldsymbol{\theta}}^T \boldsymbol{V}\boldsymbol{M}\dot{\boldsymbol{\theta}} \end{pmatrix}.$$

Since $\boldsymbol{\alpha}_1 = \gamma$, $\boldsymbol{\alpha}_2 = (\phi, \boldsymbol{\beta}^T)^T$, equation (6.72) reduces to

$$\frac{\partial \phi}{\partial \gamma^T} = -\frac{1}{n}\phi \mathbf{1}^T \boldsymbol{M}^{-1}\boldsymbol{\Psi}, \qquad \frac{\partial \boldsymbol{\beta}}{\partial \gamma^T} = \mathbf{0}.$$

It follows from $\boldsymbol{\Psi} = (\Psi_{ia}) = (\partial m_i / \partial \gamma_a)$ that

$$\begin{aligned} \frac{\partial \phi}{\partial \gamma_a} &= -\frac{1}{n}\phi \sum_{i=1}^{n} m_i^{-1} \frac{\partial m_i}{\partial \gamma_a}, \\ \phi^{-1}\frac{\partial \phi}{\partial \gamma_a} &= -\frac{1}{n}\sum_{i=1}^{n} m_i^{-1} \frac{\partial m_i}{\partial \gamma_a}, \end{aligned}$$

which leads to

$$\frac{\partial \log \phi}{\partial \gamma_a} = -\frac{1}{n}\sum_{i=1}^{n} \frac{\partial \log m_i}{\partial \gamma_a} = \frac{\partial}{\partial \gamma_a} \log\left(\prod_{i=1}^{n} m_i\right)^{-\frac{1}{n}}.$$

From this we get $\phi = \psi m_g^{-1}(\gamma)$ that results in (6.88). By direct calculation from (6.73) and (6.74), we get (6.89). ∎

From this lemma, we can get the adjusted likelihood ratio statistic.

Theorem 6.17

For normal and inverse Gaussian models, the adjusted likelihood ratio statistic for H_0 in model (6.59) can be expressed as

$$\begin{aligned} LR_A &= (1-\frac{p-2}{n})LR - \log[\det\{\boldsymbol{R}^2(\hat{\gamma})(\boldsymbol{I}_p - \boldsymbol{B}(\hat{\gamma}))m_g^{-1}(\hat{\gamma})\}] + \\ & \quad \log[\det\{\boldsymbol{R}^2(\gamma_0)(\boldsymbol{I}_p - \boldsymbol{B}(\gamma_0))\}], \end{aligned} \tag{6.90}$$

where LR is given in (6.64), $\boldsymbol{R}(\gamma)$ is given by the following Cholesky decomposition

$$\dot{\boldsymbol{\theta}}^T \boldsymbol{V}^{\frac{1}{2}} \boldsymbol{M}(\gamma) \boldsymbol{V}^{\frac{1}{2}} \dot{\boldsymbol{\theta}} = \boldsymbol{R}^T(\hat{\gamma})\boldsymbol{R}(\hat{\gamma})$$

and $\boldsymbol{B}(\gamma)$ is defined as

$$\boldsymbol{B}(\gamma) = \{[\boldsymbol{e}^T \boldsymbol{M}(\gamma)][\boldsymbol{R}^{-T}(\gamma)\ddot{\boldsymbol{\theta}}\boldsymbol{R}^{-1}(\gamma)]\}_{\tilde{\alpha}}.$$

Proof. To get ΔLR shown in (6.78), we must use (6.75). It follows from Lemma 6.5 that $\boldsymbol{\alpha}_1 = \gamma$, $\boldsymbol{\lambda} = (\psi,\ \boldsymbol{\beta}^T)^T$ and

$$\boldsymbol{\alpha}_2 = (\phi,\ \boldsymbol{\beta}^T)^T = (\psi m_g^{-1}(\gamma),\ \boldsymbol{\beta}^T)^T,$$

we have

$$\frac{\partial \boldsymbol{\alpha}_2}{\partial \boldsymbol{\lambda}^T} = \text{diag}(m_g^{-1}(\gamma),\ \boldsymbol{I}_p).$$

By direct calculation from (6.59) and (6.60), we can get $-\ddot{\boldsymbol{l}}_{\phi\phi}$, $-\ddot{\boldsymbol{l}}_{\phi\beta}$ and $-\ddot{\boldsymbol{l}}_{\beta\beta}$, which result in

$$\begin{aligned} (-\ddot{\boldsymbol{l}}_{22})_{\tilde{\alpha}} &= \begin{pmatrix} \frac{1}{2}n\phi^{-2} & \boldsymbol{0} \\ 0 & \phi\{\dot{\boldsymbol{\theta}}^T \boldsymbol{V}^{\frac{1}{2}} \boldsymbol{M}(\gamma) \boldsymbol{V}^{\frac{1}{2}} \dot{\boldsymbol{\theta}} - [\boldsymbol{e}^T \boldsymbol{M}][\ddot{\boldsymbol{\theta}}]\} \end{pmatrix}_{\tilde{\alpha}} \\ &= \begin{pmatrix} \frac{1}{2}n\phi^{-2} & \boldsymbol{0} \\ 0 & \phi \boldsymbol{R}^T(\gamma)(\boldsymbol{I}_p - \boldsymbol{B}(\gamma))\boldsymbol{R}(\gamma) \end{pmatrix}_{\tilde{\alpha}}. \end{aligned}$$

So it follows from (6.75) that

$$\begin{aligned} \det\{-\ddot{\boldsymbol{l}}^*_{\lambda\lambda}(\boldsymbol{\alpha}_1, \tilde{\boldsymbol{\lambda}}(\boldsymbol{\alpha}_1))\} &= \frac{1}{2}n\phi^{-2}m_g^{-2}(\gamma)\det\{\phi \boldsymbol{R}^T(\gamma)(\boldsymbol{I}_p - \boldsymbol{B}(\gamma))\boldsymbol{R}(\gamma)\} \\ &= \frac{1}{2}n\{\phi m_g(\gamma)\}^{p-2}\det\{\boldsymbol{R}^2(\gamma)(\boldsymbol{I}_p - \boldsymbol{B}(\gamma))m_g^{-1}(\gamma)\}. \end{aligned}$$

Substituting this result into (6.78), and taking $\boldsymbol{\alpha} = \hat{\boldsymbol{\alpha}}$ and $\boldsymbol{\alpha} = \hat{\boldsymbol{\alpha}}_0$, respectively, we get (6.90). ∎

As the special case of Theorem 6.17, (6.90) can be applied to the test for heteroscedasticity in linear and nonlinear regression. In particular, for linear regression, it is consistent with equation (8) of Simonoff and Tsai (1994), as expected.

6.6.3 Examples

We study two numerical examples as the illustration for the results of the test for varying dispersion.

Example 6.8 (European rabbit data No. 2).

These data have been studied in Example 2.4. Here we consider both the test for heteroscedasticity in normal nonlinear regression and the test for varying dispersion in inverse Gaussian nonlinear model. Ratkowsky (1983) studied these data based on the ordinary nonlinear least squares estimator with constant variance using

$$\mu_i = \alpha - \beta(x_i + \delta)^{-1}, \quad i = 1, \cdots, 71. \tag{6.91}$$

However, the plot of the studentized residual r_i against the covariate x_i (shown in Figure 6.4) displays an obvious discrepancy. The left opening megaphone of the pattern suggests that the variance decreases with x_i. This may be evidence that covariate x_i causes the nonconstant variance (McCullagh and Nelder 1989, p.392).

To test varying dispersion (here it is just the test for heteroscedasticity), we use model (6.59), that is $y_i \sim N(\mu_i, \sigma^2 m_i^{-1})$ with (6.91), $m_i = x_i^\gamma$, $i = 1, \cdots, 71$ and $H_0 : \gamma = 0$. For this test, the likelihood ratio statistic, the score statistic and their adjusted versions are computed from (6.64), (6.70), (6.87) and (6.90), respectively. The results are listed in Table 6.8.

Table 6.8 Score statistics and likelihood ratio statistics for European rabbit data

Distribution	SC	SC_A	LR	LR_A
Normal	10.2610	11.2274	10.2502	10.9168
Inverse Gaussian	37.2635	39.4696	34.7999	35.9633

$\chi^2_{0.01}(1) = 6.635 \quad \chi^2_{0.05}(1) = 3.841$

Table 6.8 and Figure 6.4 give definite evidence of heteroscedasticity. Therefore we may apply model (6.59) with (6.91) and $m_i = x_i^\gamma$ to fit the data and expect to get more preferable results. It is easy to get $\hat{\gamma} = 0.5327 \approx 0.5$, $\hat{\sigma}^{-2} = 20.566 \approx 20$ and $m_i \approx \sqrt{x_i}$ which vary from observation to observation. The variance function of y_i may be written as $\mathrm{Var}(y_i) = \sigma^2 m_i^{-1} \approx (20\sqrt{x_i})^{-1}$, which is inversely proportional to $\sqrt{x_i}$. The associated plot of studentized residuals r_i' versus the covariates x_i based on this model is given in Figure 6.5, as constant dispersion. So model (6.59) more adequately fits these data.

As an alternative, we use model (6.91) with inverse Gaussian error to fit the data. Both the test statistics (shown in Table 6.8) and the residual plot (which is quite similar to Figure 6.4, so we have omitted it here) give definite evidence of varying dispersion. So model (6.59) with $m_i = x_i^\gamma$ may more adequately fit the data. In this case we have $\hat{\gamma} = 0.92$, $\hat{\sigma}^2 = 0.0032$ and $m_i = x_i^{0.92}$ approximately equal to the covariate x_i itself, that is $m_i \approx x_i$.

Example 6.9 (Grass yield data No. 2).

These data have been discussed in Example 6.2. By the analysis shown in McCullagh and Nelder (1989, p.383), there are two problems with these data: (a) the value 2.94 for levels (0, 3, 2) might be a possible outlier and may be replaced by 2.49. The results given in Example 6.2 have confirmed this fact. In the following, we shall delete this point (case 15) or change the value 2.94 for 2.49 (for these two situations, the results are almost the same, see Table 6.9); and (b) the standard errors for $\hat{\alpha}_2$ and $\hat{\alpha}_3$ are too large, which may be caused by varying dispersion and leads to impossible negative values of α_2 and α_3 in confidence intervals. To test for a departure from constant dispersion in model (6.28), we use model (6.59), that is $y_i \sim GA(\mu_i, \sigma^2 m_i^{-1})$ with a factor function $m_i = \exp\{\sum_j z_{ij}\gamma_j\}$ and $H_0 : \gamma_j = 0$ for all js. Table 6.9 gives the results of a score test based on formulas (6.66) and (6.80) with various zs. The score statistics with $z = (x_1, x_2, x_3)$ are large, giving a definite evidence for varying dispersion. Moreover, Table 6.9 shows that the test statistics corresponding to x_2 are about 5.1 and are much larger than the critical value 3.841, but this is not true for x_1 and x_3. So the factor function $m(\cdot;\cdot)$ may be connected with x_2 and the effects of x_1 and x_3 are relatively small. Besides, as a comparison, Table 6.9 also gives the score statistics with 2.49 replacing 2.94 in case 15. The results show that the difference between these two cases is less significant.

Table 6.9 Score test for grass yield data

z	d.f.	Deleting 2.94		Changing 2.94 for 2.49	
		SC	SC_A	SC	SC_A
x_1, x_2, x_3	3	9.5192	8.8529	10.1530	9.4561
x_1, x_2	2	8.6805	7.9826	9.3250	8.5979
x_2, x_3	2	5.9399	5.8884	6.4936	6.4608
x_2	1	5.1838	5.0989	5.5881	5.5255
x_1, x_3	2	4.6763	4.0706	4.4872	3.8534
x_1	1	3.8735	3.2273	3.6092	2.9504
x_3	1	0.8560	0.8889	0.8166	0.8465

$\chi^2_{0.05}(1) = 3.841$ $\chi^2_{0.05}(2) = 5.991$ $\chi^2_{0.05}(3) = 7.815$
d.f. = degrees of freedom

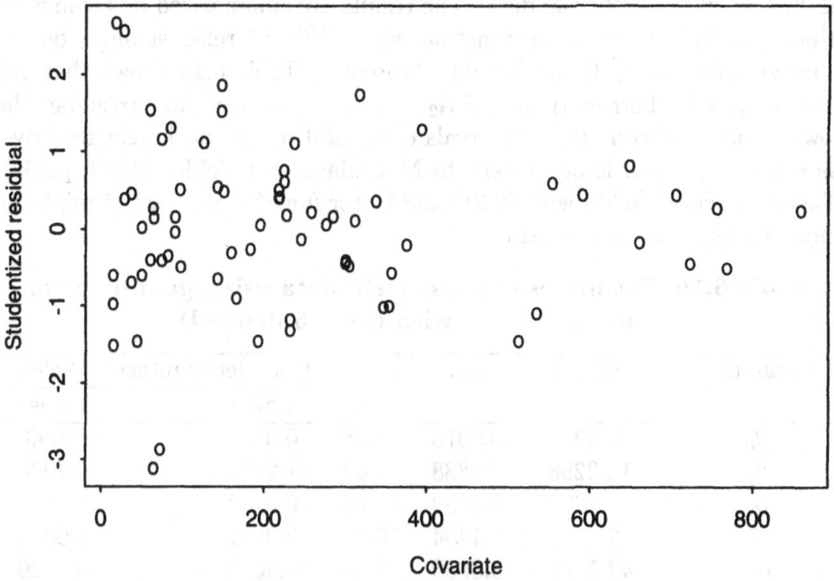

Figure 6.4 PLOT OF STUDENTIZED RESIDUAL r_i VERSUS COVARIATE x_i FOR EUROPEAN RABBIT DATA USING MODEL (6.91)

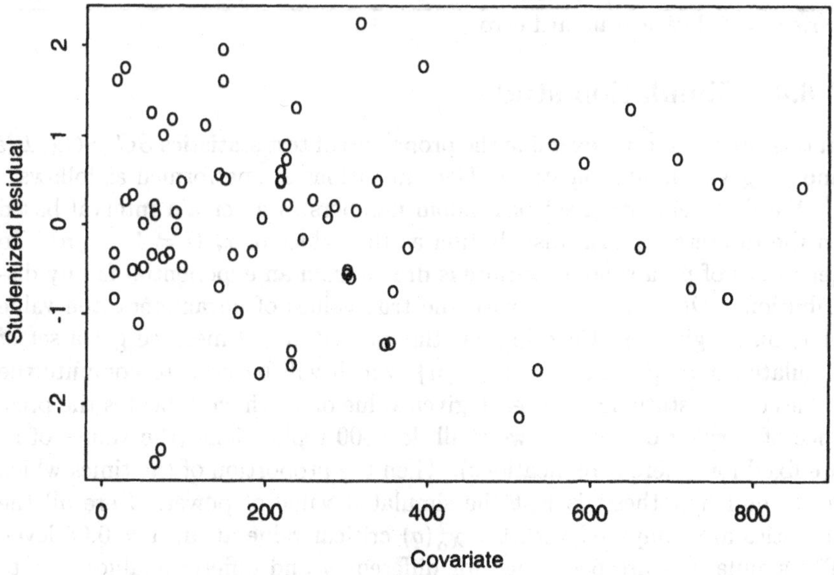

Figure 6.5 PLOT OF STUDENTIZED RESIDUAL r_i' VERSUS COVARIATE x_i FOR EUROPEAN RABBIT DATA USING MODEL (6.91) WITH (6.59)

To investigate the data further, we now apply model (6.59) with (6.28) and $m_i = e^{x_{i2}\gamma}$ to refit the data. The results are summarized in Table 6.10. Since $\hat{\gamma} = 0.0195$, the factor function $m_i = e^{0.0195x_{i2}}$ relies strongly on the nonzero value of x_2 (phosphorus). Moreover, Table 6.10 shows that the relative standard errors (r.s.e.) of $\hat{\alpha}_2$ and $\hat{\alpha}_3$ are reduced, in particular, the lower bounds of confidence intervals of α_2 and α_3 are no longer negative, as expected, which is of concern to McCullagh and Nelder (1989, p.383). Therefore model (6.59) with (6.28) and factor function $m_i = e^{x_{i2}\gamma}$ might be more adequate to fit the data.

Table 6.10 Estimates of grass yield data using gamma error and $m_i = e^{x_{i2}\gamma}$ (with case 15 deleted)

Parameter	Estimate	s.e.	r.s.e.	Confidence interval (95%)	
			(%)	Lower	Upper
β_0	0.0909	0.0078	8.58	0.0753	0.1066
β_1	13.2298	0.9838	7.44	11.2590	15.2006
β_2	0.7897	0.3999	50.64	-0.0113	1.5908
β_3	2.2978	1.1494	50.02	-0.0047	4.6003
α_1	44.5483	3.0669	6.88	38.4045	50.6920
α_2	17.4188	7.8220	44.91	1.7495	33.0881
α_3	49.2697	19.4960	39.57	10.2145	88.3249
	$\hat{\gamma} = 0.0195$			$\hat{\sigma}^2$=0.0046	

r.e.s. = relative standard error

6.6.4 Simulation study

In this subsection, we examine the properties of test statistics SC, SC_A, LR and LR_A via simulation study. The simulations are performed as follows.

We first generate a set of random numbers in a certain interval based on the discrete uniform distribution as the values of x_i $(i = 1, \cdots, n)$. To get values of y_i, a random variate is drawn from an exponential family distribution $ED(\mu_i,\ (\phi m_i)^{-1})$ with the true values of parameters, the value of x_i and a given γ. Carrying out this procedure n times, we get a set of simulated data $\{y_i,\ x_i,\ i = 1, \cdots, n\}$, which will be used to compute the values of test statistics. For each given value of γ, which indicates the presence of varying dispersion, we shall do 5000 replications (the values of x_i are fixed for different replications). Then the proportion of the times which reject null hypothesis is just the simulated value of power. Here all the statistics are compared with the $\chi^2_\alpha(q)$ critical value at an $\alpha = 0.05$ level. The simulations are performed for different n and different values of γ to get simulated power functions of various test statistics. We shall show some representative results by figures. It should be noted that the simulation study is based on a fact that the standard test statistic LR is asymptotically chi-squared. If the underlying distribution of LR is very far from the chi-squared distribution, then the simulation study will not make any sense. To guarantee this, the adequate sample size n in nonlinear models usually should be relatively larger than that in linear models. For normal, gamma and inverse Gaussian nonlinear models, the simulation results are respectively presented as follows.

(a) Normal distribution

We use the model of Ratkowsky (1983, p.108) to do the simulation study. As a usual way, the estimated values of parameters are treated as true values in simulation studies. The model is

$$\begin{cases} y_i \sim N(\mu_i,\ (\phi m_i)^{-1}), \\ \mu_i = \alpha - \beta(x_i + \delta)^{-1}, \ \text{ and } m_i = x_i^{\gamma}, \end{cases} \tag{6.92}$$

for $i = 1, \cdots, n$, and the true values are respectively $\alpha = 5.6$, $\beta = 130$, $\delta = 37.6$ and $\phi = 20$.

We first generate a set of random numbers from a discrete uniform distribution in the interval [11 900] as the values of x_i. For getting values of y_i, a random variate is drawn from the standard normal distribution, and then transformed to $N(\mu_i,\ (\phi m_i)^{-1})$ with the true values of parameters, the value of x_i and a given γ. Repeating this procedure n times, we get a set of simulated data $\{y_i,\ x_i,\ i = 1, \cdots, n\}$. The values of test statistics LR_A, SC_A, LR and SC are computed by the formulas shown in previous subsections. Then, for each given value of γ, the simulated value of power is obtained via 5000 replications. Some representative results are shown in Figure 6.6. Figure 6.6(a) displays the plots of power functions of the tests with respect to varying values of γ with sample size $n = 100$. The plots show that the powers of tests LR_A and SC_A are larger than those of LR and SC in almost every situation. It means that the likelihood ratio test and the score test based on the adjusted profile likelihood keep their size better and are more powerful than unadjusted versions for the tests of varying dispersion, as expected. This conclusion coincides with the previous theoretical results discussed in Subsection 6.6.2 and also coincides with the results of linear regression models given by Simonoff and Tsai (1994). Furthermore, Figure 6.6(a) shows that the difference between LR and SC is quite small, and also between LR_A and SC_A, but the differences between the adjusted tests and the unadjusted tests are significant. However, when n is getting larger, the difference between the adjusted one and the unadjusted one will be getting less significant. Figure 6.6(b), in which $n = 200$, justifies this fact. This fact was also mentioned by Simonoff and Tsai (1994) for linear regression models.

(b) Gamma distribution

The simulation structure is very similar to that described in part (a) except that the simulated values y_is are generated from gamma distribution. So we will skip the details. The model we used to do simulation is also similar to (6.92), that is $y_i \sim GA(\mu_i,\ (\phi m_i)^{-1})$ with $\mu_i = \alpha - \beta(x_i + \delta)^{-1}$ and $m_i = x_i^\gamma$ for $i = 1, \cdots, n$. The true values of α, β, δ and γ are the same as those given in part (a).

To compute SC and SC_A by (6.66) and (6.80), we need to compute $\ddot{s}(\phi)$, where $s(\phi) = -2\{\phi \log \phi - \log \Gamma(\phi)\}$. By Abramowitz and Stegun (1965, Ch.6) or Lawless (1981, p.512), we have

$$\begin{aligned}\frac{\partial \log \Gamma(\phi)}{\partial \phi} &= \log \phi - \frac{1}{2\phi} - \frac{1}{12\phi^2} + \frac{1}{120\phi^4} - \frac{1}{252\phi^6} + \cdots, \\ \ddot{s}(\phi) &= \frac{1}{\phi^2} + \frac{1}{3\phi^3} - \frac{1}{15\phi^5} + \frac{1}{21\phi^7} + \cdots.\end{aligned}$$

Then the test statistics SC and SC_A can be computed by using (6.66), (6.80) and the above two formulas. For different n and different values of γ, simulated power functions of SC and SC_A are obtained by 5000 replications. A representative result is shown in Figure 6.7(a) with sample size $n = 200$. Here, the plots of power functions for the tests with respect to varying values of γ are displayed. The figure shows that the powers of test SC_A are larger than that of SC in almost every situation. So the adjusted test is also better than the unadjusted test in gamma nonlinear model. In Figure 6.7(b) we give the plots of power functions corresponding to $n = 400$ which show that the difference between SC and SC_A is getting less significant. All these results are consistent with those of part (a), as expected.

(c) Inverse Gaussian distribution

Using similar procedure as described in parts (a) and (b), we now study the simulation result for inverse Gaussian nonlinear model. It should be noted that since the random variate of inverse Gaussian distribution is not easily generated from the common software packages, such as IMSL, we use the algorithm which is described in detail by Seshadri (1993, pp.203–4) to generate the values for inverse Gaussian random variables y_i. The model is $y_i \sim IG(\mu_i,\ (\phi m_i)^{-1})$ with the same expressions of μ_i and m_i as model (6.92). Notice that the variance of inverse Gaussian variable is quite sensitive to the mean value (up to cubic, see Table 1.1), so we choose $\phi = 300$. The true values of other parameters are the same as before. The simulation results are very similar to parts (a) and (b). Figure 6.8 is the plots of power functions with sample size $n = 300$. It shows that the powers of adjusted tests are also better than unadjusted tests in inverse Gaussian nonlinear model. The other conclusions are similar to the previous two parts.

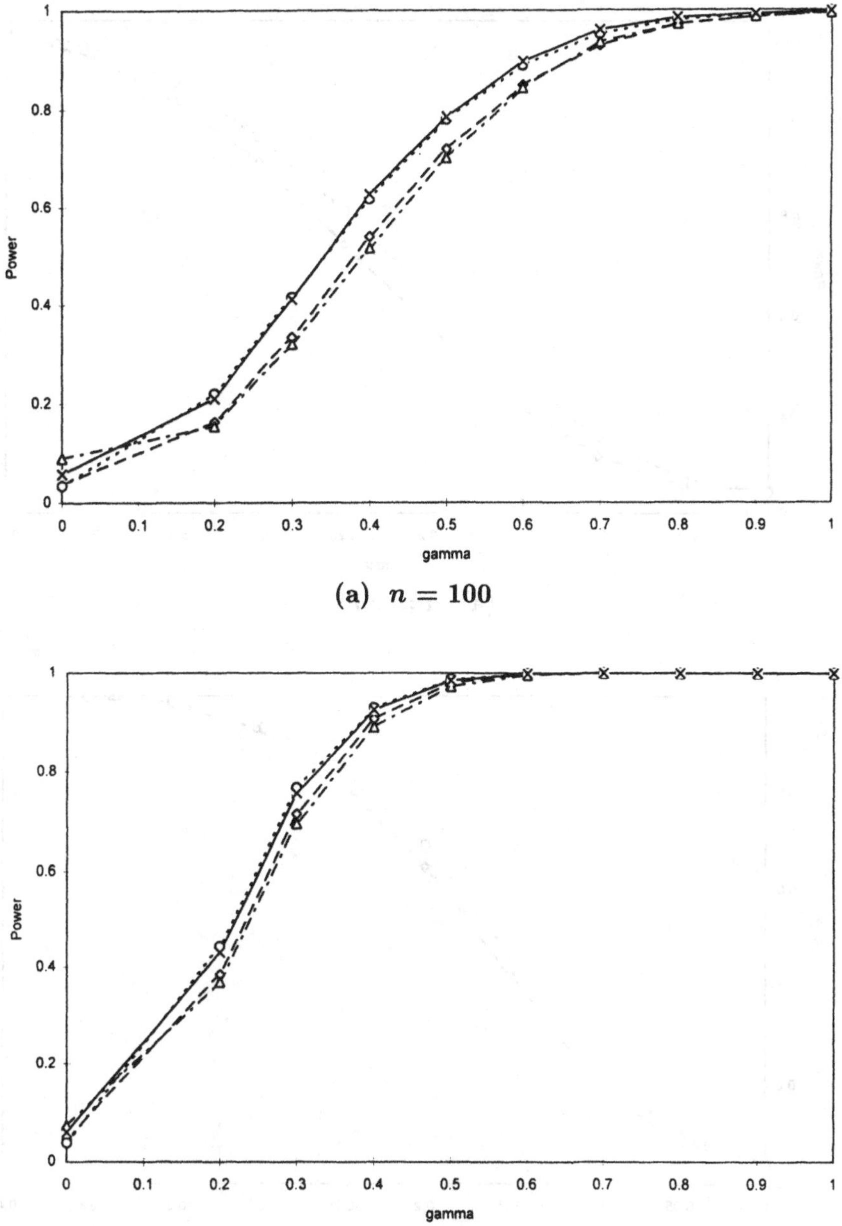

(a) $n = 100$

(b) $n = 200$

Figure 6.6 SIMULATED POWER FUNCTIONS OF TEST STATISTICS FOR NORMAL NONLINEAR MODELS WITH (a), $n = 100$ AND (b), $n = 200$.
HERE: "——" and "×" represent the adjusted likelihood ratio test; "....." and "∘" represent the adjusted score test; "— · —" and "Δ" represent the likelihood ratio test; "– – –" and "⋄" represent the score test.

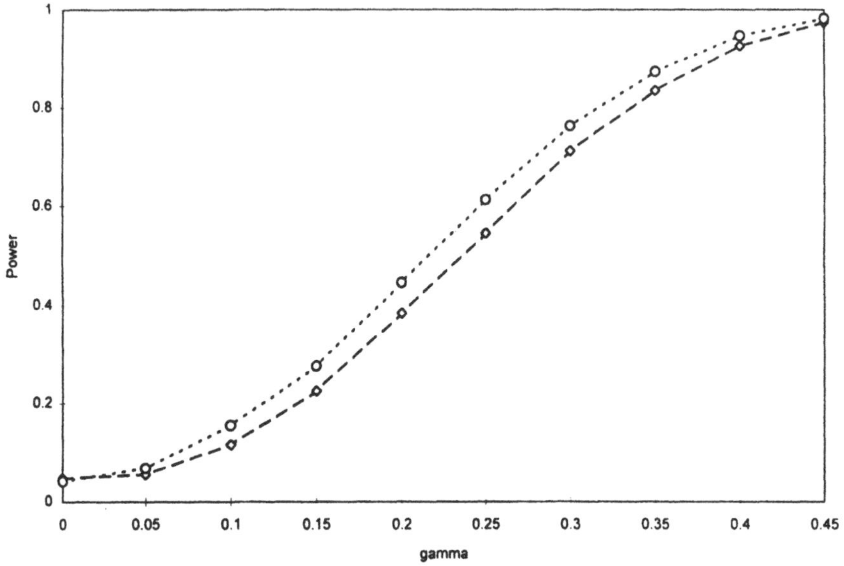

(a) $n = 200$

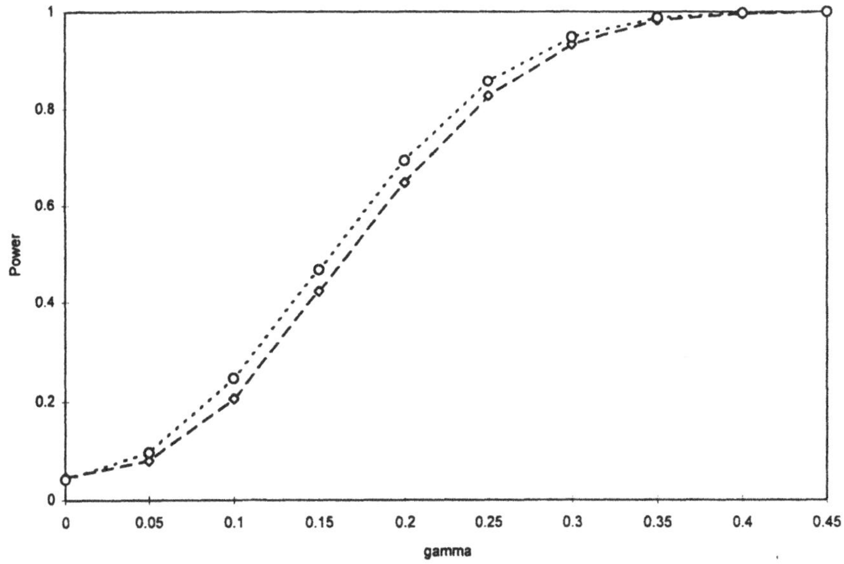

(b) $n = 400$

Figure 6.7 SIMULATED POWER FUNCTIONS OF TEST STATISTICS FOR GAMMA NONLINEAR MODELS WITH (a), $n = 200$ AND (b), $n = 400$.
HERE: "$\cdot\cdot\cdot\cdot\cdot\cdot$" and "$\circ$" represent the adjusted score test; "$- - -$" and "$\diamond$" represent the score test.

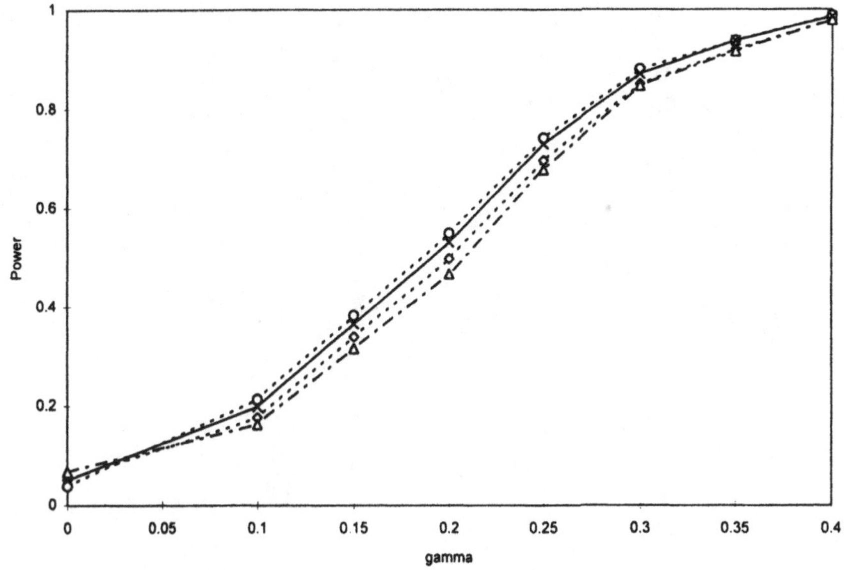

Figure 6.8 SIMULATED POWER FUNCTIONS OF TEST STATISTICS FOR INVERSE GAUSSIAN NONLINEAR MODELS WITH $n = 300$.
HERE: "——" and "×" represent the adjusted likelihood ratio test; "……" and "o" represent the adjusted score test; "— · —" and "Δ" represent the likelihood ratio test; "– – –" and "◇" represent the score test.

Chapter 7

Extension

In this chapter, we explore the possibility of extending the materials discussed in Chapters 3 to 6 to some other models. We focus on the description of the usefulness of our modified BW geometric framework to some specific models, rather than technical details because they are similar to those introduced in Chapters 3 to 6. Section 7.1 discusses curved exponential families with iid observations which were studied by Efron (1975) first and developed by Amari (1982a; 1985) by introducing a Riemannian geometric framework. As an alternative, we introduce a modified BW geometric framework as shown in Chapter 3 in Euclidean space for these models. All the results shown in Chapters 3 to 5 still hold for curved exponential families except for some minor changes (because here we treat iid observations). These results are consistent with those of Efron (1975) and Amari (1985). The multinomial nonlinear models (Rao 1973, p.392; Kass 1989; McCullagh and Nelder 1989, p211) may be regarded as specific curved exponential families, but with a constraint (because the sum of attribute frequencies is equal to unit). In Section 7.2 we deal with these models by using modified BW geometric framework. It is shown that the modified BW geometric framework can be applied to the models much more generally than normal nonlinear regression models. Section 7.3 deals with a class of rather general models, the embedded models in regular parametric families. Section 7.4 introduces our modified BW geometric framework to quasi-likelihood models (Wedderburn 1974; McCullagh 1991). As an application of quasi-likelihood models, we study covariance structure models (Lee and Jennrich 1979; Lee and Bentler 1992) by using the modified BW geometric framework. We show that there are good geometric properties in covariance structure analysis, even though the normality assumption may not hold.

7.1 Curved Exponential Families

7.1.1 Introduction and geometric framework

We have briefly introduced curved exponential families in Section 2.1. Let $\boldsymbol{Y}^m = (y_1^m, \cdots, y_n^m)^T$, $m = 1, \cdots, M$, be iid observations from family (1.1) with $\boldsymbol{\theta} = \boldsymbol{\theta}(\boldsymbol{\beta})$, where $\boldsymbol{\beta} = (\beta_1, \cdots, \beta_p)^T$ are the parameters of interest, $\boldsymbol{\theta}$ is an n-dimensional natural parameter (n is fixed here) and M is the sample size. We study statistical behavior when $M \to +\infty$. For curved exponential families (1.8) still holds for each m so that

$$\mathrm{E}(\boldsymbol{Y}^m) = \dot{b}(\boldsymbol{\theta}) = \boldsymbol{\mu}(\boldsymbol{\beta}), \quad \mathrm{Var}(\boldsymbol{Y}^m) = \sigma^2 \ddot{b}(\boldsymbol{\theta}) = \sigma^2 \boldsymbol{V}(\boldsymbol{\beta}).$$

But unlike exponential family nonlinear models, here $\boldsymbol{V}(\boldsymbol{\beta})$ is not necessarily diagonal. Moreover, the joint distribution of $\boldsymbol{Y}^1, \cdots, \boldsymbol{Y}^M$ is

$$p(\boldsymbol{y}^1, \cdots \boldsymbol{y}^M;\ \boldsymbol{\theta}, \phi) = \exp[M\phi\{\overline{\boldsymbol{y}}^T\boldsymbol{\theta} - b(\boldsymbol{\theta}) - \overline{c}\}], \quad \boldsymbol{\theta} = \boldsymbol{\theta}(\boldsymbol{\beta}), \tag{7.1}$$

where

$$\overline{\boldsymbol{y}} = \frac{1}{M}\sum_{m=1}^{M} \boldsymbol{y}^m, \quad \overline{c} = \frac{1}{M}\sum_{m=1}^{M} c(\boldsymbol{y}^m, \phi).$$

Equation (7.1) shows that the sufficient statistic $\overline{\boldsymbol{Y}}$ plays the role of observed vector and all the statistical inference can be made based on it. Since $\mathrm{E}(\overline{\boldsymbol{Y}}) = \boldsymbol{\mu}(\boldsymbol{\beta})$, by central limit theorem we have

$$\boldsymbol{e}(\boldsymbol{\beta}) \triangleq \sqrt{M}(\overline{\boldsymbol{Y}} - \boldsymbol{\mu}(\boldsymbol{\beta})) \xrightarrow{L} N(\boldsymbol{0}, \sigma^2 \boldsymbol{V}(\boldsymbol{\beta})), \quad M \to \infty \tag{7.2}$$

for any fixed ϕ and $\boldsymbol{\beta}$. Here $\boldsymbol{e}(\boldsymbol{\beta})$ can be regarded as a random error having order $\boldsymbol{e}(\boldsymbol{\beta}) = O_p(1)$.

It easily follows from (7.1) that the score function, the observed information and the Fisher information of $\boldsymbol{Y}^1, \cdots, \boldsymbol{Y}^M$ for $\boldsymbol{\beta}$ are

$$\begin{aligned} \dot{\boldsymbol{l}}(\boldsymbol{\beta}) &= \sqrt{M}\phi \boldsymbol{D}^T \boldsymbol{V}^{-1} \boldsymbol{e} = \sqrt{M}\phi \boldsymbol{D}_\theta^T \boldsymbol{e}, & (7.3)\\ -\ddot{\boldsymbol{l}}(\boldsymbol{\beta}) &= M\phi\{\boldsymbol{D}_\theta^T \boldsymbol{V} \boldsymbol{D}_\theta - [\boldsymbol{e}^T][\boldsymbol{W}_\theta]/\sqrt{M}\}, & (7.4) \end{aligned}$$

and

$$\boldsymbol{J}(\boldsymbol{\beta}) = M\phi \boldsymbol{D}_\theta^T \boldsymbol{V} \boldsymbol{D}_\theta = M\phi \boldsymbol{D}^T \boldsymbol{V}^{-1} \boldsymbol{D}, \tag{7.5}$$

where $\boldsymbol{D}$, $\boldsymbol{D}_\theta$ and $\boldsymbol{W}_\theta$ are defined in Section 2.2. Besides, the Fisher information matrices of $\boldsymbol{Y}^1, \cdots, \boldsymbol{Y}^M$ for $\boldsymbol{\theta}$ and $\boldsymbol{\mu}$ are $M\phi\boldsymbol{V}$ and $M\phi\boldsymbol{V}^{-1}$, respectively. We assume that $\boldsymbol{\theta} = \boldsymbol{\theta}(\boldsymbol{\beta})$ has finite and continuous derivatives up to the third order. From the above equations, we can directly get $\dot{\boldsymbol{l}}(\boldsymbol{\beta}) = O_p(\sqrt{M})$, $\ddot{\boldsymbol{l}}(\boldsymbol{\beta}) = O_p(M)$ and $M^{-1}\boldsymbol{J}(\boldsymbol{\beta}) = \phi \boldsymbol{D}^T \boldsymbol{V}^{-1} \boldsymbol{D} > 0$. Here most of the regularity conditions shown in Chapters 2 and 4 (such as (4.1)) are no longer needed. Equations (7.4) and (7.5) are very similar to equations (2.8), (2.9) and (2.11) except for having the sample size M. So it is expected that model (7.1) has an analogous geometric structure to exponential family nonlinear models (2.2).

The following is a brief outline of how the geometric framework can be established for model (7.1). It is easily seen from (7.4) that the maximum likelihood estimator $\hat{\boldsymbol{\beta}}$ of $\boldsymbol{\beta}$ must satisfy

$$\boldsymbol{D}^T(\hat{\boldsymbol{\beta}})\boldsymbol{V}^{-1}(\hat{\boldsymbol{\beta}})\boldsymbol{e}(\hat{\boldsymbol{\beta}}) = \boldsymbol{0}, \qquad \boldsymbol{e}(\boldsymbol{\beta}) = \sqrt{M}(\overline{\boldsymbol{Y}} - \boldsymbol{\mu}(\boldsymbol{\beta})); \tag{7.6}$$

$$\boldsymbol{D}_\theta^T(\boldsymbol{\beta})\boldsymbol{V}(\boldsymbol{\beta})\boldsymbol{e}^*(\hat{\boldsymbol{\beta}}) = \boldsymbol{0}, \qquad \boldsymbol{e}^*(\boldsymbol{\beta}) = \sqrt{M}(\overline{\boldsymbol{Y}}^* - \boldsymbol{\theta}(\boldsymbol{\beta})); \tag{7.7}$$

where $\boldsymbol{e}^*(\boldsymbol{\beta}) = \boldsymbol{V}^{-1}(\boldsymbol{\beta})\boldsymbol{e}(\boldsymbol{\beta})$ and $\overline{\boldsymbol{Y}}^* = \boldsymbol{\theta}(\boldsymbol{\beta}) + \boldsymbol{V}^{-1}(\overline{\boldsymbol{Y}} - \boldsymbol{\mu}(\boldsymbol{\beta}))$. These equations exactly coincide with equations (3.1) and (3.21). So we can introduce the geometric framework for model (7.1) both in expectation parameter space and natural parameter space, as in Sections 3.2 and 3.3. More concretely, since $\mathrm{E}\overline{\boldsymbol{Y}} = \boldsymbol{\mu}(\boldsymbol{\beta})$, $\mathrm{E}\overline{\boldsymbol{Y}}^* = \boldsymbol{\theta}(\boldsymbol{\beta})$, the solution loci $\boldsymbol{\mu} = \boldsymbol{\mu}(\boldsymbol{\beta})$, $\boldsymbol{\theta} = \boldsymbol{\theta}(\boldsymbol{\beta})$ and their tangent spaces T_β, $T(\boldsymbol{\beta})$ are the same as those defined in Sections 3.2 and 3.3. Then the definitions of curvatures given in (3.5) and (3.23) are still valid for model (7.1) without any change.

Notice that even though the geometric framework of model (7.1) seems the same as that of model (2.2), there are some substantial differences between these two models in statistical inference. The essential one is that here we deal with iid observations and the solution loci $\boldsymbol{\mu} = \boldsymbol{\mu}(\boldsymbol{\beta})$, $\boldsymbol{\theta} = \boldsymbol{\theta}(\boldsymbol{\beta})$ are independent of the sample size M while (2.2) is a regression model and all the geometric quantities are connected with the sample size n. This fact results in the following: (a) curvature arrays defined in (3.5) and (3.23) for model (7.1) are independent of the sample size M (here n is no longer the sample size, but is the dimensions of $\overline{\boldsymbol{Y}}$ and $\boldsymbol{\theta}$); (b) most of the regularity conditions added to (2.2) are no longer needed for (7.1), for example, condition (b) of Assumption C (see Chapter 4) becomes $M^{-1}\boldsymbol{J}(\boldsymbol{\beta}) = \phi \boldsymbol{D}^T\boldsymbol{V}^{-1}\boldsymbol{D}$ that always exists and is positive definite (see (7.5)); and (c) some results such as Theorem 4.1 become a well-known fact for model (7.1) since $\boldsymbol{Y}^1, \cdots, \boldsymbol{Y}^M$ are iid observations.

7.1.2 Statistical analysis

The iid observations make the problems easier to deal with, so almost all the results shown in Chapters 3 to 5 still hold for model (7.1) except for some minor changes. In the following, we describe the basic ideas.

To study asymptotic properties of $\hat{\boldsymbol{\beta}}$, we need the following lemma.

Lemma 7.1

The stochastic expansion of $\hat{\boldsymbol{\beta}}$ can be expressed as

$$\Delta\beta = \frac{1}{\sqrt{M}} L\tau + \frac{1}{M} L\{[\lambda^T][A_\theta^I]\tau - \frac{1}{2}\tau^T A^P \tau\} + O_p(M^{-3/2}), \qquad (7.8)$$

where all the notation are the same as those given in Theorem 4.2 except that in the expressions $\tau = Q^T V^{-1} e$ and $\lambda = N^T V^{-1} e$, e is defined by $\sqrt{M}(\overline{Y} - \mu(\beta))$ in (7.2), but is not equal to $Y - \mu(\beta)$ defined in (4.7).

Proof. It is easily seen that equations (4.2) to (4.4) still hold for model (7.1). From (7.2), the counterpart of (4.7) is

$$e = \hat{e} + \sqrt{M}\Delta\mu.$$

It follows from (4.2), (4.3) and $\sqrt{M}(\hat{\beta} - \beta_0) \xrightarrow{L} N(0,\ \sigma^2(D^T V^{-1} D)^{-1})$ that the counterpart of (4.8) is

$$e = N\hat{\nu} + [\hat{\nu}^T][F]\Delta\beta + \sqrt{M}\{D\Delta\beta + \frac{1}{2}\Delta\beta^T W \Delta\beta\} + O_p(M^{-1}).$$

Multiplying this equation by $D^T V^{-1}$ and $N^T V^{-1}$ respectively yields

$$\begin{aligned} D^T V^{-1} e &= [\hat{\nu}^T][D^T V^{-1} F]\Delta\beta + \sqrt{M}(D^T V^{-1} D)\Delta\beta + \\ &\quad \frac{1}{2}\sqrt{M}\Delta\beta^T [D^T V^{-1}][W]\Delta\beta + O_p(M^{-1}), \\ N^T V^{-1} e &= \hat{\nu} + [\hat{\nu}^T][N^T V^{-1} F]\Delta\beta + \\ &\quad \frac{1}{2}\sqrt{M}\Delta\beta^T [N^T V^{-1}][W]\Delta\beta + O_p(M^{-1}). \end{aligned}$$

Since $\Delta\beta = O_p(M^{-\frac{1}{2}})$ and all the derivatives are finite, we have

$$\begin{aligned} \Delta\beta &= \frac{1}{\sqrt{M}}(D^T V^{-1} D)^{-1}\{R^T \tau + [\hat{\nu}^T][R^T A_\theta^I R]\Delta\beta - \\ &\quad \frac{1}{2}\sqrt{M}\Delta\beta^T R^T [R^T][A^P] R\Delta\beta\} + O_p(M^{-1}), \qquad (7.9) \end{aligned}$$

$$\hat{\nu} = \lambda - [\hat{\nu}^T][N^T V^{-1} F]\Delta\beta - \frac{1}{2}\sqrt{M}\Delta\beta^T (R^T A^I R)\Delta\beta + O_p(M^{-1}).$$

From these we get

$$\Delta\beta = \frac{1}{\sqrt{M}} L\tau + O_p(M^{-1}), \qquad \hat{\nu} = \lambda + O_p(M^{-\frac{1}{2}}).$$

Substituting these results into (7.9) gives (7.8). ∎

The expansion (7.8) is very similar to (4.6), but here the order of the expansion is explicit. In Chapter 4, most of the results are based on the expansion (4.6), so we can get parallel results to Theorems 4.3 to 4.6 for model (7.1) based on (7.8). Now we give a theorem that has not been given by Efron (1975) and Amari (1985).

Theorem 7.1

Suppose that $\ddot{l}(\boldsymbol{\beta})$, $\boldsymbol{J}(\boldsymbol{\beta})$ are given in (7.4), (7.5) and set

$$\boldsymbol{\Lambda} = \sqrt{M}\{-\ddot{l}(\boldsymbol{\beta})\boldsymbol{J}^{-1}(\boldsymbol{\beta}) - \boldsymbol{I}_p\}_{\boldsymbol{\beta}=\hat{\boldsymbol{\beta}}}, \tag{7.10}$$

then $\text{Vec}(\boldsymbol{\Lambda})$ and $\text{tr}(\boldsymbol{\Lambda})$ are asymptotically normally distributed as $N(\boldsymbol{0}, \sigma^2\boldsymbol{\Sigma})$ and $N(0, \sigma^2 a^2)$ respectively, where $a^2 = \{\text{tr}[\boldsymbol{A}_\theta^I]\}^T\{\text{tr}[\boldsymbol{A}_\theta^I]\}$, and

$$\boldsymbol{\Sigma} = (\boldsymbol{L}\boldsymbol{L}^T \otimes \boldsymbol{I}_p)\boldsymbol{\Omega}(\boldsymbol{L}\boldsymbol{L}^T \otimes \boldsymbol{I}_p), \tag{7.11}$$

$$\boldsymbol{\Omega} = \{\text{Vec}[\boldsymbol{R}^T\boldsymbol{A}_\theta^I\boldsymbol{R}]\}\{\text{Vec}[\boldsymbol{R}^T\boldsymbol{A}_\theta^I\boldsymbol{R}]\}^T,$$

Proof. It follows from (7.4) and (7.5) that

$$\boldsymbol{\Lambda} = -[\hat{\boldsymbol{e}}^T][\hat{\boldsymbol{W}}_\theta](\hat{\boldsymbol{D}}^T\hat{\boldsymbol{V}}^{-1}\hat{\boldsymbol{D}})^{-1}.$$

By similar derivations as shown in Theorem 4.5, we get

$$\boldsymbol{\Lambda} = -[\boldsymbol{\lambda}^T][\boldsymbol{R}^T\boldsymbol{A}_\theta^I\boldsymbol{R}](\boldsymbol{L}\boldsymbol{L}^T) + O_p(M^{-\frac{1}{2}});$$

$$\text{Vec}(\boldsymbol{\Lambda}) = -(\boldsymbol{L}\boldsymbol{L}^T \otimes \boldsymbol{I}_p)\{\text{Vec}[\boldsymbol{R}^T\boldsymbol{A}_\theta^I\boldsymbol{R}]\}^T\boldsymbol{\lambda} + O_p(M^{-\frac{1}{2}}),$$

and

$$\text{tr}(\boldsymbol{\Lambda}) = -\{\text{tr}[\boldsymbol{A}_\theta^I]\}^T\boldsymbol{\lambda} + O_p(M^{-\frac{1}{2}}).$$

From these, we get the desired results. ∎

Note that there is a difference between Theorems 7.1 and 4.5. Unlike (4.1) and (4.30), here $M^{-1}\boldsymbol{J}(\boldsymbol{\beta}) = \boldsymbol{D}^T\boldsymbol{V}^{-1}\boldsymbol{D} = \boldsymbol{R}^T\boldsymbol{R}$ automatically holds, so $\boldsymbol{\Omega}$ and a^2 are not obtained by the limit procedures.

Corollary

If $\boldsymbol{\beta}$ is a scalar, then $\boldsymbol{\Lambda}$ is asymptotically normally distributed as $N(0, \gamma_\beta^2)$ where γ_β^2 is the Efron curvature given by (3.36).

Proof. We use the notation given in Subsection 3.4.3. Then it is easily seen that

$$\boldsymbol{R} = \boldsymbol{L}^{-1} = \|\dot{\boldsymbol{\theta}}_\beta\|_\theta = (\dot{\boldsymbol{\theta}}_\beta^T\boldsymbol{V}\dot{\boldsymbol{\theta}}_\beta)^{\frac{1}{2}}, \quad \boldsymbol{R}^T\boldsymbol{A}_\theta^I\boldsymbol{R} = \boldsymbol{N}_\theta^T\boldsymbol{V}\ddot{\boldsymbol{\theta}}_\beta = (\ddot{\boldsymbol{\theta}}_\beta)_N.$$

So it follows from (7.11) and (3.36) that

$$\sigma^2\boldsymbol{\Sigma} = \phi^{-1}(\dot{\boldsymbol{\theta}}_\beta^T\boldsymbol{V}\dot{\boldsymbol{\theta}}_\beta)^{-2}(\ddot{\boldsymbol{\theta}}_\beta)_N^T(\ddot{\boldsymbol{\theta}}_\beta)_N = \gamma_\beta^2. \quad ∎$$

This result was originally obtained by Eforn and Hinkley (1978), but here we derive it based on the modified BW geometric framework.

We can also obtain confidence regions in terms of the curvature for model (7.1) as discussed in Chapter 5, because the score function, the observed information and the Fisher information in model (7.1) are almost the same as those in model (2.2). As shown in Chapter 5, the key point is to make a transformation (5.5) from parameter space to tangent space. Here for model (7.1), we can make such a transformation as

$$\boldsymbol{u}(\boldsymbol{\beta}) = \sqrt{M}\boldsymbol{Q}^T\boldsymbol{V}^{-1}\{\boldsymbol{\mu}(\boldsymbol{\beta}) - \boldsymbol{\mu}(\hat{\boldsymbol{\beta}})\}, \tag{7.12}$$

where factor $\sqrt{M}$ is added so that $\boldsymbol{u}(\boldsymbol{\beta}) = O_p(1)$. Once the transformation (7.12) is established, several inference regions can be easily obtained as discussed in Chapter 5. The results and derivations are quite similar to those of Theorems 5.1 to 5.4. For example, the counterpart of (5.11) becomes

$$K(\boldsymbol{Y}) = \{\boldsymbol{\beta} : \ \boldsymbol{u}^T(\boldsymbol{\beta})(\boldsymbol{I}_p - \boldsymbol{B}_\theta/\sqrt{M})\boldsymbol{u}(\boldsymbol{\beta}) \le \rho^2(\alpha)\}.$$

7.2 Multinomial Nonlinear Models

Suppose that $\boldsymbol{Y} = (y_1, \cdots, y_n)^T$ is a multinomial random vector with the probability density function in the form

$$p(\boldsymbol{y}; \boldsymbol{\pi}(\boldsymbol{\beta})) = M! \prod_{i=1}^{n} \{\pi_i(\boldsymbol{\beta})^{y_i}/y_i!\}, \tag{7.13}$$

$$\sum_{i=1}^{n} \pi_i(\boldsymbol{\beta}) = 1, \qquad \sum_{i=1}^{n} y_i = M, \tag{7.14}$$

where $\boldsymbol{\pi}(\boldsymbol{\beta}) = (\pi_1(\boldsymbol{\beta}), \cdots, \pi_n(\boldsymbol{\beta}))^T$, $0 < \pi_i(\boldsymbol{\beta}) < 1$ and $\boldsymbol{\beta} = (\beta_1, \cdots, \beta_p)^T$ $(p < n-1)$ are the parameters of interest. This model is referred to as multinomial nonlinear model and denoted by $MN(M, \pi(\boldsymbol{\beta}))$ which was studied, for example, by Rao (1973), Kass (1989), McCullagh and Nelder (1989), and Wei (1996). From (7.13), the log-likelihood of $\boldsymbol{Y}$ may be written as

$$l(\boldsymbol{\pi}(\boldsymbol{\beta}); \boldsymbol{y}) = \sum_{i=1}^{n} \{y_i \log \pi_i(\boldsymbol{\beta}) - \log(y_i!)\} + \log(M!), \quad \sum_{i=1}^{n} \pi_i(\boldsymbol{\beta}) = 1. \tag{7.15}$$

If we set $\theta_i = \log \pi_i(\boldsymbol{\beta})$, then (7.15) has the form of curved exponential family with the constraint (7.14). Substituting $\pi_n(\boldsymbol{\beta}) = 1 - \pi_1(\boldsymbol{\beta}) - \cdots - \pi_{n-1}(\boldsymbol{\beta})$ and $y_n = M - y_1 - \cdots - y_{n-1}$ into (7.15), we can get a nonconstraint curved exponential family with $\theta_i = \log \pi_i(\boldsymbol{\beta})$, $i = 1, \cdots, n-1$. However, that is rather complicated, and one usually prefers to use (7.15) subject to the constraint (7.14) (McCullagh and Nelder 1989, p.211).

It is well known that the random vector $\boldsymbol{Y}$ may be regarded as a sum of M iid random vectors $\boldsymbol{Y}^1, \cdots, \boldsymbol{Y}^M$ with $MN(1, \boldsymbol{\pi}(\boldsymbol{\beta}))$ distributions, that is $\boldsymbol{Y} = \sum_m \boldsymbol{Y}^m$. So the central limit theorem and the law of large numbers can be applied to $\boldsymbol{Y}$ as $M \to +\infty$. By direct calculations we have

$$\mathrm{E}(\boldsymbol{Y}) = M\boldsymbol{\pi}(\boldsymbol{\beta}), \quad \mathrm{Var}(\boldsymbol{Y}) = M\boldsymbol{\Phi}, \quad \boldsymbol{\Phi} = \boldsymbol{V}(\boldsymbol{\beta}) - \boldsymbol{\pi}(\boldsymbol{\beta})\boldsymbol{\pi}^T(\boldsymbol{\beta}), \tag{7.16}$$

where

$$\boldsymbol{V}(\boldsymbol{\beta}) = \mathrm{diag}(\pi_1(\boldsymbol{\beta}), \cdots, \pi_n(\boldsymbol{\beta})).$$

Then the central limit theorem gives

$$\boldsymbol{e}(\boldsymbol{\beta}) \triangleq \sqrt{M}\{\overline{\boldsymbol{Y}} - \boldsymbol{\pi}(\boldsymbol{\beta})\} \xrightarrow{L} N(\boldsymbol{0}, \boldsymbol{\Phi}), \tag{7.17}$$

as $M \to +\infty$, where $\overline{\boldsymbol{Y}} = M^{-1}\boldsymbol{Y} = M^{-1}\sum_m \boldsymbol{Y}^m$. This result is very similar to (7.2), so we may regard $\overline{\boldsymbol{Y}}$ as an observed vector, $\boldsymbol{\pi}(\boldsymbol{\beta})$ as a mean vector and $\boldsymbol{e}(\boldsymbol{\beta})$ as a random error. But note that here $y_1, \cdots, y_n$ is restricted by $\sum_i y_i = M$.

To get likelihoods, we assume that $\boldsymbol{\pi}(\boldsymbol{\beta})$ is thrice continuously differentiable with respect to $\boldsymbol{\beta}$ in some neighborhood. The first two derivatives of $\boldsymbol{\pi}(\boldsymbol{\beta})$ are denoted by $\boldsymbol{D}(\boldsymbol{\beta})$, $\boldsymbol{W}(\boldsymbol{\beta})$ and $\boldsymbol{D}(\boldsymbol{\beta})$ is assumed to be of full rank in column. Differentiating the constraint condition $\sum_i \pi_i(\boldsymbol{\beta}) = 1$ in $\boldsymbol{\beta}$ yields

$$\boldsymbol{1}^T\boldsymbol{D} = \boldsymbol{\pi}^T\boldsymbol{V}^{-1}\boldsymbol{D} = \boldsymbol{0}, \qquad [\boldsymbol{1}^T][\boldsymbol{W}] = [\boldsymbol{\pi}^T\boldsymbol{V}^{-1}][\boldsymbol{W}] = \boldsymbol{0}. \tag{7.18}$$

These equations reflect the characteristics related to multinomial nonlinear models. Then we have the following lemma.

Lemma 7.2

Let $l(\boldsymbol{\beta}) = l(\boldsymbol{\pi}(\boldsymbol{\beta}); \boldsymbol{y})$, then the score function, the observed information and the Fisher information of $\boldsymbol{Y}$ for $\boldsymbol{\beta}$ are respectively given by

$$\dot{l}(\boldsymbol{\beta}) = \sqrt{M}\boldsymbol{D}^T\boldsymbol{V}^{-1}\boldsymbol{e}, \quad \boldsymbol{e}(\boldsymbol{\beta}) = \sqrt{M}\{\overline{\boldsymbol{Y}} - \boldsymbol{\pi}(\boldsymbol{\beta})\}; \tag{7.19}$$

$$-\ddot{l}(\boldsymbol{\beta}) = M\{\boldsymbol{D}^T\boldsymbol{V}^{-1}\boldsymbol{D} - [\boldsymbol{e}\boldsymbol{V}^{-1}][\boldsymbol{W} - \boldsymbol{\Gamma}]/\sqrt{M}\}; \tag{7.20}$$

and

$$\boldsymbol{J}(\boldsymbol{\beta}) = M\boldsymbol{D}^T\boldsymbol{V}^{-1}\boldsymbol{D}; \tag{7.21}$$

where $\boldsymbol{\Gamma} = \boldsymbol{D}^T\boldsymbol{V}^{-1}\boldsymbol{S}\boldsymbol{V}^{-1}\boldsymbol{D}$ and $\boldsymbol{S}$ is an $n \times n \times n$ array with $S_{iii} = \pi_i(\boldsymbol{\beta})$ and zeros elsewhere.

Proof. It follows from (7.14), (7.15) and (7.18) that

$$\frac{\partial l}{\partial \beta_a} = \sum_{i=1}^n \frac{y_i}{\pi_i}\frac{\partial \pi_i}{\partial \beta_a} = \sum_{i=1}^n \pi_i^{-1}(y_i - M\pi_i)\frac{\partial \pi_i}{\partial \beta_a} \tag{7.22}$$

for $a = 1, \cdots, p$. Here we use the result $\sum_i \partial\pi_i/\partial\beta_a = 0$ and $e_i = (y_i - M\pi_i)/\sqrt{M}$. From (7.17) and (7.22), we get (7.20). Taking the derivative for (7.22) in β_b $(b = 1, \cdots, p)$ and using matrix notation, we get (7.20), while (7.21) is easily obtained from (7.20) by taking expectation for $-\ddot{l}(\boldsymbol{\beta})$. ∎

This lemma is very similar to Lemma 2.3 and equations (7.4) and (7.5). So we can once again introduce the geometric framework for model (7.15). But note that the constraints (7.14) and (7.18) must be considered here, which results in some differences between models (7.15) and (7.1).

Now we first briefly introduce the geometric framework for multinomial nonlinear model (7.15). It is easily seen from (7.20) that the maximum likelihood estimator $\hat{\boldsymbol{\beta}}$ satisfies

$$\boldsymbol{D}^T(\hat{\boldsymbol{\beta}})\boldsymbol{V}^{-1}(\hat{\boldsymbol{\beta}})\boldsymbol{e}(\hat{\boldsymbol{\beta}}) = \boldsymbol{0}, \quad \boldsymbol{e}(\boldsymbol{\beta}) = \sqrt{M}tM\{\overline{\boldsymbol{Y}} - \boldsymbol{\pi}(\boldsymbol{\beta})\}.$$

This equation exactly coincides with equations (3.1) and (7.7). So we can introduce a geometric framework for (7.15) in expectation parameter space. Now take $\boldsymbol{\pi}$ as a set of coordinates in R^n since $\mathrm{E}\overline{\boldsymbol{Y}} = \boldsymbol{\pi}(\boldsymbol{\beta})$. Then the solution locus is defined as $\boldsymbol{\pi} = \boldsymbol{\pi}(\boldsymbol{\beta})$ whose tangent space at $\boldsymbol{\beta}$ is spanned by the columns of $\boldsymbol{D}(\boldsymbol{\beta})$. The inner product for any two vectors $\boldsymbol{a}, \boldsymbol{b}$ in R^n is still defined as $< \boldsymbol{a}, \boldsymbol{b} >= \boldsymbol{a}^T\boldsymbol{V}^{-1}\boldsymbol{b}$ (see (3.3)). Then the QR decomposition of $\boldsymbol{D}$ is still given in (2.4) and Definition 3.1 is valid for model (7.15).

Note that this geometric framework has some minor differences from that of curved exponential family (7.1): (a) we have $\mathrm{E}(-\partial^2 l/\partial\boldsymbol{\pi}\partial\boldsymbol{\pi}^T) = M\boldsymbol{V}^{-1}$ but it is not equal to the Fisher information matrix $\mathrm{Var}(\partial l/\partial\boldsymbol{\pi})$ for $\boldsymbol{\pi}$, since $\boldsymbol{\pi}$ is constrained by $\sum_i \pi_i = 1$; and (b) the constraint (7.14) leads to (7.18) which states that the unit vector $\boldsymbol{\pi}$ is in normal space, so we may set

$$\boldsymbol{N} = (\boldsymbol{\pi}, \ \boldsymbol{N}_1), \qquad \boldsymbol{\pi}^T\boldsymbol{V}^{-1}\boldsymbol{N}_1 = \boldsymbol{0}, \qquad \boldsymbol{A}_1^I = \boldsymbol{0}, \tag{7.23}$$

where $\boldsymbol{A}_1^I$ is the first face of $\boldsymbol{A}^I$. We shall use the convention (7.23) in later discussions.

The geometric framework introduced here is based on the expectation parameter $\boldsymbol{\pi}$. We can also introduce an associated dual geometry based on the dual parameter $\boldsymbol{\theta} = (\theta_1, \cdots, \theta_n)^T$ of $\boldsymbol{\pi}$ with $\theta_i = \log \pi_i$, $i = 1, \cdots, n$ (see (7.15)). Then $\boldsymbol{\mu}$ and $\boldsymbol{\theta}$ satisfy the following dual relationship

$$\frac{\partial\boldsymbol{\theta}}{\partial\boldsymbol{\pi}^T} = \boldsymbol{V}^{-1}, \qquad \frac{\partial\boldsymbol{\pi}}{\partial\boldsymbol{\theta}^T} = \boldsymbol{V}.$$

From this basic connection, we can defined the solution locus $\boldsymbol{\theta} = \boldsymbol{\theta}(\boldsymbol{\beta}) = \boldsymbol{\theta}(\pi(\boldsymbol{\beta}))$, then the geometric framework introduced in Section 3.3 can be applied to model (7.15). So the inner product (3.22), the curvature arrays (3.23) and Theorem 3.3 are all valid for (7.15).

About the statistical inference, model (7.15) can be regarded as a constraint curved exponential family, and the results that hold for model (7.1) usually hold for model (7.15) except for some minor changes caused by the constraint (7.14). Thus as pointed out in Section 7.1, almost all the results shown in Chapters 3 to 5 also hold for model (7.15) except for some minor changes. Here, we only give some remarks (see Wei 1996 for more details).

(a) The stochastic expansion (7.8) still holds for model (7.15), but here $\boldsymbol{\lambda} = \boldsymbol{N}^T\boldsymbol{V}^{-1}\boldsymbol{e}$ and $\boldsymbol{\tau} = \boldsymbol{Q}^T\boldsymbol{V}^{-1}\boldsymbol{e}$ must be subject to (7.17) with $\mathrm{E}(\boldsymbol{e}) = \boldsymbol{0}$, $\mathrm{Var}(\boldsymbol{e}) = \boldsymbol{\Phi} = \boldsymbol{V} - \boldsymbol{\pi}\boldsymbol{\pi}^T$ (see (7.16)), which is different from (7.2). Fortunately, from (7.18) we still have

$$\mathrm{Cov}(\boldsymbol{\lambda}, \boldsymbol{\tau}) = \boldsymbol{N}^T\boldsymbol{V}^{-1}(\boldsymbol{V} - \boldsymbol{\pi}\boldsymbol{\pi}^T)\boldsymbol{V}^{-1}\boldsymbol{Q}^T = \boldsymbol{N}^T\boldsymbol{V}^{-1}\boldsymbol{Q} = \boldsymbol{0}.$$

Similarly, we have $\mathrm{Var}(\boldsymbol{\tau}) = \boldsymbol{I}_p$, but $\mathrm{Var}(\boldsymbol{\lambda})$ has something different:

$$\mathrm{Var}(\boldsymbol{\lambda}) = \boldsymbol{N}^T\boldsymbol{V}^{-1}(\boldsymbol{V} - \boldsymbol{\pi}\boldsymbol{\pi}^T)\boldsymbol{V}^{-1}\boldsymbol{N} = \boldsymbol{I}_{n-p} - \boldsymbol{N}^T\boldsymbol{V}^{-1}\boldsymbol{\pi}\boldsymbol{\pi}^T\boldsymbol{V}^{-1}\boldsymbol{N}.$$

It follows from (7.23) that $\boldsymbol{\pi}^T\boldsymbol{V}^{-1}\boldsymbol{N} = (1, 0, \cdots, 0)^T$, which leads to

$$\mathrm{Var}(\boldsymbol{\lambda}) = \boldsymbol{I}_{n-p} - \mathrm{diag}(1, 0, \cdots, 0).$$

Thus we have

$$\boldsymbol{\lambda} = (\lambda_1,\ \boldsymbol{\lambda}_2^T)^T; \quad \lambda_1 = 0\ (a.e.), \quad \boldsymbol{\lambda}_2 \xrightarrow{L} N(\boldsymbol{0},\ \boldsymbol{I}_{n-p-1}), \tag{7.24}$$

as $M \to +\infty$. ∎

(b) All the results shown in Chapters 3 to 5 and Section 7.1 still hold for model (7.15). But $\boldsymbol{\lambda}$ must satisfy (7.24), which will cause some minor changes. For instance, if $\boldsymbol{\Lambda}$ is given in (7.10), then we have $\mathrm{tr}(\boldsymbol{\Lambda}) \xrightarrow{L} N(0,\ a^2)$ as $M \to +\infty$, with

$$a^2 = \sum_{k=2}^{n-p} (\mathrm{tr} A_{\theta k}^I)^2$$

since $\lambda_1 = 0$ $(a.e.)$, where $\boldsymbol{A}_{\theta k}^I$ is the k-th face of $\boldsymbol{A}_\theta^I$. ∎

(c) As an example, we verify a well-known result. Let

$$\kappa(\hat{\boldsymbol{\beta}}) = \sum_{i=1}^{n} \frac{\{y_i - M\pi_i(\hat{\boldsymbol{\beta}})\}^2}{M\pi_i(\hat{\boldsymbol{\beta}})},$$

that is just the chi-square test statistic for goodness of fit (Rao 1973, p.392) and it asymptotically has $\chi^2(n-p-1)$ distribution. We now can easily prove this well-known result.

Proof. It follows from (7.17) that

$$\hat{e}_i = e_i(\hat{\boldsymbol{\beta}}) = \frac{y_i - M\pi_i(\hat{\boldsymbol{\beta}})}{\sqrt{M}}.$$

Since $\boldsymbol{V} = \mathrm{diag}(\pi_1, \cdots, \pi_n)$, we have

$$\kappa(\hat{\boldsymbol{\beta}}) = \sum_{i=1}^{n} \hat{e}_i^2 \hat{V}_i^{-1} = \hat{\boldsymbol{e}}^T \hat{\boldsymbol{V}}^{-1} \hat{\boldsymbol{e}}.$$

We can get the stochastic expansion for $\hat{\boldsymbol{e}}$ as given in (4.14). Then it is easily seen from (4.14) that $\hat{\boldsymbol{e}} = \boldsymbol{N}\boldsymbol{\lambda} + O_p(M^{-\frac{1}{2}})$ and thus

$$\kappa(\hat{\boldsymbol{\beta}}) = \boldsymbol{\lambda}^T \boldsymbol{\lambda} + O_p(M^{-\frac{1}{2}}).$$

From (7.24) we get $\kappa(\hat{\boldsymbol{\beta}}) \xrightarrow{L} \chi^2(n-p-1)$ as $M \to +\infty$. ∎

7.3 Embedded Models

This section shows that the modified BW geometric framework can be used for the embedded models in regular parametric families. It is quite a general parametric model.

Suppose that the observed matrix of random vectors is denoted by $\boldsymbol{Y} = (\boldsymbol{Y}^1, \cdots, \boldsymbol{Y}^M)^T$ with $\boldsymbol{Y}^m = (y_1^m, \cdots, y_k^m)^T$, $m = 1, \cdots, M$, which is connected with an n-vector parameter $\boldsymbol{\theta}$, an open subset Θ of parameter space in R^n and log-likelihood $l(\boldsymbol{\theta})$. The first three derivatives of $l(\boldsymbol{\theta})$ are assumed to be existing and denoted by $\dot{\boldsymbol{l}}(\boldsymbol{\theta}) = (\partial l/\partial \theta_i)$, $\ddot{\boldsymbol{l}}(\boldsymbol{\theta}) = (\partial^2 l/\partial\theta_i \partial\theta_j)$ and $\boldsymbol{l}^{(3)}(\boldsymbol{\theta}) = (\partial^3 l/\partial\theta_i\partial\theta_j\partial\theta_k)$ respectively, where $i, j, k = 1, \cdots, n$. We assume that the family is regular (Cox and Hinkley 1974, p.281; Jorgensen 1983) so that

(a) $l(\boldsymbol{\theta}_1) = l(\boldsymbol{\theta}_2)$ (*a.e.*) implies $\boldsymbol{\theta}_1 = \boldsymbol{\theta}_2$ and

$$\mathrm{E}\{\dot{\boldsymbol{l}}(\boldsymbol{\theta})\} = \boldsymbol{0}, \ \mathrm{E}\{-\ddot{\boldsymbol{l}}(\boldsymbol{\theta})\} = \mathrm{Var}\{\dot{\boldsymbol{l}}(\boldsymbol{\theta})\} = M\boldsymbol{v}(\boldsymbol{\theta}), \tag{7.25}$$

$$M^{-1}|l_{ijk}^{(3)}(\boldsymbol{\theta})| \leq K(\boldsymbol{\theta}), \quad \mathrm{E}\{K(\boldsymbol{\theta})\} < \infty,$$

where $M\boldsymbol{v}(\boldsymbol{\theta}) = \boldsymbol{J}_\theta(\boldsymbol{Y})$ is the Fisher information matrix of $\boldsymbol{Y}$ for $\boldsymbol{\theta}$, assumed to be positive definite in Θ.

(b) Let $\boldsymbol{\Delta}(\boldsymbol{\theta}) = \{-\ddot{\boldsymbol{l}}(\boldsymbol{\theta}) - M\boldsymbol{v}(\boldsymbol{\theta})\}/\sqrt{M} = (\Delta_{ij})$, which is $1/\sqrt{M}$ times the difference between the observed information and the expected information for $\boldsymbol{\theta}$. We assume that $\dot{\boldsymbol{l}}(\boldsymbol{\theta})/\sqrt{M}$ and $\Delta_{ij}(\boldsymbol{\theta})$ are asymptotically normal (note that if $\boldsymbol{Y}^1, \cdots, \boldsymbol{Y}^M$ are iid observations, then these assumptions hold automatically). These assumptions imply that

$$\frac{1}{\sqrt{M}}\dot{\boldsymbol{l}}(\boldsymbol{\theta}) \to N(\boldsymbol{0}, \boldsymbol{v}(\boldsymbol{\theta})), \quad -\ddot{\boldsymbol{l}}(\boldsymbol{\theta}) = M\boldsymbol{v}(\boldsymbol{\theta}) + \sqrt{M}\boldsymbol{\Delta}(\boldsymbol{\theta}). \tag{7.26}$$

We also have $\dot{\boldsymbol{l}}(\boldsymbol{\theta}) = O_p(M^{\frac{1}{2}})$, $-\ddot{\boldsymbol{l}}(\boldsymbol{\theta}) = O_p(M)$, and $\boldsymbol{\Delta}(\boldsymbol{\theta}) = O_p(1)$.

Now we consider an embedded model dominated by $\{l(\boldsymbol{\theta}),\ \boldsymbol{\theta} = \boldsymbol{\theta}(\boldsymbol{\beta}),\ \boldsymbol{\beta} \in \mathcal{B}\}$ where $\boldsymbol{\theta} = \boldsymbol{\theta}(\boldsymbol{\beta})$ is a smooth function of $\boldsymbol{\beta}$ that is a p-vector parameter of interest taking values in an open subset $\mathcal{B}$ in R^p $(p < n)$. The first two derivatives of $l(\boldsymbol{\beta}) = l(\boldsymbol{\theta}(\boldsymbol{\beta}))$ are denoted by $\dot{\boldsymbol{l}}(\boldsymbol{\beta})$ and $\ddot{\boldsymbol{l}}(\boldsymbol{\beta})$ for short. We futher assume that

(c) The maximum likelihood estimator $\hat{\boldsymbol{\beta}}$ of $\boldsymbol{\beta}$ is an interior point of $\mathcal{B}$ satisfying $\dot{\boldsymbol{l}}(\hat{\boldsymbol{\beta}}) = \mathbf{0}$ and $-\ddot{\boldsymbol{l}}(\boldsymbol{\beta}) > \mathbf{0}$ is in some neighborhood of $\hat{\boldsymbol{\beta}}$. Further, the $n \times p$ matrix $\partial\boldsymbol{\theta}/\partial\boldsymbol{\beta}^T = \boldsymbol{D}_\theta(\boldsymbol{\beta})$ is of full rank in column, The $n \times p \times p$ array $\boldsymbol{W}_\theta(\boldsymbol{\beta}) = \partial^2\boldsymbol{\theta}/\partial\boldsymbol{\beta}\partial\boldsymbol{\beta}^T$ and the third derivatives of $\boldsymbol{\theta}(\boldsymbol{\beta})$ exist for $\boldsymbol{\beta} \in \mathcal{B}$.

The models satisfying assumptions (a) to (c) may be called the embedded models in regular parametric families or embedded models for short (see Wei 1994 for details).

To introduce the geometric framework for these models, we may use the idea described in Section 3.3. In fact, these embedded models have some similarities to exponential family nonlinear models and curved exponential families in likelihoods. We have the following lemma.

Lemma 7.3

For the embedded models stated earlier, the score function, the observed information and the Fisher information of $\boldsymbol{Y}$ for $\boldsymbol{\beta}$ can be represented as

$$\dot{\boldsymbol{l}}(\boldsymbol{\beta}) = \sqrt{M}\boldsymbol{D}_\theta^T(\boldsymbol{\beta})\boldsymbol{v}(\boldsymbol{\beta})\boldsymbol{e}(\boldsymbol{\beta}),\quad \boldsymbol{e}(\boldsymbol{\beta}) = \boldsymbol{Z} - \boldsymbol{\theta}(\boldsymbol{\beta}), \tag{7.27}$$

$$\boldsymbol{e}(\boldsymbol{\beta}) = \frac{1}{\sqrt{M}}\boldsymbol{v}^{-1}(\boldsymbol{\theta}(\boldsymbol{\beta}))\dot{\boldsymbol{l}}(\boldsymbol{\theta}(\boldsymbol{\beta})),\ \boldsymbol{Z} = \boldsymbol{\theta}(\boldsymbol{\beta}) + \boldsymbol{v}^{-1}(\boldsymbol{\theta})\dot{\boldsymbol{l}}(\boldsymbol{\theta})/\sqrt{M}; \tag{7.28}$$

$$-\ddot{\boldsymbol{l}}(\boldsymbol{\beta}) = M\boldsymbol{D}_\theta^T\boldsymbol{v}\boldsymbol{D}_\theta - \sqrt{M}[\boldsymbol{e}^T\boldsymbol{v}][\boldsymbol{W}_\theta] + \sqrt{M}\boldsymbol{D}_\theta^T\boldsymbol{\Delta}\boldsymbol{D}_\theta; \tag{7.29}$$

and

$$\boldsymbol{J}_\beta(\boldsymbol{Y}) = M\boldsymbol{D}_\theta^T\boldsymbol{v}\boldsymbol{D}_\theta. \tag{7.30}$$

Proof. By taking derivatives for $l(\boldsymbol{\theta}(\boldsymbol{\beta}))$ in $\boldsymbol{\beta}$, we have

$$\dot{\boldsymbol{l}}(\boldsymbol{\beta}) = \boldsymbol{D}_\theta^T\dot{\boldsymbol{l}}(\boldsymbol{\theta}),\quad -\ddot{\boldsymbol{l}}(\boldsymbol{\beta}) = \boldsymbol{D}_\theta^T(-\ddot{\boldsymbol{l}}(\boldsymbol{\theta}))\boldsymbol{D}_\theta + [\dot{\boldsymbol{l}}^T(\boldsymbol{\theta})][\boldsymbol{W}_\theta].$$

It follows from (7.26) and (7.28) that

$$\dot{\boldsymbol{l}}(\boldsymbol{\theta}) = \sqrt{M}\boldsymbol{v}(\boldsymbol{\theta})\boldsymbol{e},\quad -\ddot{\boldsymbol{l}}(\boldsymbol{\theta}) = M\boldsymbol{v}(\boldsymbol{\theta}) + \sqrt{M}\boldsymbol{\Delta}(\boldsymbol{\theta}).$$

Substituting these results into $\dot{\boldsymbol{l}}(\boldsymbol{\beta})$ and $\ddot{\boldsymbol{l}}(\boldsymbol{\beta})$, shown earlier, yields (7.27) and (7.29), while (7.30) is easily obtained from (7.29). ∎

The results of this lemma are very similar to Lemmas 7.2, 2.3 and equations (7.5) to (7.7). So it is expected that the geometric framework introduced in Chapter 3 and the statistical inference studied in Chapters 4 and 5 may be valid for the present models.

First we introduce the geometric framework for embedded models. Lemma 7.3 shows that $\mathrm{E}(\boldsymbol{Z}) = \boldsymbol{\theta}(\boldsymbol{\beta})$, $\mathrm{Var}(\boldsymbol{Z}) = \boldsymbol{v}^{-1}(\boldsymbol{\theta})$, so the vector $\boldsymbol{Z}$ may be regarded as an "observed vector", $\boldsymbol{e} = \boldsymbol{Z} - \boldsymbol{\theta}(\boldsymbol{\beta})$ may be regarded as a "random error" and $\boldsymbol{\theta} = \boldsymbol{\theta}(\boldsymbol{\beta})$ may define a solution locus in R^n. Further, the maximum likelihood estimator $\hat{\boldsymbol{\beta}}$ satisfies

$$\boldsymbol{D}_\theta^T(\hat{\boldsymbol{\beta}})\boldsymbol{v}(\hat{\boldsymbol{\beta}})\boldsymbol{e}(\hat{\boldsymbol{\beta}}) = \boldsymbol{0}, \quad \boldsymbol{e}(\boldsymbol{\beta}) = \boldsymbol{Z} - \boldsymbol{\theta}(\boldsymbol{\beta}).$$

This equation exactly coincides with (3.21) (see also (2.10)). So we may introduce a modified BW geometric framework as shown in Section 3.3 for embedded models as follows.

Take $\boldsymbol{\theta}$ as a coordinate in Euclidean space R^n, the solution locus π for the embedded model is defined as $\boldsymbol{\theta} = \boldsymbol{\theta}(\boldsymbol{\beta})$ which is a p-dimensional surface in R^n. The tangent space of π at $\boldsymbol{\beta}$ is spanned by the columns of $\boldsymbol{D}_\theta(\boldsymbol{\beta})$. For any two vectors $\boldsymbol{a}$ and $\boldsymbol{b}$ in R^n, we define the Fisher information (see (7.25)) inner product as $< \boldsymbol{a}, \boldsymbol{b} >_\theta = \boldsymbol{a}^T\boldsymbol{v}\boldsymbol{b}$ (see also (3.22)). Then we can make a QR decomposition for $\boldsymbol{D}_\theta$ with respect to this inner product, and define curvatures by the same procedures as shown in Section 3.3. So the Definition 3.3 and the inner product (3.23) are still valid for embedded models.

The above geometry is based on the "canonical" parameter $\boldsymbol{\theta}$ and its Fisher information matrix (note that $\boldsymbol{J}_\theta(\boldsymbol{Y}) = M\boldsymbol{v}$). We may introduce a dual parameter of $\boldsymbol{\theta}$ and its associated dual geometry for embedded models. To this aim, we need some specific conditions. As pointed out by McCullagh and Nelder (1989, p.334), if matrix $\boldsymbol{v}(\boldsymbol{\theta}) = (v_{ij})$ satisfies $\partial v_{ij}/\partial\theta_k = \partial v_{ik}/\partial\theta_j = \partial v_{kj}/\partial\theta_i$ for all i, j and k, then there exists a convex functions $b(\boldsymbol{\theta})$ such that $\partial^2 b(\boldsymbol{\theta})/\partial\boldsymbol{\theta}\partial\boldsymbol{\theta}^T = \boldsymbol{v}(\boldsymbol{\theta})$. We assume that these conditions hold and let $\boldsymbol{\mu} = \dot{\boldsymbol{b}}(\boldsymbol{\theta})$, $\boldsymbol{\mu}(\boldsymbol{\beta}) = \dot{\boldsymbol{b}}(\boldsymbol{\theta}(\boldsymbol{\beta}))$ and $\boldsymbol{D} = \partial\boldsymbol{\mu}/\partial\boldsymbol{\beta}^T$. Then we get a basic dual relationship between $\boldsymbol{\theta}$ and $\boldsymbol{\mu}$:

$$\frac{\partial\boldsymbol{\mu}}{\partial\boldsymbol{\theta}^T} = \boldsymbol{v}, \quad \frac{\partial\boldsymbol{\theta}}{\partial\boldsymbol{\mu}^T} = \boldsymbol{v}^{-1}.$$

From this we have $\boldsymbol{J}_\mu(\boldsymbol{Y}) = M\boldsymbol{v}^{-1}$, that is the Fisher information matrix of $\boldsymbol{Y}$ for $\boldsymbol{\mu}$. Further, since $\boldsymbol{D} = \boldsymbol{v}\boldsymbol{D}_\theta$, we may rewrite (7.27) in terms of the dual parameter $\boldsymbol{\mu}$ as

$$\dot{\boldsymbol{l}}(\boldsymbol{\beta}) = \sqrt{M}\boldsymbol{D}^T(\boldsymbol{\beta})\boldsymbol{v}^{-1}(\boldsymbol{\beta})\boldsymbol{e}^*(\boldsymbol{\beta}), \quad \boldsymbol{e}^*(\boldsymbol{\beta}) = \boldsymbol{Z}^* - \boldsymbol{\mu}(\boldsymbol{\beta}),$$

where $\boldsymbol{e}^*(\boldsymbol{\beta}) = \dot{\boldsymbol{l}}(\boldsymbol{\theta}(\boldsymbol{\beta}))/\sqrt{M}$ and $\boldsymbol{Z}^* = \boldsymbol{\mu}(\boldsymbol{\beta}) + \dot{\boldsymbol{l}}(\boldsymbol{\theta}(\boldsymbol{\beta}))/\sqrt{M}$ satisfying $\mathrm{E}(\boldsymbol{Z}^*) = \boldsymbol{\mu}(\boldsymbol{\beta})$ and $\mathrm{Var}(\boldsymbol{Z}^*) = \boldsymbol{v}(\boldsymbol{\beta})$. So the maximum likelihood estimator $\hat{\boldsymbol{\beta}}$ satisfies

$$\boldsymbol{D}^T(\hat{\boldsymbol{\beta}})\boldsymbol{v}^{-1}(\hat{\boldsymbol{\beta}})\boldsymbol{e}^*(\hat{\boldsymbol{\beta}}) = \boldsymbol{0}, \quad \boldsymbol{e}^*(\boldsymbol{\beta}) = \boldsymbol{Z}^* - \boldsymbol{\mu}(\boldsymbol{\beta}).$$

This equation is exactly consistent with (3.1). Hence we may introduce a geometry as given in Section 3.2 based on parameter $\boldsymbol{\mu}$ and the corresponding solution locus $\boldsymbol{\mu} = \boldsymbol{\mu}(\boldsymbol{\beta})$. Then the results given there are all valid for embedded models. But note that here parameter $\boldsymbol{\mu} = \dot{\boldsymbol{b}}(\boldsymbol{\theta})$ is usually not the expectation of the real observations.

About the statistical inference, we may partly extend the results of Chapter 4 to embedded models, but it is rather complicated and we have to omit the details. The reason is that unlike exponential families, we cannot get simple expressions of the first four moments for the "random error" $\boldsymbol{e}(\boldsymbol{\beta})$ as given in (1.9) to (1.11). However, the results of Chapter 5 can be completely extended to embedded models (Wei 1994). This is because the inference regions given in Chapter 5 are based on the modified BW geometric framework introduced in Chapter 3 and the likelihood results given in Lemma 2.3, while we have similar geometric framework and parallel likelihood results for embedded models. From Lemma 7.3 we may make a transformation

$$\boldsymbol{u}(\boldsymbol{\beta}) = \sqrt{M}\boldsymbol{Q}_\theta^T \boldsymbol{v}\{\boldsymbol{\theta}(\boldsymbol{\beta}) - \boldsymbol{\theta}(\hat{\boldsymbol{\beta}})\}.$$

From this transformation and Lemma 7.3, it is not too hard to get likelihood regions and score-based regions for parameters and parameter subsets in terms of curvatures for embedded models (see Wei 1994 for details).

7.4 Quasi-likelihood Models

If there is no sufficient information to know the family of underlying distributions of observations, the quasi-likelihood (Wedderburn 1974) can be used to play a role of likelihood function in some situations. The inference based on quasi-likelihood has been well developed in the past two decades (see McCullagh and Nelder 1989 for a comprehensive description; see also McCullagh 1983; 1991; Firth 1987; Nelder and Pregibon 1987; Hill and Tsai 1988; Nelder and Lee 1992). In this section, we first review the basis of quasi-likelihood, and then introduce the geometric aspect of quasi-likelihood nonlinear models.

Suppose that the components of the response vector $\boldsymbol{Y}$ are independent with mean vector $\boldsymbol{\mu}$ and covariance matrix $\sigma^2 \boldsymbol{V}(\boldsymbol{\mu})$, where $\sigma^2 = \phi^{-1}$ may be unknown and the variance function $\boldsymbol{V}(\boldsymbol{\mu})$ is a diagonal matrix having the form

$$\boldsymbol{V}(\boldsymbol{\mu}) = \text{diag}(V_1(\mu_1), \cdots, V_n(\mu_n)). \tag{7.31}$$

It is assumed that the parameters of interest, $\boldsymbol{\beta}$, relate to the dependence of $\boldsymbol{\mu}$ on covariate $\boldsymbol{x}$, and we may write

$$\mu_i = \mu(\boldsymbol{x}_i; \boldsymbol{\beta}) = \mu_i(\boldsymbol{\beta}),\ i = 1, \cdots, n. \tag{7.32}$$

Because we only have information about the first two moments of $\boldsymbol{Y}$, we cannot get log-likelihood and related quantities such as score function, observed information and Fisher information of $\boldsymbol{Y}$ for $\boldsymbol{\mu}$ and $\boldsymbol{\beta}$. However, as an alternative, we may introduce the associated quasi-quantities based on (7.31) and (7.32).

The quasi-likelihood or more correctly the log quasi-likelihood for $\boldsymbol{\mu}$ based on data $\boldsymbol{Y}$ is defined as

$$l_q(\boldsymbol{\mu}, \boldsymbol{y}) = \sum_{i=1}^{n} q_i(\mu_i, y_i), \quad q_i(\mu_i, y_i) = \int^{\mu_i} \frac{y_i - t}{\sigma^2 V_i(t)} dt, \tag{7.33}$$

so that the quasi-score function for $\boldsymbol{\mu}$ and $\boldsymbol{\beta}$ are respectively

$$\frac{\partial l_q}{\partial \boldsymbol{\mu}} = \phi \boldsymbol{V}^{-1}(\boldsymbol{\mu})(\boldsymbol{Y} - \boldsymbol{\mu}), \quad \frac{\partial l_q}{\partial \boldsymbol{\beta}} = \phi \boldsymbol{D}^T \boldsymbol{V}^{-1} \boldsymbol{e}, \tag{7.34}$$

where $\boldsymbol{D} = \partial \boldsymbol{\mu} / \partial \boldsymbol{\beta}^T$ and $\boldsymbol{e} = \boldsymbol{Y} - \boldsymbol{\mu}(\boldsymbol{\beta})$ as before. The maximum quasi-likelihood estimator $\hat{\boldsymbol{\beta}}$ is defined as an estimator that maximizes $l_q(\boldsymbol{\mu}(\boldsymbol{\beta}), \boldsymbol{y})$, so that it satisfies

$$\boldsymbol{D}^T(\hat{\boldsymbol{\beta}}) \boldsymbol{V}^{-1}(\hat{\boldsymbol{\beta}}) \boldsymbol{e}(\hat{\boldsymbol{\beta}}) = \boldsymbol{0}, \quad \boldsymbol{e}(\boldsymbol{\beta}) = \boldsymbol{Y} - \boldsymbol{\mu}(\boldsymbol{\beta}). \tag{7.35}$$

Then all the statistical inference and geometric interpretation will be based on equations (7.32) to (7.35).

Because the quasi-score functions $\partial l_q / \partial \boldsymbol{\mu}$ and $\partial l_q / \partial \boldsymbol{\beta}$ given in (7.34) exactly coincide with (1.18) and (2.8), we may get the same expressions of quasi-observed information matrices $-\partial^2 l_q / \partial \boldsymbol{\mu} \partial \boldsymbol{\mu}^T$, $-\partial^2 l_q / \partial \boldsymbol{\beta} \partial \boldsymbol{\beta}^T$ and their expectations, the quasi-Fisher information matrices, as given in (1.20), (2.9) and (2.11), respectively. Therefore, under some regularity conditions (see (7.36)), the inference about quasi-likelihood estimator should be quite similar to the inference of maximum likelihood estimator in exponential family linear or nonlinear models (McCullagh 1983; McCullagh and Nelder 1989; Li and McCullagh 1994). The details are omitted here, but the following important argument should be emphasized. As pointed out by McCullagh and Nelder (1989, p.334), only if the quasi-observed information matrix $-\partial^2 l_q / \partial \boldsymbol{\mu} \partial \boldsymbol{\mu}^T$ is symmetric, does it make sense to use quasi-likelihood for statistical inference. This is equivalent to require that the integral (7.33) is path-independent. To make the above integral path-independent, the sufficient conditions are

$$\frac{\partial V^{ij}}{\partial \mu_k} = \frac{\partial V^{ik}}{\partial \mu_j} = \frac{\partial V^{kj}}{\partial \mu_i} \tag{7.36}$$

for all i, j and k, where we denote V^{-1} by (V^{ij}) (McCullagh and Nelder 1989, p.334). Also notice that since quasi-likelihood is not necessary to equal the real log-likelihood (McCullagh 1983), the quasi-likelihood estimator may lose more efficiency in some cases (Firth 1987; Hill and Tsai 1988).

We now introduce the geometric framework for quasi-likelihood models. Equation (7.35) exactly coincides with (3.1), so we can introduce the same geometric framework for quasi-likelihood models as that introduced in Section 3.2 for exponential family nonlinear models. Once coordinate $\boldsymbol{\mu}$ and solution locus $\boldsymbol{\mu} = \boldsymbol{\mu}(\boldsymbol{\beta})$ are defined in R^n, the definitions of curvature arrays and directional curvatures are all valid for quasi-likelihood models. Further, if (7.36) holds, we can also introduce the dual geometry for the models as discussed in Section 3.3 (see also Section 7.3). Now, we define dual parameters $\boldsymbol{\theta} = (\theta_1, \cdots, \theta_n)^T$ of $\boldsymbol{\mu}$ as

$$\theta_i = \int^{\mu_i} V_i^{-1}(t)dt, \quad i = 1, \cdots, n. \tag{7.37}$$

Then we have

$$\frac{\partial \boldsymbol{\theta}}{\partial \boldsymbol{\mu}^T} = \boldsymbol{V}^{-1}(\boldsymbol{\mu}), \quad \frac{\partial \boldsymbol{\mu}}{\partial \boldsymbol{\theta}^T} = \boldsymbol{V}(\boldsymbol{\theta}), \quad \frac{\partial \boldsymbol{\theta}}{\partial \boldsymbol{\beta}^T} = \boldsymbol{V}^{-1}(\boldsymbol{\beta})\boldsymbol{D}(\boldsymbol{\beta}). \tag{7.38}$$

Here $\boldsymbol{\theta}$ play a role of natural parameters as discussed in Chapters 2 and 3. Once coordinate $\boldsymbol{\theta}$ and solution locus $\boldsymbol{\theta} = \boldsymbol{\theta}(\boldsymbol{\beta})$ are defined in R^n from (7.37) and (7.38), we can get all the expressions shown in Section 3.3 and the dual geometry can also be introduced for quasi-likelihood models.

About the statistical inference related to the curvature, we usually cannot get asymptotic results as shown in Chapter 4 for quasi-likelihood models. The reasons are: (a) we have no real likelihoods; and (b) we only have the first two moments of $\boldsymbol{Y}$, but the first four moments are usually needed to obtain the asymptotic results. However, under some regularity conditions, we can get quasi-likelihood ratio statistics (McCullagh 1983). So inference regions as shown in Chapter 5 can be extended to quasi-likelihood models.

Finally, we note that if the components of the response $\boldsymbol{Y}$ are not independent, that is $\boldsymbol{V}(\boldsymbol{\mu})$ is not diagonal, one can similarly define the quasi-likelihood functions (McCullagh and Nelder 1989). But in this case, the statistical inference is rather complicated even though the geometric framework is quite similar to the independent case.

7.5 Covariance Structure Models

As an application of quasi-likelihood models discussed in Section 7.4, we study covariance structure models which have wide applicability in psychology, education, marketing research, econometrics and so forth (see Lee and Bentler 1992 for a review of the theory; see also Browne 1974).

Suppose that $q \times 1$ vectors $\boldsymbol{X}_1, \cdots, \boldsymbol{X}_n, \boldsymbol{X}_{n+1}$ are iid observations with mean zero and covariance matrix $\boldsymbol{\Sigma}(\boldsymbol{\beta})$, where $\boldsymbol{\Sigma}(\boldsymbol{\beta}) = (\sigma_{ij}(\boldsymbol{\beta}))$, $(i, j =$

$1, \cdots, q$), and $\sigma_{ij}(\boldsymbol{\beta})$'s are functions of an unknown $p \times 1$ vector parameter $\boldsymbol{\beta}$ defined in an open subset $\mathcal{B}$ in R^p. In covariance structure analysis, one's interest is to investigate the statistical behavior related to the structure parameter $\boldsymbol{\beta}$ based on the sample covariance matrix $\boldsymbol{S}$, the unbiased estimator of $\boldsymbol{\Sigma}$, where

$$\boldsymbol{S} = \frac{1}{n}\sum_{i=1}^{n+1}(\boldsymbol{X}_i - \overline{\boldsymbol{X}})(\boldsymbol{X}_i - \overline{\boldsymbol{X}})^T, \quad \overline{\boldsymbol{X}} = \frac{1}{n+1}\sum_{i=1}^{n+1}\boldsymbol{X}_i.$$

Following Browne (1974), we assume that the fourth-order cumulant of $\boldsymbol{X}_1$ is zero. Then the elements of $\boldsymbol{S} = (s_{ij})$ asymptotically have normal distributions and satisfy

$$\mathrm{E}(s_{ij}) = \sigma_{ij}, \quad \mathrm{Cov}(s_{ij},\ s_{gh}) = n^{-1}(\sigma_{ig}\sigma_{jh} + \sigma_{ih}\sigma_{jg}). \tag{7.39}$$

To study estimation problems in covariance structure analysis, the following notations will be used. Suppose $\boldsymbol{A}$ is a $q \times q$ symmetric matrix, $\mathrm{vec}(\boldsymbol{A})$ represents the $q^2 \times 1$ column vector stacking the column of $\boldsymbol{A}$, while $\mathrm{vecs}(\boldsymbol{A})$ represents the column vector obtained from the $q^* = q(q+1)/2$ nonduplicated elements of $\boldsymbol{A}$. Let $\boldsymbol{K}_q$ be the q^2 by q^* transition matrix such that $\mathrm{vecs}(\boldsymbol{A}) = \boldsymbol{K}_q^T\mathrm{vec}(\boldsymbol{A})$ and $\mathrm{vec}(\boldsymbol{A}) = \boldsymbol{K}_q^{-T}\mathrm{vecs}(\boldsymbol{A})$, where $\boldsymbol{K}_q^- = (\boldsymbol{K}_q^T\boldsymbol{K}_q)^{-1}\boldsymbol{K}_q^T$ and $\boldsymbol{K}_q^{-T} = (\boldsymbol{K}_q^-)^T$. The basic properties of $\boldsymbol{K}_q$ can be found in Browne (1974). Further, we let $\boldsymbol{s} = \mathrm{vecs}(\boldsymbol{S})$, $\boldsymbol{\sigma} = \mathrm{vecs}(\boldsymbol{\Sigma})$ and $\boldsymbol{e} = \sqrt{n}(\boldsymbol{s} - \boldsymbol{\sigma})$. When $\boldsymbol{\Sigma} = \boldsymbol{\Sigma}(\boldsymbol{\beta})$, then $\boldsymbol{\sigma} = \boldsymbol{\sigma}(\boldsymbol{\beta})$ and $\boldsymbol{e} = \boldsymbol{e}(\boldsymbol{\beta})$. The first two derivatives of $\boldsymbol{\sigma}(\boldsymbol{\beta})$ in $\boldsymbol{\beta}$ are denoted by $\boldsymbol{D}(\boldsymbol{\beta})$ and $\boldsymbol{W}(\boldsymbol{\beta})$, which are $q^* \times p$ matrix and $q^* \times p \times p$ array, respectively. The Kronecker product of matrices is denoted by $\otimes$.

We shall study three kinds of estimators of $\boldsymbol{\beta}$ and unify them to quasi-likelihood estimator.

(a) Maximum likelihood estimator. We first consider the normal case in which $\boldsymbol{X}_1$ is normally distributed as $N_q(\boldsymbol{0},\ \boldsymbol{\Sigma})$. Then $\boldsymbol{S}$ has Wishart distribution $W_q(n,\ n^{-1}\boldsymbol{\Sigma})$ and satisfies

$$\mathrm{E}(\boldsymbol{s}) = \boldsymbol{\sigma}, \quad \mathrm{Var}(\boldsymbol{s}) = n^{-1}\boldsymbol{V}, \quad \boldsymbol{V} = 2\boldsymbol{K}_q^T(\boldsymbol{\Sigma}\otimes\boldsymbol{\Sigma})\boldsymbol{K}_q, \tag{7.40}$$

which can be obtained from (7.39) (see also Browne 1974). The log-likelihood of $\boldsymbol{S}$ is

$$l(\boldsymbol{\Sigma},\ \boldsymbol{S}) = -2^{-1}n\{\log|\boldsymbol{\Sigma}| + \mathrm{tr}(\boldsymbol{S}\boldsymbol{\Sigma}^{-1}) - \log|\boldsymbol{S}| - q\}. \tag{7.41}$$

Then the score functions for $\boldsymbol{\sigma}$ and $\boldsymbol{\beta}$ are easily obtained from this. In fact, we have

$$\frac{\partial l}{\partial \sigma_t} = \frac{n}{2}\mathrm{tr}\{\frac{\partial \boldsymbol{\Sigma}}{\partial \sigma_t}\ \boldsymbol{\Sigma}^{-1}(\boldsymbol{S} - \boldsymbol{\Sigma})\boldsymbol{\Sigma}^{-1}\}, \quad t = 1, \cdots, q^*, \tag{7.42}$$

which results in

$$\frac{\partial l}{\partial \sigma} = \sqrt{n}\boldsymbol{V}^{-1}e, \quad \frac{\partial l}{\partial \boldsymbol{\beta}} = \sqrt{n}\boldsymbol{D}^T\boldsymbol{V}^{-1}e, \quad e = \sqrt{n}(s - \boldsymbol{\sigma}). \tag{7.43}$$

Then $\hat{\boldsymbol{\beta}}$, the maximum likelihood estimator of $\boldsymbol{\beta}$, will be determined by the score function (7.43) and satisfies $\{\boldsymbol{D}^T(\boldsymbol{\beta})\boldsymbol{V}^{-1}(\boldsymbol{\beta})\boldsymbol{e}(\boldsymbol{\beta})\}_{\hat{\boldsymbol{\beta}}} = \mathbf{0}$, which is similar to (7.35). It is easily seen that the score function (7.43) exactly coincides with (7.34), so it may be regarded as a special case of quasi-score function.

(b) Generalized least squares estimator. If the distribution of $\boldsymbol{X}_1$ is unknown, but its fourth-order cumulant is assumed to vanish, then the generalized least squares estimator is often considered (Lee and Jennrich 1979; Lee and Bentler 1992). The generalized least squares estimator $\tilde{\boldsymbol{\beta}}$ of $\boldsymbol{\beta}$ minimizes the following objective function (Lee and Bentler 1992, p.108)

$$G(\boldsymbol{\beta}) = 2^{-1}n \ \mathrm{tr}\{(\boldsymbol{S} - \boldsymbol{\Sigma}(\boldsymbol{\beta}))\boldsymbol{Z}\}^2,$$

where $\boldsymbol{Z}$ is a positive definite matrix that converges in probability to a positive definite matrix $\boldsymbol{Z}^*$. Then $\tilde{\boldsymbol{\beta}}$ satisfies $(\partial G/\partial \boldsymbol{\beta})_{\tilde{\boldsymbol{\beta}}} = \mathbf{0}$, where

$$\frac{\partial G}{\partial \boldsymbol{\beta}} = -2\sqrt{n}\boldsymbol{D}^T\boldsymbol{V}_z^{-1}e, \quad \boldsymbol{V}_z^{-1} = \frac{1}{2}\boldsymbol{K}_q^-(\boldsymbol{Z} \otimes \boldsymbol{Z})\boldsymbol{K}_q^{-T}. \tag{7.44}$$

Ignoring the constant, (7.44) is very similar to (7.43). In particular, if we take $\boldsymbol{Z}^* = \boldsymbol{\Sigma}^{-1}$, then $\boldsymbol{V}_z^{-1}$ converges in probability to $\boldsymbol{V}^{-1}$ (see (7.40)) and $\tilde{\boldsymbol{\beta}}$ becomes the best generalized least squares estimator of $\boldsymbol{\beta}$ (Lee and Bentler 1992). In this case, (7.44) is the same as (7.43) (ignoring a constant), so generalized least squares estimator can also be regarded as a special case of quasi-likelihood estimator.

(c) Quasi-likelihood estimator. As the extensions of maximum likelihood estimator and generalized least squares estimator introduced above, we may consider the quasi-likelihood estimator based on (7.43) and (7.44). Now we also assume that the fourth-order cumulant of $\boldsymbol{X}_1$ is zero. Then following (7.33) to (7.35), the quasi-likelihood function in covariance structure analysis is defined as

$$l_q(\boldsymbol{\sigma}, \boldsymbol{S}) = \int^{\boldsymbol{\sigma}} (\sqrt{n}\boldsymbol{V}^{-1}\boldsymbol{e})^T d\boldsymbol{\sigma}, \tag{7.45}$$

where $\boldsymbol{V}$ and $\boldsymbol{e}$ are defined in (7.40) and (7.43), respectively. Then the quasi-score functions for $\boldsymbol{\sigma}$ and $\boldsymbol{\beta}$ are respectively

$$\frac{\partial l_q}{\partial \boldsymbol{\sigma}} = \sqrt{n}\boldsymbol{V}^{-1}(\boldsymbol{\sigma})e(\boldsymbol{\sigma}), \quad \frac{\partial l_q}{\partial \boldsymbol{\beta}} = \sqrt{n}\boldsymbol{D}^T\boldsymbol{V}^{-1}e. \tag{7.46}$$

The maximum quasi-likelihood estimator $\tilde{\beta}$ is then defined as an estimator that maximizes $l_q(\sigma(\beta); y)$, so that it satisfies (7.35) but with e replaced by $e(\beta) = \sqrt{n}(s - \sigma(\beta))$.

Combining these three estimators, we may unify them to quasi-likelihood estimators as discussed in Section 7.4. So the geometric framework introduced there is also valid to covariance structure models and it can be used to the statistical analysis related to maximum likelihood estimator, generalized least squares estimator and quasi-likelihood estimator introduced earlier.

As pointed out in Section 7.4, to use quasi-likelihood, $-\partial^2 l_q/\partial\sigma\partial\sigma^T$ should be symmetric or the conditions in (7.36) are satisfied. Fortunately, covariance structure models have good properties in this respect. The following two lemmas support the use of quasi-likelihood theory to covariance structure models.

Lemma 7.4

For covariance structure models, suppose that the quasi-likelihood function and the quasi-score functions are defined in (7.45) and (7.46), respectively, then the quasi-observed informations and the quasi-Fisher informations of S for σ and β are respectively given by

$$-(\ddot{l}_q)_{\sigma\sigma} = -\frac{\partial^2 l_q}{\partial\sigma\partial\sigma^T} = nV^{-1} - \sqrt{n}[e^T][\Omega]; \tag{7.47}$$

$$-(\ddot{l}_q)_{\beta\beta} = -\frac{\partial^2 l_q}{\partial\beta\partial\beta^T} = nD^T V^{-1} D - \sqrt{n}[e^T V^{-1}][\Gamma], \tag{7.48}$$

$$\Gamma = W + [V][D^T \Omega D];$$

and

$$J(\sigma) = -\mathrm{E}((\ddot{l}_q)_{\sigma\sigma}) = nV^{-1}, \quad J(\beta) = -\mathrm{E}((\ddot{l}_q)_{\beta\beta}) = nD^T V^{-1} D; \tag{7.49}$$

where $-(\ddot{l}_q)_{\sigma\sigma}$ is a symmetric matrix, $\Omega = \partial V^{-1}(\sigma)/\partial\sigma$ is a symmetric array and Ω satisfies

$$[e^T][\Omega] = -\frac{1}{2}\sqrt{n}K_q^-\{[\Sigma^{-1}(S-\Sigma)\Sigma^{-1}]\otimes\Sigma^{-1} + \Sigma^{-1}\otimes[\Sigma^{-1}(S-\Sigma)\Sigma^{-1}]\}K_q^{-T}. \tag{7.50}$$

Proof. Equations (7.48) and (7.49) can be obtained from (7.47), so we just prove (7.47) and (7.50). It follows from (7.40), (7.43) and (7.46) that

$$-(\ddot{l}_q)_{\sigma\sigma} = -\sqrt{n}\frac{\partial(V^{-1}e)}{\partial\sigma^T}$$

$$
\begin{aligned}
&= -\sqrt{n}\frac{\partial}{\partial \sigma^T}\left\{\frac{1}{2}K_q^-(\Sigma^{-1}\otimes\Sigma^{-1})K_q^{-T}\sqrt{n}\mathrm{vecs}(S-\Sigma)\right\}\\
&= -\frac{1}{2}nK_q^-\frac{\partial}{\partial\sigma^T}\{(\Sigma^{-1}\otimes\Sigma^{-1})\mathrm{vec}(S-\Sigma)\}\\
&= -\frac{1}{2}nK_q^-\frac{\partial}{\partial\sigma^T}\mathrm{vec}\{\Sigma^{-1}(S-\Sigma)\Sigma^{-1}\}. \qquad (7.51)
\end{aligned}
$$

Now for any σ_t $(t = 1,\cdots,q^*)$, we have

$$
\begin{aligned}
&-\frac{\partial}{\partial\sigma_t}\mathrm{vec}\{\Sigma^{-1}(S-\Sigma)\Sigma^{-1}\}\\
&= \mathrm{vec}\{\Sigma^{-1}\frac{\partial\Sigma}{\partial\sigma_t}\Sigma^{-1}(S-\Sigma)\Sigma^{-1}+\Sigma^{-1}(S-\Sigma)\Sigma^{-1}\frac{\partial\Sigma}{\partial\sigma_t}\Sigma^{-1}+\\
&\quad \Sigma^{-1}\frac{\partial\Sigma}{\partial\sigma_t}\Sigma^{-1}\}\\
&= \{[\Sigma^{-1}(S-\Sigma)\Sigma^{-1}]\otimes\Sigma^{-1}+\Sigma^{-1}\otimes[\Sigma^{-1}(S-\Sigma)\Sigma^{-1}]+\\
&\quad \Sigma^{-1}\otimes\Sigma^{-1}\}\mathrm{vec}\frac{\partial\Sigma}{\partial\sigma_t}\\
&= \{[\Sigma^{-1}(S-\Sigma)\Sigma^{-1}]\otimes\Sigma^{-1}+\Sigma^{-1}\otimes[\Sigma^{-1}(S-\Sigma)\Sigma^{-1}]+\\
&\quad \Sigma^{-1}\otimes\Sigma^{-1}\}K_q^{-T}\frac{\partial(\mathrm{vecs}\Sigma)}{\partial\sigma_t}.
\end{aligned}
$$

Substituting this result into (7.51) yields

$$
\begin{aligned}
-(\ddot{l}_q)_{\sigma\sigma} &= \frac{1}{2}nK_q^-(\Sigma^{-1}\otimes\Sigma^{-1})K_q^{-T}+\\
&\quad \frac{1}{2}nK_q^-\{[\Sigma^{-1}(S-\Sigma)\Sigma^{-1}]\otimes\Sigma^{-1}+\\
&\quad \Sigma^{-1}\otimes[\Sigma^{-1}(S-\Sigma)\Sigma^{-1}]\}K_q^{-T}. \qquad (7.52)
\end{aligned}
$$

On the other hand, we also have

$$
\begin{aligned}
-(\ddot{l}_q)_{\sigma\sigma} &= -\sqrt{n}\frac{\partial(V^{-1}e)}{\partial\sigma^T}\\
&= -\sqrt{n}[e^T][\frac{\partial V^{-1}}{\partial\sigma}]-\sqrt{n}V^{-1}\frac{\partial e}{\partial\sigma^T}\\
&= nV^{-1}-\sqrt{n}[e^T][\Omega]. \qquad (7.53)
\end{aligned}
$$

Combining (7.40), (7.52) and (7.53), we get (7.48) and (7.50). It is easily seen from (7.52) that $-(\ddot{l}_q)_{\sigma\sigma}$ is symmetric. We can also directly prove that $\Omega = \partial V^{-1}(\sigma)/\partial\sigma$ is a symmetric array. ∎

This lemma shows that $-(\ddot{l}_q)_{\sigma\sigma}$ is a symmetric matrix and the quasi-likelihood theory can be applied to covariance structure models (see Section 7.4). Further, the dual parameter of σ must exist (see (7.33), (7.34)), so that the dual geometry can be established in covariance structure analysis. In fact, we can get an explicit expression for the dual parameter.

Lemma 7.5

For the covariance structure models introduced earlier, there exists a parameter $\boldsymbol{\rho} = \boldsymbol{\rho}(\boldsymbol{\sigma})$ satisfying

$$\boldsymbol{\rho} = -\frac{1}{2}(\boldsymbol{K}_q^T \boldsymbol{K}_q)^{-1}\text{vecs}(\boldsymbol{\Sigma}^{-1}), \quad \frac{\partial \boldsymbol{\rho}}{\partial \boldsymbol{\sigma}^T} = \boldsymbol{V}^{-1}, \quad \frac{\partial \boldsymbol{\sigma}}{\partial \boldsymbol{\rho}^T} = \boldsymbol{V}.$$

Proof. We start with the following identity

$$\frac{\partial}{\partial \sigma_t}\text{vec}(\boldsymbol{\Sigma}\boldsymbol{\Sigma}^{-1}) = \mathbf{0}, \qquad t = 1, \cdots, q^*,$$

which leads to

$$\text{vec}\left\{\frac{\partial \boldsymbol{\Sigma}}{\partial \sigma_t}\boldsymbol{\Sigma}^{-1} + \boldsymbol{\Sigma}\frac{\partial \boldsymbol{\Sigma}^{-1}}{\partial \sigma_t}\right\} = \mathbf{0}.$$

So we have

$$\begin{aligned}(\boldsymbol{I}_q \otimes \boldsymbol{\Sigma})\frac{\partial \text{vec}(\boldsymbol{\Sigma}^{-1})}{\partial \sigma_t} &= -(\boldsymbol{\Sigma}^{-1} \otimes \boldsymbol{I}_q)\frac{\partial\ \text{vec}(\boldsymbol{\Sigma})}{\partial \sigma_t}, \\ \boldsymbol{K}_q^{-T}\frac{\partial \text{vecs}(\boldsymbol{\Sigma}^{-1})}{\partial \sigma_t} &= -(\boldsymbol{\Sigma}^{-1} \otimes \boldsymbol{\Sigma}^{-1})\boldsymbol{K}_q^{-T}\frac{\partial\ \text{vecs}(\boldsymbol{\Sigma})}{\partial \sigma_t}.\end{aligned}$$

Since $\boldsymbol{V}^{-1} = 2^{-1}\boldsymbol{K}_q^{-}(\boldsymbol{\Sigma}^{-1} \otimes \boldsymbol{\Sigma}^{-1})\boldsymbol{K}_q^{-T}$, we get

$$-\frac{1}{2}\boldsymbol{K}_q^{-}\boldsymbol{K}_q^{-T}\frac{\partial \text{vecs}(\boldsymbol{\Sigma}^{-1})}{\partial \sigma_t} = \boldsymbol{V}^{-1}\frac{\partial \text{vecs}(\boldsymbol{\Sigma})}{\partial \sigma_t},$$

which gives the desired result since $\text{vecs}(\boldsymbol{\Sigma}) = \boldsymbol{\sigma}$. ∎

About the statistical inference related to the curvatures, almost all the results given in Chapters 4 and 5 still hold in covariance structure models. The reasons for this are: (a) we can get very similar geometric framework as shown in Chapter 3 for these models; and (b) the "random error" $\boldsymbol{e} = \sqrt{n}(\boldsymbol{s} - \boldsymbol{\sigma})$ is asymptotic normal, so we can get the first four asymptotic moments of $\boldsymbol{e}$. From (a) and (b), we can get parallel results as shown in Chapters 4 and 5 in covariance structure models and the details are omitted here. Wang and Lee (1995) provided some asymptotic results for the model from the geometric viewpoint.

Appendix A

Matrix and Array

A.1 Projection in Weighted Inner Product Space

Suppose that the inner product of two vectors $\boldsymbol{a}$ and $\boldsymbol{b}$ in R^n is defined as $\langle \boldsymbol{a}, \boldsymbol{b} \rangle = \boldsymbol{a}^T \boldsymbol{V}^{-1} \boldsymbol{b}$, where $\boldsymbol{V}$ is a positive definite matrix. The norm of a vector $\boldsymbol{a}$ is defined as $\|\boldsymbol{a}\| = (\boldsymbol{a}^T \boldsymbol{V}^{-1} \boldsymbol{a})^{\frac{1}{2}}$. Given a linear subspace S in R^n, the linear subspace $S^{\perp}$ is called the orthocomplement of S if $S + S^{\perp} = R^n$ and vector $\boldsymbol{b} \in S^{\perp}$ if and only if $\boldsymbol{a}^T \boldsymbol{V}^{-1} \boldsymbol{b} = 0$ for any $\boldsymbol{a} \in S$. For any given vector $\boldsymbol{x} \in R^n$, if $\boldsymbol{x}$ is decomposed as

$$\boldsymbol{x} = \boldsymbol{a} + \boldsymbol{b}, \quad \boldsymbol{a} \in S, \quad \boldsymbol{b} \in S^{\perp}, \tag{A.1}$$

then $\boldsymbol{a}$ is called the orthogonal projection of $\boldsymbol{x}$ onto S. It is easily seen that this decomposition is unique. In fact, if $\boldsymbol{x} = \boldsymbol{a} + \boldsymbol{b} = \boldsymbol{a}' + \boldsymbol{b}'$ ($\boldsymbol{a}' \in S$, $\boldsymbol{b}' \in S^{\perp}$), then $\boldsymbol{a} - \boldsymbol{a}' = \boldsymbol{b}' - \boldsymbol{b}$. Since $\boldsymbol{a} - \boldsymbol{a}' \in S$ and $\boldsymbol{b}' - \boldsymbol{b} \in S^{\perp}$, we have $\boldsymbol{a} - \boldsymbol{a}' = \boldsymbol{b} - \boldsymbol{b}' = \boldsymbol{0}$, which results in $\boldsymbol{a} = \boldsymbol{a}'$, $\boldsymbol{b} = \boldsymbol{b}'$. If an $n \times n$ matrix $\boldsymbol{P}$ satisfies $\boldsymbol{P}\boldsymbol{x} = \boldsymbol{a}$ for any $\boldsymbol{x} \in R^n$ subject to (A.1), then $\boldsymbol{P}$ is called the orthogonal projection matrix of S. In particular, if S is spanned by an $n \times p$ matrix $\boldsymbol{A}$ ($p \leq n$), then $\boldsymbol{P}$ is denoted by $\boldsymbol{P}_A$ and is called the orthogonal projection matrix of $\boldsymbol{A}$, or the projection matrix of $\boldsymbol{A}$ in brief. In this case, the orthogonal projection matrix of $S^{\perp}$ may be denoted by $\boldsymbol{P}_B$ where $\boldsymbol{B}$ is an $n \times (n-r)$ matrix satisfying $\boldsymbol{A}^T \boldsymbol{V}^{-1} \boldsymbol{B} = \boldsymbol{0}$ and r is the rank of $\boldsymbol{A}$. In fact, we can get $\boldsymbol{B}$ as follows. The solutions of homogeneous linear equation $\boldsymbol{A}^T \boldsymbol{V}^{-1} \boldsymbol{y} = \boldsymbol{0}$ consist of $(n-r)$ linearly independent vectors, say $\boldsymbol{b}_1, \cdots, \boldsymbol{b}_{n-r}$, then we can set $\boldsymbol{B} = (\boldsymbol{b}_1, \cdots, \boldsymbol{b}_{n-r})$.

Theorem A.1

If $\boldsymbol{A}$ is an $n \times p$ matrix ($p \leq n$) and of full rank in column, then we have

$$\boldsymbol{P}_A = \boldsymbol{A}(\boldsymbol{A}^T\boldsymbol{V}^{-1}\boldsymbol{A})^{-1}\boldsymbol{A}^T\boldsymbol{V}^{-1}$$

and

$$\boldsymbol{P}_A^2 = \boldsymbol{P}_A, \qquad \boldsymbol{P}_A^T\boldsymbol{V}^{-1} = \boldsymbol{V}^{-1}\boldsymbol{P}_A. \tag{A.2}$$

Proof. Let S be a subspace of R^n spanned by $\boldsymbol{A}$ and $S^\perp$ be its orthocomplement. Suppose that $\boldsymbol{B}$ is an $n\times(n-p)$ matrix so that $\boldsymbol{A}^T\boldsymbol{V}^{-1}\boldsymbol{B} = \boldsymbol{0}$ and $S^\perp$ is spanned by $\boldsymbol{B}$. For any $\boldsymbol{x} = \boldsymbol{a} + \boldsymbol{b}$ satisfying (A.1), there exist two vectors $\boldsymbol{\alpha}$ and $\boldsymbol{\beta}$ such that $\boldsymbol{a} = \boldsymbol{A\alpha}$ and $\boldsymbol{b} = \boldsymbol{B\beta}$ hold. Thus we get $\boldsymbol{x} = \boldsymbol{A\alpha} + \boldsymbol{B\beta}$ and

$$\boldsymbol{P}_A\boldsymbol{x} = \boldsymbol{P}_A\boldsymbol{A\alpha} + \boldsymbol{P}_A\boldsymbol{B\beta} = \boldsymbol{A\alpha}.$$

This equation holds for any $\boldsymbol{x} \in R^n$, and for any $\boldsymbol{\alpha}$ and $\boldsymbol{\beta}$. Then we have

$$\boldsymbol{P}_A\boldsymbol{A} = \boldsymbol{A}, \qquad \boldsymbol{P}_A\boldsymbol{B} = \boldsymbol{0}. \tag{A.3}$$

It follows from $\boldsymbol{P}_A\boldsymbol{B} = (\boldsymbol{P}_A\boldsymbol{V})\boldsymbol{V}^{-1}\boldsymbol{B} = \boldsymbol{0}$ that all the columns of $(\boldsymbol{P}_A\boldsymbol{V})^T$ are orthogonal to $S^\perp$ and hence are located in the subspace S. Then $(\boldsymbol{P}_A\boldsymbol{V})^T$ can be represented as $(\boldsymbol{P}_A\boldsymbol{V})^T = \boldsymbol{AU}$ for some $\boldsymbol{U}$, which results in

$$\boldsymbol{P}_A = \boldsymbol{U}^T\boldsymbol{A}^T\boldsymbol{V}^{-1}. \tag{A.4}$$

Substitution of this result into (A.3) yields

$$\boldsymbol{U}^T\boldsymbol{A}^T\boldsymbol{V}^{-1}\boldsymbol{A} = \boldsymbol{A}, \qquad \boldsymbol{U} = (\boldsymbol{A}^T\boldsymbol{V}^{-1}\boldsymbol{A})^{-1}\boldsymbol{A}^T.$$

Substituting $\boldsymbol{U}^T = \boldsymbol{A}(\boldsymbol{A}^T\boldsymbol{V}^{-1}\boldsymbol{A})^{-1}$ into (A.4) gives $\boldsymbol{P}_A = \boldsymbol{A}(\boldsymbol{A}^T\boldsymbol{V}^{-1}\boldsymbol{A})^{-1}$ $\boldsymbol{A}^T\boldsymbol{V}^{-1}$ and (A.2) can be easily obtained from this. ∎

Corollary

If the columns of $\boldsymbol{Q}$ and $\boldsymbol{N}$ are respectively an orthonormal basis of S and $S^\perp$, then the projection matrices of S and $S^\perp$ are

$$\boldsymbol{P}_T = \boldsymbol{Q}\boldsymbol{Q}^T\boldsymbol{V}^{-1}, \qquad \boldsymbol{P}_N = \boldsymbol{N}\boldsymbol{N}^T\boldsymbol{V}^{-1} \tag{A.5}$$

satisfying $\boldsymbol{P}_T + \boldsymbol{P}_N = \boldsymbol{I}_n$ that is an identity matrix.

Proof. Equation (A.5) is easily obtained from $\boldsymbol{Q}^T\boldsymbol{V}^{-1}\boldsymbol{Q} = \boldsymbol{I}_p$ and $\boldsymbol{N}^T\boldsymbol{V}^{-1}\boldsymbol{N} = \boldsymbol{I}_{n-p}$. Further, for any given vector $\boldsymbol{x} \in R^n$, it can be decomposed as the form of (A.1). Then we have

$$(\boldsymbol{P}_T + \boldsymbol{P}_N)\boldsymbol{x} = (\boldsymbol{P}_T + \boldsymbol{P}_N)(\boldsymbol{a} + \boldsymbol{b}) = \boldsymbol{a} + \boldsymbol{b} = \boldsymbol{x}$$

for any $\boldsymbol{x} \in R^n$, which results in $\boldsymbol{P}_T + \boldsymbol{P}_N = \boldsymbol{I}_n$. ∎

Theorem A.2

If A is an $n \times p$ matrix ($p \leq n$) and of full rank in column, then A can be decomposed as

$$A = (Q,\ N)\begin{pmatrix} R \\ 0 \end{pmatrix} = QR, \tag{A.6}$$

which is called the QR decomposition of A, where R is an upper triangular matrix with positive diagonal elements; Q and N satisfy $Q^T V^{-1} Q = I_p$, $Q^T V^{-1} N = 0$ and $N^T V^{-1} N = I_{n-p}$, respectively.

Proof. By Cholesky decomposition (Anderson 1984, p.586), $A^T V^{-1} A$ can be decomposed as $A^T V^{-1} A = R^T R$, where R is an upper triangular matrix with positive diagonal elements. Then A can be represented as

$$A = (AR^{-1})R = QR$$

with $Q = AR^{-1}$. In fact, it is easily seen that

$$Q^T V^{-1} Q = R^{-T}(A^T V^{-1} A) R^{-1} = I_p.$$

Similarly, Let B be an $n \times (n-p)$ matrix so that $A^T V^{-1} B = 0$, then B can be decomposed as $B = NR_1$ satisfying $N^T V^{-1} N = I_{n-p}$ and $Q^T V^{-1} N = 0$ (this follows from $A^T V^{-1} B = 0$). So (A.6) is proved. ∎

Theorem A.3

Suppose that $A = (A_1, A_2)$ and the projection matrix of A_1 and A_2 are respectively denoted by P_1 and P_2. Then we have the following results.

(a) If ${A_1}^T V^{-1} A_2 = 0$, then $P_A = P_1 + P_2$.

(b) For the general case, $P_A = P_1 + P_{(I-P_1)A_2}$.

Proof. **(a)** It follows from $A = (A_1, A_2)$ and $A_1^T V^{-1} A_2 = 0$ that

$$A^T V^{-1} A = \operatorname{diag}(A_1^T V^{-1} A_1,\ A_2^T V^{-1} A_2).$$

Then by direct calculation using $P_A = A(A^T V^{-1} A)^{-1} A^T V^{-1}$, we get $P_A = P_1 + P_2$.

(b) Let S be the subspace of R^n spanned by A, then a vector $x \in S$ if and only if $x = Aa$ for some a. Let $a^T = (a_1^T,\ a_2^T)$, then we have

$$\begin{aligned} x &= A_1 a_1 + A_2 a_2 \\ &= A_1 a_1 + P_1 A_2 a_2 + (I - P_1) A_2 a_2 \end{aligned}$$

$$\begin{aligned} &= \boldsymbol{A}_1\{\boldsymbol{a}_1 + (\boldsymbol{A}_1^T \boldsymbol{V}^{-1}\boldsymbol{A}_1)^{-1}\boldsymbol{A}_1^T\boldsymbol{V}^{-1}\boldsymbol{A}_2\boldsymbol{a}_2\} + (\boldsymbol{I} - \boldsymbol{P}_1)\boldsymbol{A}_2\boldsymbol{a}_2 \\ &= \boldsymbol{A}_1\boldsymbol{a}_1' + (\boldsymbol{I} - \boldsymbol{P}_1)\boldsymbol{A}_2\boldsymbol{a}_2 = \boldsymbol{A}'\boldsymbol{a}', \end{aligned}$$

where $\boldsymbol{A}' = (\boldsymbol{A}_1,\ (\boldsymbol{I} - \boldsymbol{P}_1)\boldsymbol{A}_2)$, $(\boldsymbol{a}')^T = ((\boldsymbol{a}_1')^T,\ \boldsymbol{a}_2^T)$ and

$$\boldsymbol{a}_1{}' = \boldsymbol{a}_1 + (\boldsymbol{A}_1^T\boldsymbol{V}^{-1}\boldsymbol{A}_1)^{-1}\boldsymbol{A}_1^T\boldsymbol{V}^{-1}\boldsymbol{A}_2\boldsymbol{a}_2.$$

So S is also spanned by $\boldsymbol{A}'$, i.e. $\boldsymbol{P}_A = \boldsymbol{P}_{A'}$. Since $\boldsymbol{A}_1^T\boldsymbol{V}^{-1}(\boldsymbol{I} - \boldsymbol{P}_1)\boldsymbol{A}_2 = \boldsymbol{0}$, we get (b) from (a). ∎

A.2 Array Multiplication

Multiplication of 3-dimensional arrays was first introduced by Bates and Watts (1980). Since then, many authors have used and discussed this multiplication. Tsai (1983), Cook and Goldberg (1986) and Seber and Wild (1989) have presented good descriptions for that. Here we give more extensions and technical details (see also Wei 1989).

A 3-dimensional $n \times p \times q$ array is denoted by $\boldsymbol{X} = (X_{tij})$, where indices t, i and j indicate face, row and column, respectively. There are two ways to see an array: (a) $\boldsymbol{A} = (\boldsymbol{A}_{ij})$ and each $\boldsymbol{A}_{ij}$ is an n-vector $\boldsymbol{A}_{ij} = (A_{1ij}, \cdots, A_{nij})^T$ for any fixed i, j; and (b) $\boldsymbol{A} = (\boldsymbol{A}_t)$ and each $\boldsymbol{A}_t$ is a $p \times q$ matrix $\boldsymbol{A}_t = (A_{tij})$ for any fixed t and $\boldsymbol{A}_t$ is called the t-th face of $\boldsymbol{A}$.

Definition A.1

If $\boldsymbol{X}$ is an $n \times p \times q$ array, $\boldsymbol{A}$ and $\boldsymbol{B}$ are $r \times p$ and $q \times s$ matrices respectively, then $\boldsymbol{Y} = \boldsymbol{AXB}$ is defined as an $n \times r \times s$ array with elements:

$$Y_{tkl} = \sum_{i=1}^{p}\sum_{j=1}^{q} A_{ki}X_{tij}B_{jl}.$$

Definition A.2

If $\boldsymbol{X}$ is an $n \times p \times q$ array, $\boldsymbol{A}$ is an $m \times n$ matrix, then $\boldsymbol{Y} = [\boldsymbol{A}][\boldsymbol{X}]$ is called the bracket product of $\boldsymbol{A}$ and $\boldsymbol{X}$, that is an $m \times p \times q$ array with elements

$$Y_{tij} = \sum_{k=1}^{n} A_{tk}X_{kij}.$$

Definition A.3

If $\boldsymbol{X}$ is an $n \times p \times p$ array, then the trace of $\boldsymbol{X}$ is denoted by $\mathrm{tr}[\boldsymbol{X}]$ and defined as an n-vector with the element $\mathrm{tr}(\boldsymbol{X}_i)$ where $\boldsymbol{X}_i$ is the i-th face of

$\boldsymbol{X}$, $i = 1, \cdots, n$. If $\boldsymbol{Y}$ is an $n \times p \times q$ array, then the vectorization of $\boldsymbol{Y}$ is defined as a $pq \times n$ matrix Vec[$\boldsymbol{Y}$] whose t-th column is Vec($\boldsymbol{Y}_t$), i.e. Y_{tij} is the element of Vec[$\boldsymbol{Y}$] at $\{(j-1)p+i, t\}$.

The following properties can be obtained directly from the above definitions. Now let $\boldsymbol{X}, \boldsymbol{Y} \cdots$ denote arrays; $\boldsymbol{A}$, $\boldsymbol{B}$, $\boldsymbol{L}$, $\boldsymbol{M}, \cdots$ denote matrices, $\boldsymbol{a}$, $\boldsymbol{b}$, $\boldsymbol{c}$, $\cdots$ denote vectors and λ, μ denote real numbers. Then we have

(1) $[\boldsymbol{I}_n][\boldsymbol{X}] = \boldsymbol{X}$ ($\boldsymbol{I}_n$ is an identity matrix).

(2) $[\lambda \boldsymbol{A} + \mu \boldsymbol{B}][\boldsymbol{X}] = \lambda[\boldsymbol{A}][\boldsymbol{X}] + \mu[\boldsymbol{B}][\boldsymbol{X}]$.

(3) $[\boldsymbol{A}][\lambda \boldsymbol{X} + \mu \boldsymbol{Y}] = \lambda[\boldsymbol{A}][\boldsymbol{X}] + \mu[\boldsymbol{A}][\boldsymbol{Y}]$.

(4) $[\boldsymbol{A}][\boldsymbol{L}\boldsymbol{X}\boldsymbol{M}] = \boldsymbol{L}[\boldsymbol{A}][\boldsymbol{X}]\boldsymbol{M}$.

(5) $[\boldsymbol{A}\boldsymbol{B}][\boldsymbol{X}] = [\boldsymbol{A}][[\boldsymbol{B}][\boldsymbol{X}]]$.

(6) $\boldsymbol{A}(\boldsymbol{a}^T\boldsymbol{X}\boldsymbol{a}) = [\boldsymbol{A}][\boldsymbol{a}^T\boldsymbol{X}\boldsymbol{a}] = \boldsymbol{a}^T[\boldsymbol{A}][\boldsymbol{X}]\boldsymbol{a}$.

(7) $\boldsymbol{a}^T\boldsymbol{X}\boldsymbol{a} = (\boldsymbol{X}\boldsymbol{a})\boldsymbol{a} = \sum_{i=1}^{p}\sum_{j=1}^{p} \boldsymbol{X}_{ij}a_i a_j$.

(8) $[\boldsymbol{A}\boldsymbol{a}][(\boldsymbol{B}\boldsymbol{b})(\boldsymbol{C}\boldsymbol{c})^T] = \boldsymbol{B}[\boldsymbol{A}][[\boldsymbol{a}][\boldsymbol{b}\boldsymbol{c}^T]]\boldsymbol{C}^T$.

(9) tr$[\boldsymbol{A}\boldsymbol{X}]$ =tr$[\boldsymbol{X}\boldsymbol{A}]$.

(10) $\boldsymbol{A}$ tr $[\boldsymbol{X}]$ =tr$[[\boldsymbol{A}][\boldsymbol{X}]]$.

(11) Vec$[[\boldsymbol{A}][\boldsymbol{X}]]$ =(Vec$[\boldsymbol{X}]$)$\boldsymbol{A}^T$.

(12) Vec$[\boldsymbol{A}\boldsymbol{X}\boldsymbol{B}] = (\boldsymbol{B}^T \otimes \boldsymbol{A})Vec[\boldsymbol{X}]$ (where "$\otimes$" denotes the Kronecker product).

Proof. Some of the above properties are quite simple, and we will prove (4), (5), (6), (7), (10) and (11).

(4). Without loss of generality, we just prove $[\boldsymbol{A}][\boldsymbol{L}\boldsymbol{X}] = \boldsymbol{L}[\boldsymbol{A}][\boldsymbol{X}]$. To do so, let $\boldsymbol{Y} = \boldsymbol{L}\boldsymbol{X}$, $\boldsymbol{Z} = [\boldsymbol{A}][\boldsymbol{L}\boldsymbol{X}]$, $\boldsymbol{W} = [\boldsymbol{A}][\boldsymbol{X}]$, and $\boldsymbol{U} = \boldsymbol{L}[\boldsymbol{A}][\boldsymbol{X}]$, then we have

$$Y_{tij} = \sum_e L_{ie}X_{tej},$$

$$Z_{sij} = \sum_t A_{st}Y_{tij} = \sum_t\sum_e A_{st}L_{ie}X_{tej},$$

$$W_{sij} = \sum_t A_{st} X_{tij};$$

hence

$$U_{sij} = \sum_e L_{ie} W_{sej} = \sum_e \sum_t L_{ie} A_{st} X_{tej} = Z_{sij}. \quad \blacksquare$$

(5). Let $\boldsymbol{C} = \boldsymbol{AB}$, $\boldsymbol{Y}$=$[\boldsymbol{AB}][\boldsymbol{X}]$, $[\boldsymbol{B}][\boldsymbol{X}] = \boldsymbol{Z}$, $[\boldsymbol{A}][\boldsymbol{Z}] = \boldsymbol{W}$, then we have

$$Y_{tij} = \sum_s C_{ts} X_{sij} = \sum_s \sum_e A_{te} B_{es} X_{sij};$$

hence

$$W_{tij} = \sum_s A_{ts} Z_{sij} = \sum_s \sum_e A_{ts} B_{se} X_{eij} = Y_{tij}. \quad \blacksquare$$

(6). Let $(\boldsymbol{a}^T \boldsymbol{X} \boldsymbol{a}) = \boldsymbol{b}$, $\boldsymbol{A}(\boldsymbol{a}^T \boldsymbol{X} \boldsymbol{a}) = \boldsymbol{c}$, $[\boldsymbol{A}][\boldsymbol{a}^T \boldsymbol{X} \boldsymbol{a}] = \boldsymbol{d}$, $[\boldsymbol{A}][\boldsymbol{X}] = \boldsymbol{Y}$ and $\boldsymbol{a}^T[\boldsymbol{A}][\boldsymbol{X}]\boldsymbol{a} = \boldsymbol{e}$, then we have

$$b_t = \sum_{i,j} X_{tij} a_i a_j; \quad c_s = \sum_t \sum_{i,j} A_{st} X_{tij} a_i a_j;$$

hence

$$d_s = \sum_t A_{st} b_t = \sum_s \sum_{i,j} A_{st} X_{tij} a_i a_j = c_s;$$

$$e_s = \sum_{i,j} Y_{sij} a_i a_j = \sum_{i,j} \sum_t A_{st} X_{tij} a_i a_j = c_s. \quad \blacksquare$$

(7). Let $\boldsymbol{Y} = \boldsymbol{X}\boldsymbol{a}$, $\boldsymbol{a}^T \boldsymbol{X} \boldsymbol{a} = \boldsymbol{b}$, and $\boldsymbol{c} = (\boldsymbol{X}\boldsymbol{a})\boldsymbol{a}$, then we have

$$Y_{ti} = \sum_j X_{tij} a_j;$$

hence

$$c_t = \sum_i Y_{ti} a_i = \sum_i \sum_e X_{tij} a_i a_j = \boldsymbol{a}^T \boldsymbol{X}_t \boldsymbol{a} = b_t. \quad \blacksquare$$

(10). Let $\boldsymbol{Y} = [\boldsymbol{A}][\boldsymbol{X}]$, $\boldsymbol{A}\mathrm{tr}[\boldsymbol{X}] = \boldsymbol{a}$, and $\mathrm{tr}[[\boldsymbol{A}][\boldsymbol{X}]] = \boldsymbol{b}$, then we have

$$a_s = \sum_t A_{st} \mathrm{tr}(\boldsymbol{X}_t) = \sum_t \sum_i A_{st} X_{tii};$$

hence

$$b_s = \mathrm{tr}(\boldsymbol{Y}_s) = \sum_i Y_{sii} = \sum_i \sum_t A_{st} X_{tii} = a_s. \quad \blacksquare$$

(11). Let $[\boldsymbol{A}][\boldsymbol{X}] = \boldsymbol{Y}$, Vec$[\boldsymbol{Y}] = \boldsymbol{L}$, Vec$[\boldsymbol{X}] = \boldsymbol{M}$, and $\boldsymbol{M}\boldsymbol{A}^T = \boldsymbol{B}$. Any positive integral s can be uniquely decomposed as $s = (j-1)p + i$, then we have

$$L_{st} = Y_{tij} = \sum_e A_{te} X_{eij};$$

hence

$$B_{st} = \sum_e M_{se} A_{te} = \sum_e X_{eij} A_{te} = L_{st}.$$

∎

A.3 Differentiation of Vector and Matrix

Let y be a scalar and $\boldsymbol{\eta}$, $\boldsymbol{\beta}$ and $\boldsymbol{\gamma}$ be $n \times 1$, $p \times 1$ and $q \times 1$ vectors, respectively. If y is the function of $\boldsymbol{\beta}$, then the derivatives of y with respect to $\boldsymbol{\beta}$ are defined as

$$\frac{\partial y}{\partial \boldsymbol{\beta}} = (\frac{\partial y}{\partial \beta_a}), \qquad \frac{\partial^2 y}{\partial \boldsymbol{\beta} \partial \boldsymbol{\beta}^T} = (\frac{\partial^2 y}{\partial \beta_a \partial \beta_b}),$$

where $a,\ b = 1, \cdots, p$; $\partial y/\partial \boldsymbol{\beta}$ is a $p \times 1$ vector and $\partial^2 y/\partial \boldsymbol{\beta}\partial\boldsymbol{\beta}^T$ is a $p \times p$ matrix, respectively. If vector $\boldsymbol{\eta}$ is the function of $\boldsymbol{\beta}$, then the derivatives of $\boldsymbol{\eta}$ with respect to $\boldsymbol{\beta}$ are defined as

$$\frac{\partial \boldsymbol{\eta}}{\partial \boldsymbol{\beta}^T} = (\frac{\partial \eta_i}{\partial \beta_a}), \qquad \frac{\partial^2 \boldsymbol{\eta}}{\partial \boldsymbol{\beta} \partial \boldsymbol{\beta}^T} = (\frac{\partial^2 \eta_i}{\partial \beta_a \partial \beta_b}),$$

where $i = 1, \cdots, n$; $a, b = 1, \cdots, p$; $\partial \boldsymbol{\eta}/\partial \boldsymbol{\beta}^T$ is an $n \times p$ matrix and $\partial^2 \boldsymbol{\eta}/\partial \boldsymbol{\beta}\partial\boldsymbol{\beta}^T$ is an $n \times p \times p$ array, respectively.

If $\boldsymbol{\beta}$ is the function of $\boldsymbol{\gamma}$, then the chain rules of the derivatives of $\boldsymbol{\eta}(\boldsymbol{\beta}(\boldsymbol{\gamma}))$ are given by the following.

(1) Chain rule of the first order derivative for a vector

$$\frac{\partial \boldsymbol{\eta}}{\partial \boldsymbol{\gamma}^T} = \frac{\partial \boldsymbol{\eta}}{\partial \boldsymbol{\beta}^T} \frac{\partial \boldsymbol{\beta}}{\partial \boldsymbol{\gamma}^T}. \tag{A.7}$$

(2) Chain rule of the second order derivative for a vector

$$\frac{\partial^2 \boldsymbol{\eta}}{\partial \boldsymbol{\gamma} \partial \boldsymbol{\gamma}^T} = (\frac{\partial \boldsymbol{\beta}}{\partial \boldsymbol{\gamma}^T})^T (\frac{\partial^2 \boldsymbol{\eta}}{\partial \boldsymbol{\beta} \partial \boldsymbol{\beta}^T}) (\frac{\partial \boldsymbol{\beta}}{\partial \boldsymbol{\gamma}^T}) + [\frac{\partial \boldsymbol{\eta}}{\partial \boldsymbol{\beta}^T}][\frac{\partial^2 \boldsymbol{\beta}}{\partial \boldsymbol{\gamma} \partial \boldsymbol{\gamma}^T}]. \tag{A.8}$$

Proof. It follows from the chain rule of the differentiation that

$$\frac{\partial \eta_i}{\partial \gamma_k} = \sum_{a=1}^{p} \frac{\partial \eta_i}{\partial \beta_a} \frac{\partial \beta_a}{\partial \gamma_k} \quad (i = 1, \cdots, n;\ k = 1, \cdots, q);$$

$$\frac{\partial^2 \eta_i}{\partial \gamma_k \partial \gamma_j} = \sum_{a=1}^{p}\sum_{b=1}^{p} \frac{\partial^2 \eta_i}{\partial \beta_a \partial \beta_b} \frac{\partial \beta_a}{\partial \gamma_k} \frac{\partial \beta_b}{\partial \gamma_j} + \sum_{a=1}^{p} \frac{\partial \eta_i}{\partial \beta_a} \frac{\partial^2 \beta_a}{\partial \gamma_k \partial \gamma_j}$$

$$(k, j = 1, \cdots, q);$$

which result in (A.7) and (A.8). ∎

Obviously, formulas (A.7) and (A.8) can be applied to the derivatives of the scalar y as follows.

(3) Chain rule of the first order derivative for a scalar

$$\frac{\partial y}{\partial \boldsymbol{\gamma}} = (\frac{\partial \boldsymbol{\beta}}{\partial \boldsymbol{\gamma}^T})^T (\frac{\partial y}{\partial \boldsymbol{\beta}}). \tag{A.9}$$

(4) Chain rule of the second order derivative for a scalar

$$\frac{\partial^2 y}{\partial \boldsymbol{\gamma} \partial \boldsymbol{\gamma}^T} = (\frac{\partial \boldsymbol{\beta}}{\partial \boldsymbol{\gamma}^T})^T (\frac{\partial^2 y}{\partial \boldsymbol{\beta} \partial \boldsymbol{\beta}^T})(\frac{\partial \boldsymbol{\beta}}{\partial \boldsymbol{\gamma}^T}) + [(\frac{\partial y}{\partial \boldsymbol{\beta}})^T][\frac{\partial^2 \boldsymbol{\beta}}{\partial \boldsymbol{\gamma} \partial \boldsymbol{\gamma}^T}]. \tag{A.10}$$

If $\boldsymbol{a}$ is a constant vector and $\boldsymbol{A}$ is a constant symmetric matrix, than we have the following.

(5) The derivative of a linear form

$$\frac{\partial (\boldsymbol{a}^T \boldsymbol{\beta})}{\partial \boldsymbol{\beta}} = \boldsymbol{a}.$$

(6) The derivative of a quadratic form

$$\frac{\partial (\boldsymbol{\beta}^T \boldsymbol{A} \boldsymbol{\beta})}{\partial \boldsymbol{\beta}} = 2\boldsymbol{A}\boldsymbol{\beta}.$$

Let the $n \times n$ matrix $\boldsymbol{X} = (x_{ij})$ be the function of a real value t, then the derivative $\partial \boldsymbol{X} / \partial t$ is defined as a matrix $\partial \boldsymbol{X} / \partial t = (\partial x_{ij} / \partial t)$. Further, the determinant of $\boldsymbol{X}$ is denoted by $\det(\boldsymbol{X})$, then we have the following.

(7) The derivative of an inverse matrix

$$\frac{\partial \boldsymbol{X}^{-1}}{\partial t} = -\boldsymbol{X}^{-1} \frac{\partial \boldsymbol{X}}{\partial t} \boldsymbol{X}^{-1}. \tag{A.11}$$

(8) The derivative of a determinant

$$\frac{\partial \det(\boldsymbol{X})}{\partial t} = \det(\boldsymbol{X}) \mathrm{tr}(\boldsymbol{X}^{-1} \frac{\partial \boldsymbol{X}}{\partial t}). \tag{A.12}$$

Proof.

(7) Differentiating $\boldsymbol{X}\boldsymbol{X}^{-1} = \boldsymbol{I}_n$ gives

$$\frac{\partial \boldsymbol{X}}{\partial t}\boldsymbol{X}^{-1} + \boldsymbol{X}\frac{\partial \boldsymbol{X}^{-1}}{\partial t} = \boldsymbol{0},$$

which results in (A.11). ∎

(8) Let the algebraic complement of x_{ij} of $\boldsymbol{X}$ be $\boldsymbol{X}_{ij}$ and $\boldsymbol{X}^{-1} = \boldsymbol{Y} = (y_{ij})$, then $y_{ij} = \boldsymbol{X}_{ji}/\det(\boldsymbol{X})$ and

$$\det(\boldsymbol{X}) = \sum_{i=1}^{n} x_{ij}\boldsymbol{X}_{ij}, \quad (j = 1, \cdots, n).$$

Since $\boldsymbol{X}_{ij}$ no longer depends on x_{ij}, we have

$$\begin{aligned}\frac{\partial \det(\boldsymbol{X})}{\partial t} &= \sum_{i,j} \frac{\partial \det(\boldsymbol{X})}{\partial x_{ij}} \frac{\partial x_{ij}}{\partial t} \\ &= \sum_{i,j} \det(\boldsymbol{X}) y_{ji} \frac{\partial x_{ij}}{\partial t} \\ &= \det(\boldsymbol{X}) \mathrm{tr}(\boldsymbol{X}^{-1}\frac{\partial \boldsymbol{X}}{\partial t}),\end{aligned}$$

which gives (A.12). ∎

Appendix B

Some Asymptotic Results

This appendix deals with some asymptotic properties of maximum likelihood estimation in exponential family nonlinear models (2.2). We shall provide: (a) the necessary condition of weak consistency of an estimator (Theorems B.1 and B.2); (b) the strong consistency and the asymptotic normality of maximum likelihood estimators (i.e. the proofs of Theorem 4.1 and Lemma 4.1); and (c) the asymptotic numerical stability of the Gauss-Newton iteration procedure for maximum likelihood estimators (i.e. the proof of Theorem 2.1).

Note that we shall use some assumptions of the text in this appendix. Assumptions A and B are given in Sections 2.1 and 2.2 respectively, and Assumption C is given in Section 4.1.

B.1 Notation and Preparation

We will first introduce some notation and basic results, to be used later. To stress the quantities depending on the sample size n, we shall add the subscripts to such quantities in this appendix. For instance, $\hat{\boldsymbol{\beta}}$, l, $\dot{\boldsymbol{l}}$ and $\ddot{\boldsymbol{l}}$ are denoted by $\hat{\boldsymbol{\beta}}_n$, l_n, $\dot{\boldsymbol{l}}_n$ and $\ddot{\boldsymbol{l}}_n$, respectively. Similar notation will be used for other quantities. In particular, the Fisher information matrix $\boldsymbol{J}_\beta(\boldsymbol{Y})$ given in (2.11) will be denoted by $\boldsymbol{J}_n$ or $\boldsymbol{J}_n(\boldsymbol{\beta})$ for short. To distinguish the true value and other values of the parameter $\boldsymbol{\beta}$ in a neighborhood, we denote the true value by $\boldsymbol{\beta}_0$ and mostly drop the argument $\boldsymbol{\beta}_0$ in $\dot{\boldsymbol{l}}(\boldsymbol{\beta}_0)$, $\boldsymbol{D}(\boldsymbol{\beta}_0)$, $\boldsymbol{J}_n(\boldsymbol{\beta}_0)$, etc. and just write $\dot{\boldsymbol{l}}$, $\boldsymbol{D}$, $\boldsymbol{J}_n$, etc., respectively. For a given $\delta > 0$, the neighborhood of $\boldsymbol{\beta}_0$ with radius δ is denoted by $\mathcal{N}(\delta) = \{\boldsymbol{\beta} : ||\boldsymbol{\beta} - \boldsymbol{\beta}_0|| < \delta\}$, $\overline{\mathcal{N}}(\delta) = \{\boldsymbol{\beta} : ||\boldsymbol{\beta} - \boldsymbol{\beta}_0|| \le \delta\}$ and $\partial\mathcal{N}(\delta) = \{\boldsymbol{\beta} : ||\boldsymbol{\beta} - \boldsymbol{\beta}_0|| = \delta\}$, respectively.

Given a symmetric matrix $\boldsymbol{A}$, the smallest eigenvalue of $\boldsymbol{A}$ is denoted by $\lambda_{min}(\boldsymbol{A})$ which satisfies $\lambda_{min}(\boldsymbol{A}) \le \boldsymbol{\lambda}^T \boldsymbol{A} \boldsymbol{\lambda}$ for any unit vector $\boldsymbol{\lambda}$. If $\boldsymbol{A}$ is a positive definite matrix, then it can be decomposed as $\boldsymbol{A} = \boldsymbol{A}^{\frac{1}{2}} \boldsymbol{A}^{\frac{T}{2}}$,

$\boldsymbol{A}^{-1} = \boldsymbol{A}^{-\frac{T}{2}}\boldsymbol{A}^{-\frac{1}{2}}$ (such as Cholesky decomposition). In particular, $\boldsymbol{J}_n = \boldsymbol{J}_n^{\frac{1}{2}}\boldsymbol{J}_n^{\frac{T}{2}}$, $\boldsymbol{K} = \boldsymbol{K}^{\frac{1}{2}}\boldsymbol{K}^{\frac{T}{2}}$ (see (4.1)) will be used. Since $n^{-1}\boldsymbol{J}_n \to \phi\boldsymbol{K}$ as $n \to \infty$ (see (4.1)), we have $(n^{-1}\boldsymbol{J}_n)^{\frac{1}{2}} \to \phi^{\frac{1}{2}}\boldsymbol{K}^{\frac{1}{2}}$, $(n^{-1}\boldsymbol{J}_n)^{-1} \to \phi^{-1}\boldsymbol{K}^{-1}$ and $(n^{-1}\boldsymbol{J}_n)^{-\frac{1}{2}} \to \phi^{-\frac{1}{2}}\boldsymbol{K}^{-\frac{1}{2}}$, respectively.

Finally, we notice a fact that will be used frequently in this appendix. From Assumption A, functions $f(\boldsymbol{x}, \boldsymbol{\beta})$, $g(\boldsymbol{\mu})$ and $b(\boldsymbol{\theta})$ are continuous and $\mathcal{X}$ is a compact subset, so all the quantities discussed, such as $\theta_i(\boldsymbol{\beta}) = \theta(\boldsymbol{x}_i, \boldsymbol{\beta})$, $\mu_i(\boldsymbol{\beta}) = \mu(\boldsymbol{x}_i, \boldsymbol{\beta})$, $b(\theta_i) = b(\theta(\boldsymbol{x}_i, \boldsymbol{\beta}))$ and their derivatives (up to the third order), are always bounded in $\mathcal{X} \times \overline{\mathcal{N}}(\delta)$ (Fahrmeir and Kaufmann 1985). In particular, the quantities such as $n^{-1}\sum_i\{\mu(\boldsymbol{x}_i, \boldsymbol{\beta}) - \mu(\boldsymbol{x}_i, \boldsymbol{\beta}')\}$, $n^{-1}\sum_i \theta_{iA}V_i\theta_{iB}$, etc. are also bounded in the region $\mathcal{X} \times \overline{\mathcal{N}}(\delta)$.

B.2 Necessary Condition of Consistency

Theorem B.1

Suppose that the Assumption A (see Chapter 2) holds in model (2.2). If there exists a consistent estimator $\overline{\boldsymbol{\beta}}_n(\boldsymbol{Y})$ satisfying

$$\overline{\boldsymbol{\beta}}_n(Y) \to \boldsymbol{\beta} \quad \text{in probability for all} \quad \boldsymbol{\beta} \in \mathcal{B}, \tag{B.1}$$

then there exists a neighborhood $\mathcal{N}(\delta) = \{\boldsymbol{\beta}' : \|\boldsymbol{\beta}' - \boldsymbol{\beta}\| < \delta\}$ such that

$$A_n(\boldsymbol{\beta}, \boldsymbol{\beta}') = \sum_{i=1}^{n} V_i(\boldsymbol{\beta})\{\theta_i(\boldsymbol{\beta}) - \theta_i(\boldsymbol{\beta}')\}^2 \to +\infty, \tag{B.2}$$

as $n \to \infty$ for all $\boldsymbol{\beta}' \neq \boldsymbol{\beta}$ in $\mathcal{N}(\delta)$.

Proof. By Prakasa Rao (1987, p.364), a necessary condition for (B.1) is

$$K_n(\boldsymbol{\beta}, \boldsymbol{\beta}') = \mathrm{E}_{\boldsymbol{\beta}}\{l_n(\boldsymbol{\beta}) - l_n(\boldsymbol{\beta}')\} \to +\infty, \quad \text{as} \quad n \to \infty,$$

for any $\boldsymbol{\beta}' \neq \boldsymbol{\beta}$ in $\mathcal{B}$, where $K_n(\boldsymbol{\beta}, \boldsymbol{\beta}')$ is the Kullback-Leibler distance (see Section 1.3). Let $\theta_i = \theta_i(\boldsymbol{\beta})$, $\theta_i' = \theta_i(\boldsymbol{\beta}')$ and $\Delta\theta_i = \theta_i' - \theta_i$, then it follows from (2.5) that

$$\begin{aligned} K_n(\boldsymbol{\beta}, \boldsymbol{\beta}') &= \phi\sum_{i=1}^{n} \mathrm{E}_{\boldsymbol{\beta}}[\{y_i\theta_i - b(\theta_i)\} - \{y_i\theta_i' - b(\theta_i')\}] \\ &= \phi\sum_{i=1}^{n}\{\dot{b}(\theta_i)(\theta_i - \theta_i') + b(\theta_i') - b(\theta_i)\} \\ &= \phi\sum_{i=1}^{n}\{\frac{1}{2}\ddot{b}(\theta_i)\Delta\theta_i^2 + \frac{1}{6}b^{(3)}(\xi_i)\Delta\theta_i^3\} \end{aligned}$$

$$= \frac{1}{2}\phi \sum_{i=1}^{n} \{V(\theta_i) + \frac{1}{3}b^{(3)}(\xi_i)\Delta\theta_i\}\Delta\theta_i^2,$$

where $\xi_i = t_i\theta_i + (1-t_i)\theta_i'$ for some $0 \le t_i \le 1$ and $V(\theta_i) = \ddot{b}(\theta_i)$. By Assumption A, we may set

$$\inf_{\Theta} \ddot{b}(\theta(\boldsymbol{x}, \boldsymbol{\beta})) = u, \quad \sup_{\Theta} |b^{(3)}(\theta(\boldsymbol{x}, \boldsymbol{\beta}))| = m,$$

where $u > 0$, $m < +\infty$ in a neighborhood $\overline{\mathcal{N}}(\delta_1)$ of $\boldsymbol{\beta}$. Since $\theta(\boldsymbol{x}, \boldsymbol{\beta}) = \dot{b}^{-1} \circ g^{-1} \circ f(\boldsymbol{x}, \boldsymbol{\beta})$ is uniformly continuous in $\mathcal{X} \times \overline{\mathcal{N}}(\delta_1)$, there exists $0 < \delta \le \delta_1$ such that

$$|\Delta\theta_i| = |\theta(\boldsymbol{x}_i, \boldsymbol{\beta}) - \theta(\boldsymbol{x}_i, \boldsymbol{\beta}')| \le \frac{3u}{2m}$$

for any $\boldsymbol{x}_i$ and $\boldsymbol{\beta}' \ne \boldsymbol{\beta}$ in $\mathcal{N}(\delta)$, which results in

$$|\frac{1}{3}b^{(3)}(\xi_i)\Delta\theta_i| \le \frac{m}{3} \cdot \frac{3u}{2m} = \frac{u}{2},$$

$$V(\theta_i) + \frac{1}{3}b^{(3)}(\xi_i)\Delta\theta_i \le 2V(\theta_i)$$

for any $\boldsymbol{x}_i$ and $\boldsymbol{\beta}' \ne \boldsymbol{\beta}$ in $\mathcal{N}(\delta)$. Therefore we have

$$K_n(\boldsymbol{\beta}, \boldsymbol{\beta}') \le \phi \sum_{i=1}^{n} V(\theta_i)\{\theta_i(\boldsymbol{\beta}) - \theta_i(\boldsymbol{\beta}')\}^2.$$

Since $K_n(\boldsymbol{\beta}, \boldsymbol{\beta}') \to +\infty$, as $n \to \infty$, we obtain (B.2). ∎

It follows from (B.2) that

$$\sum_{i=1}^{n} V_i(\boldsymbol{\beta})\{\theta_i(\boldsymbol{\beta}) - \theta_i(\boldsymbol{\beta}')\}^2 \le \left\{\sup_{\mathcal{X}\times\mathcal{N}(\delta)} \ddot{b}(\theta(\boldsymbol{x}, \boldsymbol{\beta}))\right\} \sum_{i=1}^{n} \left\{\theta_i(\boldsymbol{\beta}) - \theta_i(\boldsymbol{\beta}')\right\}^2.$$

Then we have

$$\sum_{i=1}^{n} \{\theta_i(\boldsymbol{\beta}) - \theta_i(\boldsymbol{\beta}')\}^2 \to +\infty,$$

if (B.1) holds. This result has been given by Wu (1981) for nonlinear regression models which may be regarded as the special cases of model (2.2).

Theorem B.2

For generalized linear models, i.e. $\theta_i = \theta(\boldsymbol{x}_i^T\boldsymbol{\beta}) = \dot{b}^{-1} \circ g^{-1} \circ f(\boldsymbol{x}_i^T\boldsymbol{\beta})$ in (2.2), if Assumption A holds and the derivatives $\dot{\theta}(\boldsymbol{x}^T\boldsymbol{\beta})$ has a positive lower bound in $\mathcal{X} \times \mathcal{N}(\delta)$ for some δ, then a necessary condition of the existing estimator $\overline{\boldsymbol{\beta}}_n(Y)$ satisfying (B.1) is

$$\lambda_{min}(J_n(\boldsymbol{\beta})) \to +\infty \quad \text{for any} \quad \boldsymbol{\beta} \in \mathcal{B}. \tag{B.3}$$

Proof. It is easily seen that $\partial\theta(\boldsymbol{x}_i{}^T\boldsymbol{\beta})/\partial\boldsymbol{\beta}^T = \dot{\theta}(\boldsymbol{x}_i^T\boldsymbol{\beta})\boldsymbol{x}_i^T$. Then the Taylor series expansion gives

$$\theta_i(\boldsymbol{\beta}) - \theta_i(\boldsymbol{\beta}') = -\dot{\theta}(\boldsymbol{x}_i^T\tilde{\boldsymbol{\beta}}_i)\boldsymbol{x}_i^T(\boldsymbol{\beta}' - \boldsymbol{\beta}),$$

where $\tilde{\boldsymbol{\beta}}_i = \tilde{t}_i\boldsymbol{\beta} + (1 - \tilde{t}_i)\boldsymbol{\beta}'$ for some $0 \le \tilde{t}_i \le 1$. Setting $\boldsymbol{\lambda} = (\boldsymbol{\beta}' - \boldsymbol{\beta})/\|\Delta\boldsymbol{\beta}\|$, $\Delta\boldsymbol{\beta} = \boldsymbol{\beta}' - \boldsymbol{\beta}$ and substituting the obtained results into (B.2) yield

$$A_n(\boldsymbol{\beta}, \boldsymbol{\beta}') = \|\Delta\boldsymbol{\beta}\|^2 \sum_{i=1}^{n} V_i(\boldsymbol{\beta})\dot{\theta}^2(\boldsymbol{x}_i^T\tilde{\boldsymbol{\beta}}_i)\boldsymbol{\lambda}^T\boldsymbol{x}_i\boldsymbol{x}_i^T\boldsymbol{\lambda} \to +\infty. \tag{B.4}$$

It follows from assumptions that $\dot{\theta}(\boldsymbol{x}^T\boldsymbol{\beta})$ is uniformly continuous and has positive lower bound in $\mathcal{X} \times \mathcal{N}(\delta)$. Therefore there exists a neighborhood $\mathcal{N}(\delta_1)$ with $0 < \delta_1 \le \delta$ such that

$$\begin{aligned}
|\dot{\theta}(\boldsymbol{x}_i^T\tilde{\boldsymbol{\beta}}_i)| &\le |\dot{\theta}(\boldsymbol{x}_i^T\boldsymbol{\beta})| + |\dot{\theta}(\boldsymbol{x}_i^T\tilde{\boldsymbol{\beta}}_i) - \dot{\theta}(\boldsymbol{x}_i^T\boldsymbol{\beta})| \\
&\le |\dot{\theta}(\boldsymbol{x}_i^T\boldsymbol{\beta})| + \varepsilon \inf_{\mathcal{X}\times\mathcal{N}(\delta_1)} |\dot{\theta}(\boldsymbol{x}^T\boldsymbol{\beta})| \\
&\le |\dot{\theta}(\boldsymbol{x}_i^T\boldsymbol{\beta})| + \varepsilon|\dot{\theta}(\boldsymbol{x}_i^T\boldsymbol{\beta})| = (1+\varepsilon)|\dot{\theta}(\boldsymbol{x}_i^T\boldsymbol{\beta})|.
\end{aligned}$$

Substituting this result into (B.4) gives

$$\begin{aligned}
A_n(\boldsymbol{\beta}, \boldsymbol{\beta}') &\le \|\Delta\boldsymbol{\beta}\|^2(1+\varepsilon)^2 \sum_{i=1}^{n} V_i(\boldsymbol{\beta})\dot{\theta}^2(\boldsymbol{x}_i^T\boldsymbol{\beta})\boldsymbol{\lambda}^T\boldsymbol{x}_i\boldsymbol{x}_i^T\boldsymbol{\lambda} \\
&= \|\Delta\boldsymbol{\beta}\|^2(1+\varepsilon)^2\boldsymbol{\lambda}^T\Big\{\sum_{i=1}^{n} V_i(\boldsymbol{\beta})\dot{\theta}^2(\boldsymbol{x}_i^T\boldsymbol{\beta})\boldsymbol{x}_i\boldsymbol{x}_i^T\Big\}\boldsymbol{\lambda}.
\end{aligned}$$

On the other hand, from (2.11) we know that

$$\boldsymbol{J}_n(\boldsymbol{\beta}) = \phi\boldsymbol{D}_\theta^T(\boldsymbol{\beta})\boldsymbol{V}(\boldsymbol{\beta})\boldsymbol{D}_\theta(\boldsymbol{\beta}) = \phi\sum_{i=1}^{n}\dot{\theta}(\boldsymbol{x}_i^T\boldsymbol{\beta})\boldsymbol{x}_iV_i(\boldsymbol{\beta})\boldsymbol{x}_i^T\dot{\theta}(\boldsymbol{x}_i^T\boldsymbol{\beta}).$$

Then from (B.4) we get $\boldsymbol{\lambda}^T\boldsymbol{J}_n(\boldsymbol{\beta})\boldsymbol{\lambda} \to +\infty$ $(n \to \infty)$, which leads to (B.3) since $\boldsymbol{\beta}'$ is an arbitrary point in $\mathcal{N}(\delta_1)$ and $\boldsymbol{\lambda}$ an arbitrary unit vector. ∎

Fahrmeir and Kaufmann (1985) pointed out that (B.3) is a basic condition to assure the consistency and the asymptotic normality of the maximum likelihood estimator in regression problem. For generalized linear models, they added some mild conditions together with (B.3) to obtain the strong consistency and the asymptotic normality of the maximum likelihood estimator. However, we have not seen the statement which regards (B.3) as a necessary condition, except the classical linear regression models with independent and identically distributed errors (Lai, Robbins and Wei 1979).

B.3 Consistency and Asymptotic Normality

To investigate the asymptotic behavior related to the maximum likelihood estimator, we first give several lemmas.

Lemma B.1

Let $\{r_i\}$ be a sequence of independent random variables with $\mathrm{E}(r_i) = 0$, $\mathrm{Var}\,(r_i) = \sigma_i^2 > 0$. If

$$\lim\ \sup_{n\to\infty}\frac{1}{n}\sum_{i=1}^{n}\sigma_i^2 < +\infty.$$

then

$$\frac{1}{n}\sum_{i=1}^{n} r_i \to 0 \quad (a.e.) \quad (\text{as } n\to\infty).$$

Proof. This lemma is a direct result of Lemma 2 of Wu (1981, p.504) with $A_n = n$ and $\delta = \frac{1}{2}$. ∎

Corollary

If Assumption A holds in model (2.2), then

$$\frac{1}{n}\sum_{i=1}^{n} e_i(\boldsymbol{\beta}_0)\theta_{iA}(\boldsymbol{\beta}') \to 0, \quad (a.e.)$$

where $e_i(\boldsymbol{\beta}_0) = y_i - \mu(\boldsymbol{x}_i, \boldsymbol{\beta}_0)$ and θ_{iA} is the derivative of $\theta(\boldsymbol{x}_i, \boldsymbol{\beta})$ defined in the beginning of Section 4.1.

Proof. Note that

$$\frac{1}{n}\sum_{i=1}^{n}\mathrm{Var}\{e_i(\boldsymbol{\beta}_0)\theta_{iA}(\boldsymbol{\beta}')\} = \frac{1}{n}\sum_{i=1}^{n}\sigma^2 V_i(\boldsymbol{\beta}_0)\theta_{iA}^2(\boldsymbol{\beta}')$$
$$\leq \sigma^2 \sup_{\mathcal{X}}\{\ddot{b}(\theta(\boldsymbol{x}, \boldsymbol{\beta}_0))\theta_{iA}^2(\boldsymbol{x}, \boldsymbol{\beta}')\} < +\infty,$$

which leads to the desired result from Lemma B.1. ∎

Lemma B.2

Suppose that $\{r_i\}$ is a sequence given in Lemma B.1, $\{\boldsymbol{x}_i\}$ is a known sequence in a compact subset $\mathcal{X}$ and $h(\boldsymbol{x}, \boldsymbol{\beta})$ is a continuous function in $\mathcal{X} \times \mathcal{B}_1$, where $\mathcal{B}_1$ is a compact subset in $\mathcal{B}$, then

$$\frac{1}{n}\sum_{i=1}^{n} r_i h(\boldsymbol{x}_i, \boldsymbol{\beta}) \to 0 \quad (a.e.) \quad \text{uniformly in } \mathcal{B}_1. \tag{B.5}$$

In particular, if $\boldsymbol{\beta}_n \to \boldsymbol{\beta}_0$ in $\overline{\mathcal{N}}(\delta)$, then

$$\frac{1}{n}\sum_{i=1}^{n} r_i h(\boldsymbol{x}_i, \boldsymbol{\beta}_n) \to 0. \quad (a.e.) \tag{B.6}$$

Proof. For any given $\boldsymbol{\beta} \in \mathcal{B}_1$,

$$\frac{1}{n}\sum_{i=1}^{n} \mathrm{Var}\{r_i h(\boldsymbol{x}_i, \boldsymbol{\beta})\} = \frac{1}{n}\sum_{i=1}^{n} h^2(\boldsymbol{x}_i, \boldsymbol{\beta})\sigma_i^2$$

$$\le \{\sup_{\mathcal{X}} h^2(\boldsymbol{x}, \boldsymbol{\beta})\}\frac{1}{n}\sum_{i=1}^{n} \sigma_i^2 < +\infty.$$

From Lemma B.1 we have $n^{-1}\sum_i r_i h(\boldsymbol{x}_i, \boldsymbol{\beta}) \to 0$ $(a.e)$ for this $\boldsymbol{\beta}$. Now consider a point $\boldsymbol{\beta}' \neq \boldsymbol{\beta}$ in a neighborhood $\mathcal{N}(\delta)$ of $\boldsymbol{\beta}$. It is easily seen that

$$\frac{1}{n}\Big|\sum_{i=1}^{n} r_i h(\boldsymbol{x}_i, \boldsymbol{\beta}')\Big| \le \frac{1}{n}\Big|\sum_{i=1}^{n} r_i\{h(\boldsymbol{x}_i, \boldsymbol{\beta}') - h(\boldsymbol{x}_i, \boldsymbol{\beta})\}\Big|$$

$$+\Big|\frac{1}{n}\sum_{i=1}^{n} r_i h(\boldsymbol{x}_i, \boldsymbol{\beta})\Big|. \tag{B.7}$$

The second term of (B.7) tends to zero $(a.e.)$ while the first term satisfies

$$\frac{1}{n}\Big|\sum_{i=1}^{n} r_i\{h(\boldsymbol{x}_i, \boldsymbol{\beta}') - h(\boldsymbol{x}_i, \boldsymbol{\beta})\}\Big|$$

$$\le \sup_{\mathcal{X}}\Big|h(\boldsymbol{x}, \boldsymbol{\beta}') - h(\boldsymbol{x}, \boldsymbol{\beta})\Big|\Big\{\frac{1}{n}\sum_{i=1}^{n}\Big|r_i\Big|\Big\}. \tag{B.8}$$

Since $h(\boldsymbol{x}, \boldsymbol{\beta})$ is uniformly continuous in $\mathcal{X} \times \mathcal{B}_1$, we have

$$\sup_{\mathcal{X}} |h(\boldsymbol{x}, \boldsymbol{\beta}') - h(\boldsymbol{x}, \boldsymbol{\beta})| \to 0 \quad \text{if} \quad \boldsymbol{\beta}' \to \boldsymbol{\beta}. \tag{B.9}$$

On the other hand, we can prove that $n^{-1}\Sigma_i|r_i|$ is finite. In fact, we have

$$\frac{1}{n}\sum_{i=1}^{n}|r_i| = \frac{1}{n}\sum_{i=1}^{n}\{|r_i| - \mathrm{E}|r_i|\} + \frac{1}{n}\sum_{i=1}^{n}\mathrm{E}|r_i|.$$

Since

$$\frac{1}{n}\sum_{i=1}^{n}\mathrm{Var}\{|r_i| - \mathrm{E}|r_i|\} \le \frac{1}{n}\sum_{i=1}^{n}\mathrm{Var}\,(r_i) = \frac{1}{n}\sum_{i=1}^{n}\sigma_i^2,$$

it follows from Lemma B.1 that

$$\frac{1}{n}\sum_{i=1}^{n}\{|r_i| - \mathrm{E}|r_i|\} \to 0 \quad (a.e.).$$

Then

$$\begin{aligned}\frac{1}{n}\sum_{i=1}^{n}\mathrm{E}|r_i| &\le \frac{1}{n}\sum_{i=1}^{n}\mathrm{E}(1+r_i^2)\\ &\le 1+\frac{1}{n}\sum_{i=1}^{n}\sigma_i^2 < +\infty \ (n\to\infty).\end{aligned}$$

Hence we get

$$\frac{1}{n}\sum_{i=1}^{n}|r_i| < +\infty \quad (a.e.)\ (n\to\infty), \tag{B.10}$$

Which shows that the second term of (B.7) tends to zero. Combining equations (B.7) to (B.10), we have the following conclusion. For any given $\varepsilon > 0$, there exists a neighborhood $\mathcal{N}(\delta_1)$ of $\boldsymbol{\beta}$ with $0 < \delta_1 \le \delta$ and an integer n_1, such that

$$P\Big\{\Big|\frac{1}{n}\sum_{i=1}^{n} r_i h(\boldsymbol{x}_i,\boldsymbol{\beta}')\Big| < \varepsilon, \quad n > n_1\Big\} = 1$$

for any $\boldsymbol{\beta}' \in \mathcal{N}(\delta_1)$. Since $\mathcal{B}_1$ is compact, it follows from the finitely covered theorem that there exists $n_2 > n_1$ such that

$$P\Big\{\Big|\frac{1}{n}\sum_{i=1}^{n} r_i h(\boldsymbol{x}_i,\boldsymbol{\beta})\Big| < \varepsilon, \quad n > n_2\Big\} = 1$$

for any $\boldsymbol{\beta} \in \mathcal{B}_1$. Then (B.5) can be proved from this and (B.6) is directly obtained from (B.5). ∎

Corollary

If Assumption A holds in model (2.2), then

$$\frac{1}{n}\sum_{i=1}^{n} e_i(\boldsymbol{\beta}_0)\theta_{iA}(\boldsymbol{\beta}_n) \to 0, \quad (a.e.)$$

where $\boldsymbol{\beta}_n \to \boldsymbol{\beta}_0$ in $\overline{\mathcal{N}}(\delta)$ for some δ. ∎

Lemma B.3

If Assumption A holds in model (2.2), and $-\ddot{\boldsymbol{l}}_n(\boldsymbol{\beta})$ given in (2.9) is denoted by $-\ddot{\boldsymbol{l}}_n(\boldsymbol{\beta}) = \boldsymbol{J}_n(\boldsymbol{\beta}) - \boldsymbol{R}_n(\boldsymbol{\beta})$ with $\boldsymbol{R}_n(\boldsymbol{\beta}) = \phi[\boldsymbol{e}^T(\boldsymbol{\beta})][\boldsymbol{W}_\theta(\boldsymbol{\beta})]$, then for any given $\varepsilon > 0$, there exist $\delta_0 > 0$ and $n_0 > 0$ such that

$$P\Big\{ \sup_{\boldsymbol{\beta}\in\overline{\mathcal{N}}(\delta_0)} n^{-1}\|\boldsymbol{R}_n(\boldsymbol{\beta})\| < \varepsilon, \quad n > n_0\Big\} = 1. \tag{B.11}$$

In particular, if $\boldsymbol{\beta}_n \to \boldsymbol{\beta}_0$ in $\overline{\mathcal{N}}(\delta)$ for some $\delta > 0$, then

$$n^{-1}\boldsymbol{R}_n(\boldsymbol{\beta}_n) \to 0. \quad (a.e.) \tag{B.12}$$

Proof. The component of $n^{-1}\boldsymbol{R}_n(\boldsymbol{\beta})$ at (a, b) may be written as

$$\{n^{-1}\boldsymbol{R}_n(\boldsymbol{\beta})\}_{ab} = n^{-1}\phi\sum_{i=1}^{n}\{y_i - \mu(\boldsymbol{x}_i, \boldsymbol{\beta})\}\ddot{\theta}_{iab}(\boldsymbol{x}_i, \boldsymbol{\beta}) = A_n + B_n;$$

where

$$\begin{aligned} A_n &= n^{-1}\phi\sum_{i=1}^{n} e_i(\boldsymbol{\beta}_0)\ddot{\theta}_{iab}(\boldsymbol{x}_i, \boldsymbol{\beta}), \\ B_n &= n^{-1}\phi\sum_{i=1}^{n}\{\mu(\boldsymbol{x}_i, \boldsymbol{\beta}_0) - \mu(\boldsymbol{x}_i, \boldsymbol{\beta})\}\ddot{\theta}_{iab}(\boldsymbol{x}_i, \boldsymbol{\beta}). \end{aligned}$$

By Lemma B.2, there exist $\delta_1 > 0$ and $n_1 > 0$ such that

$$P\Big\{|A_n| < \frac{\varepsilon}{2}, \quad n > n_1, \quad \boldsymbol{\beta} \in \overline{\mathcal{N}}(\delta_1)\Big\} = 1. \tag{B.13}$$

Further, since

$$|B_n| \le \phi\sup_{\mathcal{X}}|\mu(\boldsymbol{x}, \boldsymbol{\beta}_0) - \mu(\boldsymbol{x}, \boldsymbol{\beta})| \sup_{\mathcal{X}\times\mathcal{N}(\delta_1)}|\ddot{\theta}_{iab}(\boldsymbol{x}_i, \boldsymbol{\beta})|,$$

and $\mu(\boldsymbol{x}, \boldsymbol{\beta})$ is uniformly continuous in $\mathcal{X} \times \overline{\mathcal{N}}(\delta_1)$, there exists $0 < \delta_2 < \delta_1$ such that

$$P\Big\{\Big|B_n\Big| < \frac{\varepsilon}{2}, \qquad \boldsymbol{\beta} \in \overline{\mathcal{N}}(\delta_2)\Big\} = 1.$$

From this and (B.13), we obtain (B.11), and (B.12) is immediately obtained from (B.11). ∎

Lemma B.4

If Assumptions A to C hold in model (2.2), then

$$\boldsymbol{J}_n^{-\frac{1}{2}}\dot{\boldsymbol{l}}_n \xrightarrow{L} N(\boldsymbol{0}, \boldsymbol{I}_p) \qquad \text{as } n \to \infty,$$

where $\boldsymbol{J}_n = \boldsymbol{J}_n(\boldsymbol{\beta}_0) = \boldsymbol{J}_n^{\frac{1}{2}}\boldsymbol{J}_n^{\frac{T}{2}}$, and $\dot{\boldsymbol{l}}_n = \dot{\boldsymbol{l}}_n(\boldsymbol{\beta}_0)$.

Proof. To get this lemma, it is enough to show that

$$Z_n = \boldsymbol{\lambda}^T \boldsymbol{J}_n^{-\frac{1}{2}} \dot{\boldsymbol{l}}_n \xrightarrow{L} N(0,1)$$

holds for any $\boldsymbol{\lambda}$ satisfying $\boldsymbol{\lambda}^T\boldsymbol{\lambda} = 1$. From (2.8) and (2.11), $\dot{\boldsymbol{l}}_n = \phi \boldsymbol{D}^T \boldsymbol{V}^{-1}\boldsymbol{e}$, $\boldsymbol{J}_n = \phi \boldsymbol{D}^T\boldsymbol{V}^{-1}\boldsymbol{D}$ and Z_n may be written as

$$Z_n = \sum_{i=1}^n \phi \boldsymbol{\lambda}^T \boldsymbol{J}_n^{-\frac{1}{2}} \boldsymbol{d}_i V_i^{-1} e_i = \sum_{i=1}^n \alpha_i \varepsilon_i, \tag{B.14}$$

where

$$\alpha_i = \phi^{\frac{1}{2}} \boldsymbol{\lambda}^T \boldsymbol{J}_n^{-\frac{1}{2}} \boldsymbol{d}_i V_i^{-\frac{1}{2}}, \qquad \varepsilon_i = \phi^{\frac{1}{2}} V_i^{-\frac{1}{2}} e_i,$$

$\boldsymbol{D} = (\boldsymbol{d}_1, \cdots, \boldsymbol{d}_n)^T$ and each $\boldsymbol{d}_i = \partial\mu(\boldsymbol{x}_i, \boldsymbol{\beta})/\partial\boldsymbol{\beta}$ is a p-vector ($i = 1, \cdots, n$). Then we have $\mathrm{E}(\varepsilon_i) = 0$ and Var $(\varepsilon_i) = 1$. Now we use the Lindeberg-Feller central limit theorem for the summation (B.14). It is easily seen that $\mathrm{E}(\alpha_i\varepsilon_i) = 0$ and

$$\begin{aligned}
\mathrm{Var}(\sum_{i=1}^n \alpha_i \varepsilon_i) &= \sum_{i=1}^n \alpha_i^2 \\
&= \sum_{i=1}^n \phi \boldsymbol{\lambda}^T \boldsymbol{J}_n^{-\frac{1}{2}} \boldsymbol{d}_i V_i^{-1} \boldsymbol{d}_i^T \boldsymbol{J}_n^{-\frac{T}{2}} \boldsymbol{\lambda} \\
&= \boldsymbol{\lambda}^T \boldsymbol{J}_n^{-\frac{1}{2}} (\sum_{i=1}^n \phi \boldsymbol{d}_i V_i^{-1} \boldsymbol{d}_i^T) \boldsymbol{J}_n^{-\frac{T}{2}} \boldsymbol{\lambda} \\
&= \boldsymbol{\lambda}^T \boldsymbol{J}_n^{-\frac{1}{2}} \boldsymbol{J}_n \boldsymbol{J}_n^{-\frac{T}{2}} \boldsymbol{\lambda} = \boldsymbol{\lambda}^T \boldsymbol{\lambda} = 1.
\end{aligned}$$

We shall show that the Lindeberg condition is satisfied for (B.14), that is for any $\delta > 0$

$$g_n(\delta) = \sum_{i=1}^n \int_{|\alpha_i^2 z^2| > \delta^2} \alpha_i^2 z^2 dF(z|\boldsymbol{x}_i) \to 0 \quad (n \to \infty),$$

where $F(z|\boldsymbol{x}_i)$ is the distribution function of ε_i. Following Fahrmeir and Kaufmann (1985), $g_n(\delta)$ may be written as

$$g_n(\delta) = \sum_{i=1}^n \alpha_i^2 \int_{|z|^2 > \delta^2 \alpha_i^{-2}} z^2 dF(z|\boldsymbol{x}_i) \le \sum_{i=1}^n \alpha_i^2 h_n(\boldsymbol{x}_i), \tag{B.15}$$

where

$$h_n(\boldsymbol{x}_i) = \int_{|z|^2 > \delta^2 a_n^{-1}} z^2 dF(z|\boldsymbol{x}_i), \tag{B.16}$$

and

$$a_n = \max_{1\le i\le n} \alpha_i^2.$$

We can prove that $a_n \to 0$ (i.e. $\delta^2 a_n^{-1} \to \infty$) as $n \to \infty$. In fact, by Cauchy-Schwarz inequality, we get

$$\begin{aligned}
\alpha_i^2 &\le \phi V_i^{-1}(\boldsymbol{\lambda}^T\boldsymbol{\lambda})\boldsymbol{d}_i^T \boldsymbol{J}_n^{-\frac{T}{2}} \boldsymbol{J}_n^{-\frac{1}{2}} \boldsymbol{d}_i \\
&= \phi V_i^{-1} \boldsymbol{d}_i^T \boldsymbol{J}_n^{-1} \boldsymbol{d}_i \\
&= n^{-1}\phi V_i^{-1} \boldsymbol{d}_i^T (n^{-1}\boldsymbol{J}_n)^{-1} \boldsymbol{d}_i \\
&= n^{-1}\phi V_i^{-1} \|\boldsymbol{d}_i\|^2 \boldsymbol{\lambda}_i^T (n^{-1}\boldsymbol{J}_n)^{-1} \boldsymbol{\lambda}_i,
\end{aligned}$$

where $\boldsymbol{\lambda}_i = \boldsymbol{d}_i/\|\boldsymbol{d}_i\|$ is a unit vector. From Assumption A, $V_i^{-1} = \ddot{b}(\theta(\boldsymbol{x}_i, \boldsymbol{\beta}_0))$ and $\|\boldsymbol{d}_i\|^2 = \|\partial\mu(\boldsymbol{x}_i, \boldsymbol{\beta}_0)/\partial\boldsymbol{\beta}\|^2$ are bounded in $\mathcal{X}$. From Assumption C, $(n^{-1}\boldsymbol{J}_n)^{-1} \to \boldsymbol{K}^{-1}(\boldsymbol{\beta}_0)$ as $n \to \infty$. So $\boldsymbol{\lambda}_i^T (n^{-1}\boldsymbol{J}_n)^{-1}\boldsymbol{\lambda}_i$ is also bounded in $\mathcal{X}$. Then we have $a_n \to 0$ as $n \to \infty$. Hence it follows from (B.16) that $h_n(\boldsymbol{x}_i) \to 0$ for any $\boldsymbol{x}_i$ as $n \to \infty$. Then from (B.15) we get

$$g_n(\delta) \le \left\{ \max_{1\le i\le n} h_n(\boldsymbol{x}_i)\right\}\left(\sum_{i=1}^n \alpha_i^2\right) \le \sup_{\boldsymbol{x}\in\mathcal{X}} h_n(\boldsymbol{x}). \tag{B.17}$$

The function $h_n(\boldsymbol{x})$ has the following properties: $h_n(\boldsymbol{x}) \to 0$ (as $n \to \infty$) pointwise for any $\boldsymbol{x} \in \mathcal{X}$; $h_n(\boldsymbol{x})$ is monotonously decreasing in n; and $h_n(\boldsymbol{x})$ is continuous in $\mathcal{X}$ (by the Helly-Bray Lemma and continuity properties of exponential families). Due to these properties and the compactness of $\mathcal{X}$, $h_n(\boldsymbol{x}) \to 0$ uniformly in $\mathcal{X}$ as $n \to \infty$, that is $\sup_{\mathcal{X}} h_n(\boldsymbol{x}) \to 0$. Then from (B.17) we get $g_n(\delta) \to 0$ as $n \to \infty$, which results in $Z_n \xrightarrow{L} N(0,1)$ for any $\boldsymbol{\lambda}$ which meets $\boldsymbol{\lambda}^T\boldsymbol{\lambda} = 1$ and we complete the proof of the lemma. ∎

Proof of Theorem 4.1

(a) Strong consistency of $\hat{\boldsymbol{\beta}}_n$

It will be shown that there exist $\delta^* > 0$ and $n^* > 0$ such that for any $0 < \delta \le \delta^*$

$$P\{l_n(\boldsymbol{\beta}) - l_n(\boldsymbol{\beta}_0) < 0, \quad \boldsymbol{\beta} \in \partial\mathcal{N}(\delta), \quad n > n^*\} = 1. \tag{B.18}$$

This means that $\hat{\boldsymbol{\beta}}_n$ which maximizes $l_n(\boldsymbol{\beta})$ must be located inside of $\overline{\mathcal{N}}(\delta)$. Since $\delta \le \delta^*$ and δ is arbitrary, we have $\hat{\boldsymbol{\beta}}_n \to \boldsymbol{\beta}_0$ (*a.e.*) (see also Fahrmeir and Kaufmann 1985).

To prove (B.18), let $\boldsymbol{\lambda} = (\boldsymbol{\beta} - \boldsymbol{\beta}_0)/\delta$, then the Taylor series expansion gives

$$l_n(\boldsymbol{\beta}) - l_n(\boldsymbol{\beta}_0) = \delta\boldsymbol{\lambda}^T \dot{l}_n(\boldsymbol{\beta}_0) + \frac{1}{2}\delta^2\boldsymbol{\lambda}^T \ddot{l}_n(\tilde{\boldsymbol{\beta}}_n)\boldsymbol{\lambda},$$

where $\tilde{\beta}_n = \tilde{t}_n\beta_0 + (1-\tilde{t}_n)\beta$ for some $0 \le \tilde{t}_n \le 1$. Then (B.18) is equivalent to

$$P\Big\{\frac{1}{n}\lambda^T \dot{l}_n(\beta_0) < \frac{1}{2n}\delta\lambda^T(-\ddot{l}_n(\tilde{\beta}_n))\lambda, \quad \beta \in \partial\mathcal{N}(\delta), \quad n > n^*\Big\} = 1. \quad \text{(B.19)}$$

From (2.9), $-\ddot{l}_n(\tilde{\beta}_n) = \boldsymbol{J}_n(\tilde{\beta}_n) - \boldsymbol{R}_n(\tilde{\beta}_n)$, then the above equation is equivalent to

$$P\Big\{\frac{1}{n}\lambda^T \dot{l}_n(\beta_0) < \frac{1}{2n}\delta\lambda^T \boldsymbol{J}_n(\tilde{\beta}_n)\lambda - \frac{1}{2n}\delta\lambda^T \boldsymbol{R}_n(\tilde{\beta}_n)\lambda,$$
$$\beta \in \partial\mathcal{N}(\delta), \quad n > n^*\Big\} = 1.$$

To prove this, we notice the following facts:

(i) By the continuity of $\boldsymbol{J}_n(\beta)$ at β_0 and the assumption (4.1), there exist n_1 and δ_1 such that

$$\Big|\frac{1}{n}\lambda^T \boldsymbol{J}_n(\beta)\lambda - \lambda^T \boldsymbol{K}(\beta_0)\lambda\Big| \le \Big|\frac{1}{n}\lambda^T \boldsymbol{J}_n(\beta)\lambda - \lambda^T \boldsymbol{K}(\beta)\lambda\Big|$$
$$+\Big|\lambda^T \boldsymbol{K}(\beta)\lambda - \lambda^T \boldsymbol{K}(\beta_0)\lambda\Big| < \frac{\varepsilon}{2}$$

for any $\beta \in \mathcal{N}(\delta_1)$ and $n > n_1$. Let $\lambda_{min}(\boldsymbol{K}(\beta_0)) = c$, then we have

$$\frac{1}{n}\lambda^T \boldsymbol{J}_n(\beta)\lambda > c - \frac{\varepsilon}{2} \quad (a.e.) \quad \text{for any} \quad \beta \in \mathcal{N}(\delta_1),\ n > n_1.$$

(ii) By Lemma B.3, there exist $n_2 \ge n_1$ and $\delta^* \le \delta_1$ such that

$$\frac{1}{n}\lambda^T \boldsymbol{R}_n(\beta)\lambda < \frac{\varepsilon}{2} \quad (a.e.) \quad \text{when} \quad \beta \in \mathcal{N}(\delta^*),\ n > n_2.$$

(iii) It follows from (2.8) that the a-th component of $n^{-1}\dot{l}_n(\beta_0) = n^{-1}\phi \boldsymbol{D}_\theta^T(\beta_0)\boldsymbol{e}(\beta_0)$ is

$$n^{-1}\dot{l}_{na}(\beta_0) = \frac{1}{n}\phi\sum_{i=1}^{n}\dot{\theta}_{ia}(\beta_0)e_i(\beta_0), \quad a = 1, \cdots, p.$$

It follows from the corollary of Lemma B.2 that $n^{-1}\dot{l}_{na}(\beta_0) \to 0$ $(a.e.)$ and hence $n^{-1}\|\dot{l}_n(\beta_0)\| \to 0$ $(a.e.)$. Since

$$\lambda^T \dot{l}_n(\beta_0) \le (\lambda^T\lambda)(\dot{l}_n^T \dot{l}_n) = \|\dot{l}_n\|^2,$$

we get $n^{-1}\lambda^T \dot{l}_n(\beta_0) \to 0$ $(a.e.)$ for any $\lambda^T\lambda = 1$.

Now, combining (i) and (ii), there exist $\delta^* > 0$ and n_2 such that

$$\frac{1}{n}\boldsymbol{\lambda}^T\{-\ddot{\boldsymbol{l}}(\boldsymbol{\beta})\}\boldsymbol{\lambda} = \frac{1}{n}\boldsymbol{\lambda}^T\boldsymbol{J}_n(\boldsymbol{\beta})\boldsymbol{\lambda} - \frac{1}{n}\boldsymbol{\lambda}^T\boldsymbol{R}_n(\boldsymbol{\beta})\boldsymbol{\lambda} > c - \varepsilon \quad (a.e.)$$

for any $\boldsymbol{\beta} \in \mathcal{N}(\delta^*)$ and $n > n_2$. This also results in the existence of $\hat{\boldsymbol{\beta}}_n$ in $\mathcal{N}(\delta^*)$ when $n > n_2$.

For any given $0 < \delta \le \delta^*$, it follows from (iii) that there exists $n^* \ge n_2$ such that

$$\frac{1}{n}\boldsymbol{\lambda}^T\dot{\boldsymbol{l}}_n(\boldsymbol{\beta}_0) < \frac{1}{2}\delta(c-\varepsilon) \qquad (a.e.)$$

for any $\boldsymbol{\beta} \in \mathcal{N}(\delta)$ and $n > n^*$. So we obtain (B.19) and then (B.18) is proved. ∎

(b) Asymptotic normality of $\hat{\boldsymbol{\beta}}_n$

We shall apply Lemma B.4 to prove the asymptotic normality of $\hat{\boldsymbol{\beta}}_n$. The Taylor series expansion of $\dot{\boldsymbol{l}}_n(\boldsymbol{\beta}_0)$ at $\hat{\boldsymbol{\beta}}_n$ yields

$$\begin{aligned}\dot{\boldsymbol{l}}_n(\boldsymbol{\beta}_0) &= \dot{\boldsymbol{l}}_n(\hat{\boldsymbol{\beta}}_n) + \ddot{\boldsymbol{l}}_n(\boldsymbol{\beta}_n^*)(\boldsymbol{\beta}_0 - \hat{\boldsymbol{\beta}}_n) \\ &= -\ddot{\boldsymbol{l}}_n(\boldsymbol{\beta}_n^*)(\hat{\boldsymbol{\beta}}_n - \boldsymbol{\beta}_0),\end{aligned}$$

where $\boldsymbol{\beta}_n^* = t_n^*\boldsymbol{\beta}_0 + (1-t_n^*)\hat{\boldsymbol{\beta}}_n$ for some $0 \le t_n^* \le 1$ and $\boldsymbol{\beta}_n^* \to \boldsymbol{\beta}_0$ as $n \to \infty$ since $\hat{\boldsymbol{\beta}}_n \to \boldsymbol{\beta}_0$. To apply Lemma B.4, we rewrite the above equation as

$$\boldsymbol{J}_n^{-\frac{1}{2}}\dot{\boldsymbol{l}}_n = \boldsymbol{J}_n^{-\frac{1}{2}}\{-\ddot{\boldsymbol{l}}_n(\boldsymbol{\beta}_n^*)\}\boldsymbol{J}_n^{-\frac{T}{2}}\boldsymbol{J}_n^{\frac{T}{2}}(\hat{\boldsymbol{\beta}}_n - \boldsymbol{\beta}_0).$$

Then we have

$$\sqrt{n}(\hat{\boldsymbol{\beta}}_n - \boldsymbol{\beta}_0) = (n^{-1}\boldsymbol{J}_n)^{-\frac{T}{2}}\boldsymbol{G}_n^{-1}\boldsymbol{J}_n^{-\frac{1}{2}}\dot{\boldsymbol{l}}_n, \tag{B.20}$$

where

$$\boldsymbol{G}_n = \boldsymbol{J}_n^{-\frac{1}{2}}\{-\ddot{\boldsymbol{l}}_n(\boldsymbol{\beta}_n^*)\}\boldsymbol{J}_n^{-\frac{T}{2}}.$$

It can be shown that $\boldsymbol{G}_n \to \boldsymbol{I}_p$ (a.e.) $(n \to \infty)$. In fact, $\boldsymbol{G}_n$ can be expressed as

$$\boldsymbol{G}_n = \left(\frac{\boldsymbol{J}_n}{n}\right)^{-\frac{1}{2}}\left\{n^{-1}\boldsymbol{J}_n(\boldsymbol{\beta}_n^*) - n^{-1}\boldsymbol{R}_n(\boldsymbol{\beta}_n^*)\right\}\left(\frac{\boldsymbol{J}_n}{n}\right)^{-\frac{T}{2}}.$$

It follows from Lemma B.3 that $n^{-1}\boldsymbol{R}_n(\boldsymbol{\beta}_n^*) \to 0$ (a.e.). It follows from Assumption C that $n^{-1}\boldsymbol{J}_n \to \phi\boldsymbol{K}(\boldsymbol{\beta}_0)$, $(n^{-1}\boldsymbol{J}_n)^{-\frac{T}{2}} \to (\phi\boldsymbol{K})^{-\frac{T}{2}}$. Since $\hat{\boldsymbol{\beta}}_n \to \boldsymbol{\beta}_0$ (a.e.), we have $n^{-1}\boldsymbol{J}_n(\boldsymbol{\beta}_n^*) \to \phi\boldsymbol{K}(\boldsymbol{\beta}_0) = \phi\boldsymbol{K}^{\frac{1}{2}}\boldsymbol{K}^{\frac{T}{2}}$ (a.e.). Then $\boldsymbol{G}_n \to \boldsymbol{I}_p$ (a.e.) can be verified from the above results.

Using Lemma B.4 and the above results to (B.20), we get

$$\sqrt{n}(\hat{\boldsymbol{\beta}}_n - \boldsymbol{\beta}_0) \xrightarrow{L} N(0,\ \sigma^2\boldsymbol{K}^{-1}(\boldsymbol{\beta}_0)).$$

∎

Proof of Lemma 4.1

The proof is quite similar to that of Lemma B.4. Now let

$$\eta_n = \sigma^{-1}\boldsymbol{a}^T\boldsymbol{V}^{-1}\boldsymbol{e} = \sum_{i=1}^{n}\sigma^{-1}a_iV_i^{-1}e_i = \sum_{i=1}^{n}\xi_i\varepsilon_i, \tag{B.21}$$

where

$$\xi_i = a_iV_i^{-\frac{1}{2}}, \qquad \varepsilon_i = \sigma^{-1}V_i^{-\frac{1}{2}}e_i,$$

then $\mathrm{E}(\varepsilon_i) = 0$, $\mathrm{Var}(\varepsilon_i) = 1$. We shall show that the Lindeberg condition is satisfied for the summation (B.21), that is for any $\delta > 0$,

$$g_n(\delta) = B_n^{-2}\sum_{i=1}^{n}\int_{C_{ni}}\xi_i^2z^2dF(z|\boldsymbol{x}_i) \to 0 \quad \text{as} \quad n\to\infty,$$

where $B_n^2 = \sum_i \xi_i^2 = \boldsymbol{a}^T\boldsymbol{V}^{-1}\boldsymbol{a}$, $C_{ni} = \{z : z^2 > \xi_i^{-2}B_n^2\delta^2\}$ and $F(z|\boldsymbol{x}_i)$ is the distribution function of ε_i. Since $n^{-1}(\boldsymbol{a}^T\boldsymbol{V}^{-1}\boldsymbol{a}) \to a_0$, it follows from Lemma 3 of Wu (1981) that

$$\max_{1\le i\le n}\{a_iV_i^{-\frac{1}{2}}(\boldsymbol{a}^TV^{-1}\boldsymbol{a})^{-1}V_i^{-\frac{1}{2}}a_i\} \to 0 \quad \text{as} \quad n\to\infty,$$

which results in

$$M_n = \min_{1\le i\le n}(\xi_i^{-2}B_n^2) \to +\infty.$$

Then we have

$$h_n(\boldsymbol{x}_i) = \int_{z^2>M_n\delta^2} z^2dF(z^2|\boldsymbol{x}_i) \to 0.$$

Further, it is easily seen that

$$\int_{C_{ni}} z^2dF(z|\boldsymbol{x}_i) \le h_n(\boldsymbol{x}_i).$$

Then we get

$$\begin{aligned} g_n(\delta) &\le B_n^{-2}\sum_{i=1}^{n}\xi_i^2h_n(\boldsymbol{x}_i) \\ &\le B_n^{-2}\Big(\sum_{i=1}^{n}\xi_i^2\Big)\max_{1\le i\le n}h_n(\boldsymbol{x}_i) \\ &\le \sup_{\boldsymbol{x}\in\mathcal{X}}h_n(\boldsymbol{x}). \end{aligned}$$

According to the similar analysis given in Lemma B.4, we have $\sup_{\mathcal{X}} h_n(\boldsymbol{x}) \to 0$, which leads to $g_n(\delta) \to 0$ as $n\to\infty$. Thus Lindeberg-Feller central limit theorem results in

$$B_n^{-1}\eta_n = (\boldsymbol{a}^T\boldsymbol{V}^{-1}\boldsymbol{a})^{-\frac{1}{2}}\sigma^{-1}\boldsymbol{a}^T\boldsymbol{V}^{-1}\boldsymbol{e} \xrightarrow{L} N(0,1).$$

Then we get

$$(n^{-1}\boldsymbol{a}^T\boldsymbol{V}^{-1}\boldsymbol{a})^{-\frac{1}{2}}(\sqrt{n})^{-1}\boldsymbol{a}^T\boldsymbol{V}^{-1}\boldsymbol{e} \xrightarrow{L} N(0,\ \sigma^2),$$

which leads to

$$(\sqrt{n})^{-1}\boldsymbol{a}^T\boldsymbol{V}^{-1}\boldsymbol{e} \xrightarrow{L} N(0,\ \sigma^2 a_0).$$

∎

B.4 Asymptotic Numerical Stability of Iterations

Proof of Theorem 2.1

We shall use Jennrich's (1969) approach to prove this theorem. Now we define a mapping $\boldsymbol{\beta} \mapsto \boldsymbol{h}_n(\boldsymbol{\beta})$ as

$$\boldsymbol{h}_n(\boldsymbol{\beta}) = \boldsymbol{\beta} + (\boldsymbol{D}^T\boldsymbol{V}^{-1}\boldsymbol{D})^{-1}\boldsymbol{D}^T\boldsymbol{V}^{-1}\boldsymbol{e}.$$

Then the Gauss-Newton iteration procedure (2.16) becomes $\boldsymbol{\beta}^{i+1} = \boldsymbol{h}_n(\boldsymbol{\beta}^i)$. Since $\hat{\boldsymbol{\beta}}_n(\boldsymbol{Y})$ is the maximum likelihood estimator of $\boldsymbol{\beta}$, it follows from Theorem 4.1 that there exists a neighborhood $\mathcal{N}(\delta_1)$ of $\boldsymbol{\beta}_0$ and $n_1 > 0$ such that $\hat{\boldsymbol{\beta}}_n(\boldsymbol{Y}) \in \mathcal{N}(\delta_1)$ and $\dot{\boldsymbol{l}}_n(\hat{\boldsymbol{\beta}}_n) = 0$ for $n \geq n_1$ and almost every $\boldsymbol{Y}$. Hence $\boldsymbol{h}(\hat{\boldsymbol{\beta}}_n) = \hat{\boldsymbol{\beta}}_n$ holds for $n > n_1$ since $(\phi\boldsymbol{D}^T\boldsymbol{V}^{-1}\boldsymbol{e})_{\hat{\beta}} = \dot{\boldsymbol{l}}_n(\hat{\boldsymbol{\beta}}) = \boldsymbol{0}$. This means that $\hat{\boldsymbol{\beta}}_n$ is a fixed point of the mapping $\boldsymbol{\beta} \mapsto \boldsymbol{h}_n(\boldsymbol{\beta})$ for $n > n_1$ and almost every $\boldsymbol{Y}$. It will be shown that there exist $n_0 \geq n_1$ and $\mathcal{N}\ (\delta_0)$ with $\delta_0 \leq \delta_1$ such that

$$\|\boldsymbol{g}_n(\boldsymbol{\beta})\| = \|\frac{\partial \boldsymbol{h}_n(\boldsymbol{\beta})}{\partial \boldsymbol{\beta}^T}\| \leq c < 1. \qquad (a.e.) \tag{B.22}$$

This is a sufficient condition of asymptotic numerical stability of the iteration $\boldsymbol{\beta}^{i+1} = \boldsymbol{h}_n(\boldsymbol{\beta}^i)$ in $\mathcal{N}(\delta_0)$ (Jennrich 1969). In fact, (B.22) will lead to the uniqueness of the fixed point of the mapping $\boldsymbol{\beta} \mapsto \boldsymbol{h}_n(\boldsymbol{\beta})$ and the convergence of the sequence $\{\boldsymbol{\beta}^i\}$. Moreover, $\boldsymbol{\beta}^i$ must tend to the fixed point as $i \to \infty$, that is $\boldsymbol{\beta}^i \to \hat{\boldsymbol{\beta}}(\boldsymbol{Y})$ (a.e.) as $i \to \infty$, which is the desired result. Now let us prove (B.22). Since $\boldsymbol{D} = \boldsymbol{V}\boldsymbol{D}_\theta$ (see (3.26)), $\boldsymbol{h}_n(\boldsymbol{\beta})$ may be rewritten as

$$\boldsymbol{h}_n(\boldsymbol{\beta}) = \boldsymbol{\beta} + (\boldsymbol{D}_\theta^T\boldsymbol{V}\boldsymbol{D}_\theta)^{-1}\boldsymbol{D}_\theta^T\boldsymbol{e},$$

where $\boldsymbol{D}_\theta = \partial\boldsymbol{\theta}/\partial\boldsymbol{\beta}^T = (\dot{\theta}_{ia})$. Then $\boldsymbol{g}_n(\boldsymbol{\beta}) = \partial\boldsymbol{h}_n(\boldsymbol{\beta})/\partial\boldsymbol{\beta}^T$ is equal to

$$\begin{aligned} \boldsymbol{g}_n(\boldsymbol{\beta}) &= \boldsymbol{I}_p - (\boldsymbol{D}_\theta^T\boldsymbol{V}\boldsymbol{D}_\theta)^{-1}\boldsymbol{D}_\theta{}^T\boldsymbol{D} + (\boldsymbol{D}_\theta^T\boldsymbol{V}\boldsymbol{D}_\theta)^{-1}[\boldsymbol{e}^T][\boldsymbol{W}_\theta] + \\ &\quad \left\{\frac{\partial(\boldsymbol{D}_\theta^T\boldsymbol{V}\boldsymbol{D}_\theta)^{-1}}{\partial\boldsymbol{\beta}}\right\}\boldsymbol{D}_\theta^T\boldsymbol{e}. \end{aligned}$$

Using ${\boldsymbol D_\theta}^T \boldsymbol D = \boldsymbol D_\theta^T \boldsymbol V \boldsymbol D_\theta$ and (A.11), we get

$$\begin{aligned}\boldsymbol g_n(\boldsymbol\beta) &= (\boldsymbol D_\theta^T \boldsymbol V \boldsymbol D_\theta)^{-1}[\boldsymbol e^T][\boldsymbol W_\theta] - \\ &\quad (\boldsymbol D_\theta^T \boldsymbol V \boldsymbol D_\theta)^{-1}\Big\{\frac{\partial(\boldsymbol D_\theta^T \boldsymbol V \boldsymbol D_\theta)}{\partial\boldsymbol\beta}\Big\}(\boldsymbol D_\theta^T \boldsymbol V \boldsymbol D_\theta)^{-1}\boldsymbol D_\theta^T \boldsymbol e.\end{aligned}$$

By Assumption A, $\boldsymbol g_n(\boldsymbol\beta)$ is continuous in $\mathcal X \times \mathcal B$ for almost every $\boldsymbol Y$, so there exists a neighborhood $\mathcal N(\delta_0)$ with $\delta_0 \le \delta_1$ such that

$$\|\boldsymbol g_n(\boldsymbol\beta)\| \le \|\boldsymbol g_n(\boldsymbol\beta_0)\| + \|\boldsymbol g_n(\boldsymbol\beta) - \boldsymbol g_n(\boldsymbol\beta_0)\| \le \|\boldsymbol g_n(\boldsymbol\beta_0)\| + \frac{1}{4} \tag{B.23}$$

for $\boldsymbol\beta \in \mathcal N(\delta_0)$. On the other hand, $\boldsymbol g_n(\boldsymbol\beta_0)$ may be rewritten as

$$\begin{aligned}\boldsymbol g_n(\boldsymbol\beta_0) &= \Big(\frac{1}{n}\boldsymbol D_\theta^T \boldsymbol V \boldsymbol D_\theta\Big)^{-1}_{\beta_0}\Big\{\frac{1}{n}\Big[\boldsymbol e^T(\boldsymbol\beta_0)\Big]\Big[\boldsymbol W_\theta(\boldsymbol\beta_0)\Big]\Big\} - \\ &\quad \Big(\frac{1}{n}\boldsymbol D_\theta^T \boldsymbol V \boldsymbol D_\theta\Big)^{-1}_{\beta_0}\Big\{\frac{1}{n}\frac{\partial(\boldsymbol D_\theta^T \boldsymbol V \boldsymbol D_\theta)}{\partial\boldsymbol\beta}\Big\}_{\beta_0}\Big(\frac{1}{n}\boldsymbol D_\theta^T \boldsymbol V \boldsymbol D_\theta\Big)^{-1}_{\beta_0} \\ &\quad \Big\{\frac{1}{n}\boldsymbol D_\theta^T(\boldsymbol\beta_0)\boldsymbol e(\boldsymbol\beta_0)\Big\}.\end{aligned}$$

By Assumptions A, C and the compactness of $\mathcal X$, we have

$$\Big(\frac{1}{n}\boldsymbol D_\theta^T \boldsymbol V \boldsymbol D_\theta\Big)^{-1}_{\beta_0} = O(1), \qquad \Big\{\frac{1}{n}\frac{\partial(\boldsymbol D_\theta^T \boldsymbol V \boldsymbol D_\theta)}{\partial\boldsymbol\beta}\Big\} = O(1)$$

as $n \to +\infty$. By the corollary of Lemma B.2, we have

$$\frac{1}{n}[\boldsymbol e^T(\boldsymbol\beta_0)][\boldsymbol W_\theta(\boldsymbol\beta_0)] \to 0, \quad (a.e.)$$

$$\frac{1}{n}\boldsymbol D_\theta^T(\boldsymbol\beta_0)\boldsymbol e(\boldsymbol\beta_0) \to 0, \quad (a.e.)$$

as $n \to +\infty$. Therefore $\boldsymbol g_n(\boldsymbol\beta_0) \to 0$ $(a.e.)$ as $n \to +\infty$ and hence there exists $n_0 \ge n_1$ such that $\|\boldsymbol g_n(\boldsymbol\beta_0)\| \le \frac{1}{2}$. Then from (B.23) we have

$$\|\boldsymbol g_n(\boldsymbol\beta)\| \le \|\boldsymbol g_n(\boldsymbol\beta_0)\| + \frac{1}{4} \le \frac{3}{4} < 1 \quad (a.e.)$$

for any $\boldsymbol\beta \in \mathcal N(\delta_0)$ and $n \ge n_0$, which leads to (B.22). By Jennrich's (1969) argument, we complete the proof of the theorem. ∎

Bibliography

[1] Abramowitz, M. and Stegun, I.A. (1965). *Handbook of Mathematical Functions*. New York: Dover.

[2] Amari, S. (1982a). Differential geometry of curved exponential family-curvatures and information loss. *Ann. Statist.* **10**, 357–385.

[3] Amari, S. (1982b). Geometrical theory of asymptotic ancillarity and conditional inference. *Biometrika* **69**, 1–17.

[4] Amari, S. (1985). Differential Geometrical Methods in Statistics. *Lecture Notes in Statistics* **28**. Berlin: Springer.

[5] Amari, S. (1989). Fisher information under restriction of Shannon information in multi-terminal situations. *Ann. Inst. Statist. Math.* **41**, 623–648.

[6] Amari, S., Barndorff-Nielsen, O.E., Kass, R.E., Lauritzen, S.L. and Rao, C.R. (1987). *Differential Geometry in Statistical Inference*. Hayward, CA: IMS.

[7] Amari, S. and Kumon, M. (1983). Differential geometry of Edgeworth expansions in curved exponential family. *Ann. Inst. Statist. Math.* **35**, 1–24.

[8] Amari, S. and Kumon, M. (1988). Estimation in the presence of infinitely many nuisance parameters–geometry of estimating functions. *Ann. Statist.* **16**, 1044–1068.

[9] Andersen, E.B. (1992). Diagnostics in categorical data analysis. *J. Roy. Statist. Soc. Ser.* **B54**, 781–791.

[10] Anderson, T.W. (1984). *An Introduction to Multivariate Statistical Analysis* (second edition). New York: Wiley

[11] Atkinson, A.C. (1985). *Plots, Transformation and Regression*. Oxford: Oxford Press.

[12] Barndorff-Nielsen, O.E. (1978). *Information and Exponential Families in Statistical Theory.* New York: Wiley.

[13] Barndorff-Nielsen, O.E. (1983). On a formula for the distribution of a maximum likelihood estimator. *Biometrika* **70**, 343–365.

[14] Barndorff-Nielsen, O.E. (1986). Likelihood and observed geometries. *Ann. Statist.* **14**, 856–873.

[15] Barndorff-Nielsen, O.E. (1988). Parameter Statistical Models and Likelihood. *Lecture Notes in Statistics* **50**. Berlin: Springer.

[16] Barndorff-Nielsen, O.E., Cox, D.R. and Reid, N. (1986). The role of differential geometry in statistical theory. *Internat. Statist. Rev.* **54**, 83–96.

[17] Bates, D.M., Hamilton, D.C. and Watts, D.G. (1983). Calculation of intrinsic and parameter-effects curvatures for nonlinear regression models. *Commun. Statist.-Simula. Computa.* **12**, 469–477.

[18] Bates, D.M. and Watts, D.G. (1980). Relative curvature measures of nonlinearity. *J. Roy. Statist. Soc. Ser.* **B42**, 1–25.

[19] Bates, D.M. and Watts, D.G. (1981). Parameter transformations for improved approximate confidence region in nonlinear least squares. *Ann. Statist.* **9**, 1152–1167.

[20] Bates, D.M. and Watts, D.G. (1988). *Nonlinear Regression Analysis and Its Applications.* New York: Wiley.

[21] Beale, E.M.L. (1960). Confidence regions in non-linear estimation (with Discussion). *J. Roy. Statist. Soc. Ser.* **B22**, 41–88.

[22] Beckman, R.J. and Cook, R.D. (1983). Outlier...s. *Technometrics* **25**, 119–149.

[23] Belsley, D.A., Kuh, E. and Welsch, R.E. (1980). *Regression Diagnostics.* New York:Wiley.

[24] Breslow, N.E. (1990). Tests of hypotheses in over-dispersed Poisson regression and other quasi-likelihood models. *J. Amer. Statist. Assoc.* **85**, 565–571.

[25] Browne, M.W. (1974). Generalized least squares estimators in the analysis of covariance structures. *South Afric. Statist. J.* **8**, 1–24.

[26] Chatterjee, S. and Hadi, A.S. (1986). Influential observations, high leverage points and outliers in linear regression (with discussion). *Statist. Sci.* **1**, 379–416.

[27] Clarke, G.P.Y. (1980). Moments of the least squares estimators in a nonlinear regression model. *J. Roy. Statist. Soc. Ser.* **B42** 227–237.

[28] Clarke, G.P.Y. (1987a). Approximate confidence limits for a parameter function in nonlinear regression. *J. Amer. Statist. Assoc.* **82**, 221–230.

[29] Clarke, G.P.Y. (1987b). Marginal curvatures and their usefulness in the analysis of nonlinear regression models. *J. Amer. Statist. Assoc.* **82**, 844–850.

[30] Cook, R.D. (1986). Assessment of local influence. *J. Roy. Statist. Soc. Ser.* **B48**, 133–169.

[31] Cook, R.D. (1987). Parameter plots in nonlinear regression. *Biometrika* **74**, 669–677.

[32] Cook, R.D. and Goldberg, M.L. (1986). Curvature for parameter subsets in nonlinear regression. *Ann. Statist.* **14**, 1399–1418.

[33] Cook, R.D. and Tsai, C.L. (1985) Residuals in nonlinear regression. *Biometrika* **72**, 23–29.

[34] Cook, R.D. and Tsai, C.L. (1990). Diagnostics for assessing the accuracy of normal approximations in exponential family nonlinear models. *J. Amer. Statist. Assoc.* **85**, 770–777.

[35] Cook, R.D., Tsai, C.L. and Wei, B.C. (1986). Bias in nonlinear regression. *Biometrika* **73**, 615–623.

[36] Cook, R.D. and Weisberg, S. (1982). ***Residuals and Influence in Regression.*** London: Chapman and Hall.

[37] Cook, R.D. and Weisberg, S. (1983). Diagnostics for heteroscedasticity in regression. *Biometrika* **70**, 1–10.

[38] Cook, R.D. and Weisberg, S. (1990). Confidence curves in nonlinear regression. *J. Amer. Statist. Assoc.* **85**, 544–551.

[39] Cordeiro, G.M. and McCullagh, P. (1991). Bias correction in generalized linear models *J. Roy. Statist. Soc. Ser.* **B53**, 629–634.

[40] Cordeiro, G.M. and Paula, G.A. (1989). Improved likelihood ratio statistics for exponential family nonlinear models. *Biometrika* **76**, 93–100.

[41] Cox, D.R. (1983). Some remarks on overdispersion. *Biometrika* **70**, 269–274.

[42] Cox, D.R. and Hinkley, D.V. (1974). *Theoretical Statistics*. London: Chapman and Hall.

[43] Cox, D.R. and Reid, N. (1987). Parameter orthogonality and approximate conditional inference. *J. Roy. Statist. Soc. Ser.* **B49**, 1–39.

[44] Cox, D.R. and Reid, N. (1993). A note on the calculation of adjusted profile likelihood. *J. Roy. Statist. Soc. Ser.* **B55**, 467–471.

[45] Critchley, F., Marriott, P. and Salmon, M. (1993). Preferred point geometry and statistical manifolds. *Ann. Statist.* **21**, 1197–1242.

[46] Critchley, F., Marriott, P. and Salmon, M. (1994). Preferred point geometry and the local differential geometry of the Kullback-Leiber divergence. *Ann. Statist.* **22**, 1587–1602.

[47] Davison, A.C. and Gigli, A. (1989). Deviance residuals and normal scores plots. *Biometrika* **76**, 211–221.

[48] Davison, A.C. and Tsai, C.L. (1992). Regression model diagnostics. *Internal. Statist. Rev.* **60**, 337–355.

[49] Dean, C. (1992). Testing for overdispersion in Poisson and binomial regression models. *J. Amer. Statist. Assoc.* **87**, 451–457.

[50] Dean, C. and Lawless, J.F. (1989). Tests for detecting over-dispersion in Poisson regression models. *J. Amer. Statist. Assoc.* **84**, 467–472.

[51] del Pino, G. (1989). The unifying role of iterative generalized least squares in statistical algorithms. *Statist. Sci.* **4**, 394–408.

[52] Draper, N.R. and Smith, H. (1981). *Applied Regression Analysis* (second edition). New York: Wiley.

[53] Dudzinski, M.L. and Mykytowycz, R. (1961). The eye lens as an indicator of age in the wild rabbit in Australia. *CSIRO Wildl. Rev.* **6**, 156–159.

[54] Dzieciolowski, K. and Ross, W.H. (1990). Assessing case influence on confidence intervals in nonlinear regression. *Canad. J. Statist.* **18**, 127–139.

[55] Efron, B. (1975). Defining the curvature of a statistical problem (with application to second order efficiency). *Ann. Statist.* **3**, 1189–1242.

[56] Efron, B. (1978). The geometry of exponential families. *Ann. Statist.* **6**, 362–376.

[57] Efron, B. (1986). Double Exponential families and their use in generalized linear regression. *J. Amer. Statist. Assoc.* **81**, 709–721.

[58] Efron, B. (1991). Regression percentiles using asymmetric squared error loss. *Statist. Sinica* **1**, 93–125.

[59] Efron, B. (1992). Poisson overdispersion estimates based on the method of asymmetric maximum likelihood. *J. Amer. Statist. Assoc.* **87**, 98–107.

[60] Efron, B. and Hinkley, D.V. (1978). Assessing the accuracy of the maximum likelihood estimator: Observed versus expected Fisher information. *Biometrika* **65**, 457–487.

[61] Efron, B. and Johnstone, I.M. (1990). Fisher's information in terms of the hazard rate. *Ann. Statist.* **18**, 38–62.

[62] Eguchi, S. (1983). Second order efficiency of minimum contrast estimators in a curved exponential family. *Ann. Statist.* **11**, 793–803.

[63] Emerson, J.D., Hoaglin, D.C. and Kempthorne, P.J. (1984). Leverage in least squares additive-plus-multiplicative fits for two-way tables. *J. Amer. Statist. Assoc.* **79**, 329–335.

[64] Escobar, L.A. and Meeker, W.Q. (1992). Assessing influence in regression analysis with censored data. *Biometrics* **48**, 507–528.

[65] Fahrmeir, L. and Kaufmann, H. (1985). Consistency and asymptotic normality of the maximum likelihood estimator in generalized linear models. *Ann. Statist.* **13**, 342–368.

[66] Fisher, R.A. (1925). Theory of statistical estimation *Proc. Cambridge Philos. Soc.* **122**, 700–725.

[67] Fisher, R.A. (1950). The significance of deviations from expectation in a Poisson series. *Biometrics* **6**, 17–24.

[68] Firth, D. (1987). On the efficiency of quasi-likelihood estimation. *Biometrika* **74**, 233–45.

[69] Firth, D. (1991). Generalized linear models. In *Statistical Theory and Modeling*, eds Hinkley, D.V., Reid, N. and Snell, E.J. 55–82. London: Chapman and Hall.

[70] Gallant, A.R. (1987). *Nonlinear Statistical Models.* New York: Wiley.

[71] Ganio, L.M. and Schaffe, D.W. (1992). Diagnostics for overdispersion. *J. Amer. Statist. Assoc.* **87**, 795–804.

[72] Gay, D.M. and Welsch, R.E. (1988). Maximum likelihood and quasi-likelihood for nonlinear exponential family regression models. *J. Amer. Statist. Assoc.* **83**, 990–998.

[73] Gelfand, A. and Dalal, S. (1990). A note on overdispersed exponential families. *Biometrika* **77**, 55–64.

[74] Hamilton, D.C. (1986). Confidence regions for parameter subsets in nonlinear regression. *Biometrika* **73**, 57–64.

[75] Hamilton, D.C. and Watts, D.G. (1985). A quadratic design criterion for precise estimation in nonlinear regression models. *Technometrics* **27**, 241–250.

[76] Hamilton, D.C., Watts, D.G. and Bates, D.M. (1982). Accounting for intrinsic nonlinearity in nonlinear regression parameter inference regions. *Ann. Statist.* **10**, 386–393.

[77] Hartley, H.O. (1961). The modified Gauss-Newton method for the fitting nonlinear regression function. *Technometrics* **3**, 269–280.

[78] Hill, J.R. and Tsai, C.L. (1988). Calculating the efficiency of maximum quasi-likelihood estimation. *Applied Statist.* **37**, 219–230.

[79] Hoaglin, D.C. and Welsch, R.E. (1978). The hat matrix in regression and ANOVA. *Amer. Statist.* **32**, 17–22.

[80] Hosoya, Y. (1988). The second-order Fisher information. *Biometrika* **75**, 265–274.

[81] Hosoya, Y. (1990). Information amount and higher-order efficiency in estimation. *Ann. Inst Statist. Math.* **42** 37–49

[82] Hougaard, P. (1982). Parameterizations of non-linear models. *J. Roy. Statist. Soc. Ser.* **B44**, 244–252.

[83] Hougaard P. (1985). The appropriateness of the asymptotic distribution in nonlinear regression model in relation to curvature. *J. Roy. Statist. Soc. Ser.* **B47**, 102–114.

[84] Hougaard, P. (1986). Covariance stabilizing transformation in nonlinear regression. *Scand. J. Statist.* **13**, 207–210

[85] Jennrich, R.I. (1969). Asymptotic properties of nonlinear least squares estimators. *Ann. Math. Statist.* **40**, 633–643.

[86] Jorgensen, B. (1983). Maximum likelihood estimation and large-sample inference for generalized linear and nonlinear regression models. *Biometrika* **70**, 19–28.

[87] Jorgensen, B. (1987). Exponential dispersion models. *J. Roy. Statist. Soc. Ser.* **B49**, 127–162.

[88] Jorgensen, B. (1992a). Exponential dispersion models and extensions: a review. *Internal. Statist Rev.* **60**, 5–20.

[89] Jorgensen, B. (1992b). Finding rank leverage subsets in regression. *Scand. J. Statist.* **19**, 139–156.

[90] Kass, R.E. (1984). Canonical parameterizations and zero parameter effects curvature. *J. Roy Statist. Soc. Ser.* **B46**, 86–92.

[91] Kass, R.E. (1989). The geometry of asymptotic inference. *Statist. Sci.* **4**, 188–219.

[92] Kass, R.E. and Elizabeth, H.S. (1994). Some diagnostics of maximum likelihood and posterior nonnormality. *Ann. Statist.* **22**, 668–695.

[93] Kumon, M. and Amari, S. (1983). Geometrical theory of higher-order asymptotics of test, interval estimator and conditional inference. *Proc. Roy. Soc.* London, **A387**, 420–458.

[94] Kumon, M. and Amari, S. (1984). Estimation of a structure parameter in the presence of a large number of nuisance parameters. *Biometrika* **71**, 445–459.

[95] Lai, T.L., Robbins, H. and Wei, C.Z. (1979). Strong consistency of least squares estimator in multiple regression. *J. Multi. Analy.* **9**, 343–361.

[96] Lawless, J.F. (1982). *Statistical Models and Methods for Lifetime Data.* New York: Wiley.

[97] Lee, A.H. (1987). Diagnostic displays for assessing leverage and influence in generalized linear models. *Austral. J. Statist.* **29**, 233–243.

[98] Lee, A.H. (1988). Assessing partial influence in generalized linear models. *Biometrics* **44**, 71–77.

[99] Lee, S.Y. and Bentler, P.M. (1992). Some aspects of covariance structure analysis. In *The Development of Statistics: Recent Contributions From China*, eds Chen, X.R., Fang, K.T. and Yang, C.C., 107–124. Essex: Longman Sci. & Tech.

[100] Lee, S.Y. and Jennrich, R.I. (1979). A study of algorithm for covariance structure analysis with specific comparisons using factor. *Psychometrika* **44**, 99–113.

[101] Li, B. and McCullagh, P. (1994). Potential functions and conservative estimating functions. *Ann. Statist.* **22**, 340–356.

[102] Malinvaud, E. (1970). The consistency of nonlinear regression. *Ann. Statist.* **41**, 956–969.

[103] McCullagh, P. (1983). Quasi-likelihood functions. *Ann. Statist.* **11**, 59–67.

[104] McCullagh, P. (1987). *Tensor Method in Statistics.* London: Chapman and Hall.

[105] McCullagh, P. (1991). Quasi-likelihood and estimating equations. In *Statistical Theory and Modeling*, eds Hinkley, D.V., Reid, N. and Snell, E.J. 265–286. London: Chapman and Hall.

[106] McCullagh, P. and Nelder, J.A. (1989). *Generalized Linear Models* (second edition). London: Chapman and Hall.

[107] Millman, R.S. and Parker, G.D. (1977). *Elements of Differential Geometry.* Englewood Cliffs, NJ: Prentice Hall.

[108] Mitchell, A.F.S. (1988). Statistical manifold of univariate elliptic distributions. *Internal. Statist. Rev.* **56**, 1–16.

[109] Moolgavkar, S.H. and Venzon, D.J. (1987). Confidence regions in curved exponential families: application to matched case-control and survival studies with general relative risk function. *Ann. Statist.* **15**, 346–352.

[110] Morris, C.N. (1982). Natural exponential families with quadratic variance functions. *Ann. Statist.* **10**, 65–80.

[111] Morris, C.N. (1983). Natural exponential families with quadratic variance functions: statistical theory. *Ann. Statist.* **11**, 515–529.

[112] Morton, R. (1987). Asymmetry of estimators in nonlinear regression. *Biometrika* **74**, 679–685.

[113] Murray, M.K. and Rice, J.W. (1993). *Differential Geometry and Statistics.* London: Chapman and Hall.

[114] Nelder, J.A. and Lee, Y. (1992). Likelihood, quasi-likelihood and pseudo-likelihood: Some comparisons. *J. Roy. Statist. Soc. Ser.* **B54**, 273–284.

[115] Nelder, J.A. and Pregibon, D. (1987). An extended quasi-likelihood function. *Biometrika* **74**, 221–232.

[116] Okamoto, L., Amari, S. and Takeuchi, K. (1991). Asymptotic theory of sequential estimation: differential geometrical approach. *Ann. Statist.* **19**, 961–981.

[117] Pazman, A. (1987). On formulas for the distribution of nonlinear L.S. estimates. *Statistics* **18**, 3–15.

[118] Pazman, A. (1990). Small-sample distributional properties of nonlinear regression estimators (a geometric approach). *Statistics* **21**, 323–367.

[119] Pazman, A. (1991). Curvatures and the distribution of the maximum likelihood estimator in nonlinear exponential models. *Barzilian J. Prob. and Statist.* **5**, 43–63.

[120] Pierce, D.A. (1975). Discussion to Efron (1975). *Ann. Statist.* **3**, 1219–1221.

[121] Powsner, L. (1935). The effects of temperature on the durations of the developmental stages of Drosophila Melanogaster. *Physiological Zoology* **8**, 474–520.

[122] Prakasa Rao, B.L.S. (1987). *Asymptotic Theory of Statistical Inference.* New York: Wiley.

[123] Pregibon, D. (1981). Logistic regression diagnostics. *Ann. Statist.* **9**, 705–724.

[124] Rao, C.R. (1961). Asymptotic efficiency and limiting information. *Proc. Fourth Berkeley Symp. Math. Statist. Prob.* **1**, 531–545, Univ. of California Press.

[125] Rao, C.R. (1962). Efficient estimates and optimum inference procedures in large samples. *J. Roy. Statist. Soc. Ser.* **B24**, 46–72.

[126] Rao, C.R. (1963). Criteria of estimation in large samples. *Sankhya* **25**, 189–206.

[127] Rao, C.R. (1973). *Linear Statistical Inference and Its Applications* (second edition). New York: Wiley.

[128] Ratkowsky, D.A. (1983). *Nonlinear Regression in Modeling: A unified practical approach.* New York: Marcal Dekker.

[129] Ratkowsky, D.A. (1990). *Handbook of Nonlinear Regression Models.* New York: Marcel Dekker.

[130] Ravishanker, N., Melnick, E.L. and Tsai, C.L. (1990). Differential geometry of ARMA models. *J. Time Series Analy.* **11**, 259–274.

[131] Ross, W.H. (1987a). The geometry of case deletion and the assessment of influence in nonlinear regression. *Canad. J. Statist.* **15**, 91–103.

[132] Ross, W.H. (1987b). The expectation of the likelihood ratio criterion. *Internal. Statist. Rev.* **55**, 315–330.

[133] Schall, R. and Dunne, T.T. (1992). A note on the relationship between parameter collinearity and local influence. *Biometrika* **79**, 399–404.

[134] Seber, G.A.F. and Wild, C.J. (1989). *Nonlinear Regression.* New York: Wiley.

[135] Seshadri, V. (1993) *The Inverse Gaussian Distribution.* New York: Oxford Univ. Press Inc.

[136] Simonoff, J.S. and Tsai, C.L. (1994). Improved tests for nonconstant variance in regression based on the modified profile likelihood. *Appl. Statist.* **43**, 357–370.

[137] Skovgaard, I.M. (1985). A second-order investigation of asymptotic ancillary. *Ann. Statist.* **13**, 535–551.

[138] Smith, P.J. and Heitjan, D.F. (1993). Testing and adjusting for departures from nominal dispersion in generalized linear models. *Appl. Statist.* **42**, 31–41.

[139] Smyth, G.K. (1989). Generalized linear models with varying dispersion. *J. Roy. Statist. Soc. Ser.* **B51**, 47–60

[140] St. Laurent, R.T. and Cook, R.D. (1992). Leverage and superleverage in nonlinear regression. *J. Amer. Statist. Assoc.* **87**, 985–990.

[141] St. Laurent, R.T. and Cook, R.D. (1993). Leverage, local influence and curvature in nonlinear regression. *Biometrika* **80**, 99–106.

[142] Storer, B.E. and Crowley, J. (1985). A diagnostic for Cox regression and general conditional likelihoods. *J. Amer. Statist. Assoc.* **80**, 139–147.

[143] Thomas, W. (1990). Influence on confidence region for regression coefficients in generalized linear models. *J. Amer. Statist. Assoc.* **85**, 393–397.

[144] Thomas, W. and Cook, R.D. (1989). Assessing influence on regression coefficients in generalized linear models. *Biometrika* **76**, 741–749.

[145] Tsai, C.L. (1983). Contribution to the design and analysis of nonlinear models. Unpublished Ph.D. dissertation. Univ. Minnesota.

[146] Vos, P.W. (1989). Fundamental equations for statistical submanifold with application to the Bartlett correction. *Ann. Inst. Statist. Math.* **41**, 429–450.

[147] Vos, P.W. (1991). A geometric approach to detecting influential cases. *Ann. Statist.* **19**, 1570–1581.

[148] Wang, S.J. and Lee. S.Y. (1995). A geometric approach of the generalized least squares estimation in analysis of covariance structures. *Statist. Prob. Letters* **24**, 39–47.

[149] Wedderburn, R.W.M. (1974). Quasi-likelihood functions, generalized linear models, and the Gauss-Newton method. *Biometrika* **61**, 439–447.

[150] Wei, B.C. (1986). A solution of Amari's conjecture. *Chin. Sci. Bull.* **31**, 1448–1450.

[151] Wei, B.C. (1988). *Differential Geometry and Statistical Inference* (in Chinese). Beijing: Qinghua Univ. Press.

[152] Wei, B.C. (1989). *Modern Nonlinear Regression Analysis* (in Chinese). Nanjing: SE Univ. Press.

[153] Wei, B.C. (1991). Some second order asymptotics in nonlinear regression. *Aust. J. Statist.* **33**, 75–84.

[154] Wei, B.C. (1994). On confidence regions of embedded models in regular parametric families (A geometric approach). *Aust. J. Statist.* **36**, 327–338.

[155] Wei, B.C. (1995). Geometry of exponential family nonlinear models and asymptotic inference. *Science in China Ser. A* **38**, 298–308.

[156] Wei, B.C. (1996). Some asymptotic properties in multinomial nonlinear models (A geometric approach). *Appl. Math., J. Chin. Univs. Ser. B* **11**, 273–284.

[157] Wei, B.C and Hickernell, F.J. (1996). Regression transformation diagnostics for explanatory variables. *Statist. Sinica* **6**, 433–452.

[158] Wei, B.C., Hu, Y.Q. and Fung, W.K. (1997). Generalized leverage and its applications. *Scand. J. Statist.* (at press).

[159] Wei, B.C., Lu, G. and Shi, J.Q. (1991). *Introduction to Statistical Diagnostics* (in Chinese). Nanjing: SE Univ. Press.

[160] Wei, B.C. and Shi, J.Q. (1994). On statistical models in regression diagnostics. *Ann. Inst. Statist. Math.* **46**, 267–278.

[161] Wei, B.C., Shi, J.Q., Fung, W.K. and Hu, Y.Q. (1998). Testing for varying dispersion in exponential family nonlinear models. *Ann. Inst. Statist. Math.* (at press).

[162] Wei, B.C. and Zhao, X. (1987). A limit theorem related to the Fisher information and Amari curvature (in Chinese). *Acta Math. Appl. Sinica* **10**, 74–80.

[163] Weissfeld, L.L. (1990). Influence diagnostics for the proportional hazards models. *Statist. Prob. Letters* **10**, 59–65.

[164] Welch, L.F., Adams, W.E. and Corman, J.L. (1963). Yield response surfaces, isoquants and economic fertilizer optima for coastal Bermuda grass. *Agron. J.* **55**, 63–67.

[165] Whitmore, G.A. (1986). Inverse Gaussian ratio estimation. *Appl. Statist.* **35**, 8–15

[166] Williams, D.A. (1987). Generalized linear model diagnostics using the deviance single case deletions. *Appl. Statist.* **36** , 181–191.

[167] Wu, C.F. (1981). Asymptotic theory of nonlinear least squares estimation. *Ann. Statist.* **9**, 501–513.

[168] Wu, X. and Luo, Z. (1993a). Second-order approach to local influence. *J. Roy. Statist. Soc. Ser.* **B55**, 929–936.

[169] Wu, X. and Luo, Z. (1993b). Residual sum of squares and multiple potential diagnostics by a second order local approach. *Statist. Prob. Letters* **16**, 289–296.

[170] Wu, X. and Wan, F. (1994). A perturbation scheme for nonlinear models. *Statist. Prob. Letters* **20**, 197–202.

[171] Yoshizoe, Y. (1991). Leverage points in nonlinear regression models. *J. Japan Statist. Soc.* **21**, 1–11.

[172] Zelterman, D. and Chen, C.F. (1988). Homogeneity tests against central mixture alternatives. *J. Amer. Statist. Assoc.* **83**, 179–182

[173] Zhu, H.T. and Wei, B.C. (1997). Some notes on preferred point $\alpha-$geometry and $\alpha-$divergence function. *Statist. Prob. Letters* (at press).

Author Index

Abramowitz, M, 160
Adams, W.E., 120
Amari, S., 16, 29–31, 35, 44, 47, 55, 56, 58, 60, 62, 67, 69, 72, 74, 76, 80, 81, 165, 169
Andersen, E.B., 25, 101, 116, 117
Anderson, T.W., 187
Atkinson, A.C., 101

Barndorff-Nielsen, O.E., 1, 2, 7, 22, 29, 30, 44, 148
Bates, D.M., 14, 29, 30, 32, 34, 35, 37–39, 42, 43, 49, 64, 67, 83, 92, 93, 125, 188
Beale, E.M.L., 83
Beckman, R.J., 106
Belsley, D.A., 105
Bentler, P.M., 165, 179, 181
Breslow, N.E., 141
Browne, M.W., 179, 180

Chatterjee, S., 101, 136
Chen, C.F., 141
Clarke, G.P.Y., 30, 56, 64, 67, 69
Cook, R.D., 13, 14, 16, 28, 30, 56, 64, 67, 83, 84, 90, 93, 101, 104, 105, 123–126, 132, 134, 136, 139, 141, 142, 146, 188
Cordeiro, G.M., 2, 13, 14, 16, 22, 67
Corman, J.L., 120
Cox, D.R., 8, 9, 21, 23, 29, 60, 84, 96, 141, 147–149, 152, 174
Critchley, F., 30
Crowley, J., 106

Dalal, S., 141
Davison, A.C., 101, 114, 116, 119
Dean, C., 141
del Pino, G., 24
Draper, N.R., 90
Dudzinski, M.L., 27
Dunne, T.T., 30
Dzieciolowski, K., 106, 108

Efron, B., 16, 29–31, 35, 44, 47, 51, 52, 54–56, 58, 67, 69, 71, 72, 74, 76, 77, 80, 81, 141, 142, 165, 169, 170
Eguchi, S., 30
Elizabeth, H.S., 30, 217
Emerson, J.D., 136
Escobar, L.A., 30, 123–125

Fahrmeir, L., 15, 57, 58, 196, 198, 203, 204
Firth, D., 14, 177, 178
Fisher, R.A., 71, 76, 141
Fung, W.K., 139

Gallant, A.R., 24, 26, 27, 56, 57, 63
Ganio, L.M., 142
Gay, D.M., 102
Gelfand, A., 141
Gigli, A., 119
Goldberg, M.L., 30, 188

Hadi, A.S., 101, 136
Hamilton, D.C., 30, 83, 85, 89–91, 93, 96, 98–100
Hartley, H.O., 26
Heitjan, D.F., 141
Hickernell, F.J., 30, 123
Hill, J.R., 177, 178
Hinkley, D.V., 9, 23, 55, 60, 76, 77, 80, 81, 84, 96, 170, 174
Hoaglin, D.C., 136
Hosoya, Y., 72, 76
Hougaard, P., 30

Hu, Y.Q., 139

Jennrich, R.I., 24, 57–59, 165, 181, 208, 209
Johnstone, I.M., 30
Jorgensen, B., 1–3, 7, 13, 16, 22, 31, 44, 84, 174

Kass, R.E., 29, 30, 60, 72, 75, 76, 80, 165, 170, 217
Kaufmann, H., 15, 57, 58, 196, 198, 203, 204
Kempthorne, P.J., 136
Kuh, E., 105
Kumon, M., 30

Lai, T.L., 198
Lawless, J.F., 141, 160
Lee, A.H., 28
Lee, S.Y., 179, 181, 184
Lee, Y., 177
Li, B., 178
Lu, G., 106, 123
Luo, Z., 30, 123

Malinvaud, E., 58
Marriott, P., 30
McCullagh, P., 2, 6, 7, 12, 13, 16, 22, 31, 49, 57, 63, 67, 76, 80, 84, 101, 104, 111, 119, 120, 128, 129, 141, 155, 156, 158, 165, 170, 176–179
Meeker, W.Q., 30, 123–125
Melnick, E.L., 30
Millman, R.S., 29, 30, 34
Mitchell, A.F.S., 30, 218
Moolgavkar, S.H., 30
Morris, C.N., 1, 2, 7, 31
Morton, R., 56, 57
Murray, M.K., 29
Mykytowycz, R., 27

Nelder, J.A., 2, 7, 12, 16, 31, 49, 84, 101, 104, 111, 119, 120, 128, 129, 141, 155, 156, 158, 165, 170, 176–179

Okamoto, L., 30, 72, 76

Parker, G.D., 29, 30, 34
Paula, G.A., 13, 14, 16
Pazman, A., 13, 14, 30, 36
Pierce, D.A., 76, 80
Powsner, L., 111
Prakasa Rao, B.L.S., 196
Pregibon, D., 26, 101, 105, 116, 117, 119, 123, 177

Rao, C.R., 40, 71, 72, 103, 165, 170, 173
Ratkowsky, D.A., 14, 27, 30, 155, 159
Ravishanker, N., 30
Reid, N., 8, 21, 29, 147–149, 152
Rice, J.W., 29
Robbins, H., 198
Ross, W.H., 30, 106, 108

Salmon, M., 30
Schaffe, D.W., 141, 142
Schall, R., 30
Seber, G.A.F., 5, 6, 14, 24, 29, 30, 56, 64, 67, 69, 70, 188
Seshadri, V., 160
Shi, J.Q., 13, 102, 106, 107, 109, 123
Simonoff, J.S., 142, 152, 154, 159
Skovgaard, I.M., 76, 80, 81
Smith, H., 90
Smith, P.J., 141
Smyth, G.K., 2, 6, 11, 12, 142
St. Laurent, R.T., 30, 136, 139
Stegun, I.A., 160
Storer, B.E., 106

Takeuchi, K., 30, 72, 76
Thomas, W., 30, 90, 123

Tsai, C.L., 13, 14, 16, 30, 56, 64, 67, 83, 84, 93, 101, 114, 116, 142, 152, 154, 159, 177, 178, 188

Venzon, D.J., 30
Vos, P.W., 30

Wan, F., 123, 128, 132, 139
Wang, S.J., 184
Watts, D.G., 14, 29, 30, 32, 34, 35, 37–39, 42, 43, 49, 64, 67, 83, 90, 92, 93, 125, 188
Wedderburn, R.W.M., 165, 177
Wei, B.C., 13, 30, 56, 57, 64, 67, 69, 70, 72, 74, 76, 80, 81, 83, 102, 106, 107, 109, 123, 139, 141, 170, 173, 175, 177, 188
Wei, C.Z., 198
Weisberg, S., 28, 30, 90, 101, 104–106, 124, 134, 136, 142, 146
Weissfeld, L.L., 30
Welch, L.F., 120
Welsch, R.E., 102, 105, 136
Whitmore, G.A., 27
Wild, C.J., 5, 6, 14, 24, 29, 30, 56, 64, 67, 69, 70, 188
Williams, D.A., 106, 108
Wu, C.F., 58, 197, 207
Wu, X., 30, 123, 128, 132

Yoshizoe, Y., 136, 139

Zelterman, D., 141
Zhao, X., 30, 76, 80
Zhu, H.T., 30

Subject Index

M-estimator, 109, 139

Adjusted
  likelihood ratio statistic, 147, 148, 152, 154, 159
  profile likelihood, 147, 148, 159
  score statistic, 149, 150, 159
AP statistic, 118
Approximate
  bias, 66, 70
  variance, 67, 70, 74
Array multiplication, 5, 34, 37, 41, 188
Asymptotic
  χ^2 distribution, 9, 23, 84, 96, 173
  ancillary statistic, 60, 62, 76
  normality, 58, 77, 79, 169, 199, 206
  numerical stability, 24, 208
  sufficient statistic, 80
Asymptotics, 55, 64, 195

Bias, 64, 66, 70

Canonical link, 15, 50, 90, 119, 139
Central limit theorem, 166, 171, 203, 207
Conditional variance, 81
Confidence region, 83, 84, 93, 98, 170, 177
  likelihood based, 84, 88, 94
  parameter subset, 93
  score based, 96, 98
Connection, 48
Consistency, 58, 196, 198, 199, 204
Cook distance, 103, 104, 117
Covariance structure models, 179, 181, 183
Cramér-Rao lower bound, 64, 69

Curvature array, 32, 33, 45, 47, 167, 172, 179
Curved exponential family, 16, 55, 69, 76, 80, 166, 170

Determinant, 90, 118, 192
Deviance, 11, 17, 21, 109, 119, 146
  and Kullback-Leibler distance, 10
  and likelihood ratio, 9
  definition, 9
  for diagnostics, 119
  in diagnostic models, 109
  mean of, 17
  variance of, 17
Diagamma function, 18, 22
Diagnostics, 26, 101, 102, 104, 114, 116, 120, 126, 136, 141
  based on case deletion, 102, 116, 117, 121
  based on deviance, 119
  based on weighted least squares, 116
  models, 105, 106
Directional curvature, 36, 41, 46, 50, 52, 92, 179
Dispersion parameter, 1, 2, 8, 18, 21, 22, 56, 109, 139, 142
Dual
  geometry, 44, 47, 172, 179, 183
  parameter, 172, 176, 179, 183
Duality, 44–46, 52, 93, 176

Edgeworth expansion, 72, 76
Effective residual curvature matrix, 89, 138
Efron curvature, 51, 52, 54, 55, 75, 76, 80, 169
Ellipsoid, 85, 89–92, 95
Embedded models, 174, 177

Eular constant, 18
Expectation parameter, 3, 31, 33, 44, 93, 167, 172
Exponential family
 cumulants, 3
 definition, 1
 deviance, 9, 12
 likelihood, 7
 linear, 2, 3
 moment generating function, 3, 6
 moments, 3, 5
 variance function, 3
 weighted, 12

Fisher information, 7, 8, 21, 44, 72, 76, 96, 104, 113, 131, 139, 144, 145, 147, 153, 166, 171, 172, 174, 175, 178
 inner product, 31, 33, 35, 45, 47, 49, 176

Gamma
 distribution, 14, 111, 160
 nonlinear model, 18, 22, 120, 146, 152, 160
Gauss-Newton method, 24, 208
 modified, 26
Generalized least squares estimator, 25, 181
Generalized leverage
 and leverage, 136
 and local influence, 139
 and observed information, 138
 definition, 136
 for exponential family nonlinear models, 138
 formula, 137
 of M-estimator, 139
 of Bayesian estimator, 139
Generalized linear models, 14, 26, 49, 90, 108, 110, 119, 139, 141, 197, 198
Generalized nonlinear models, 14
Geometric framework, 29, 32, 35, 44, 47, 55, 85, 165, 167, 172, 174, 176, 179, 181, 184

Hat matrix, 117, 136, 139
Heteroscedasticity, 141, 142, 146, 152, 154, 155

Influence
 curvature, 126, 130, 132
 graph, 125, 130, 131
 local, 123
 matrix, 124, 126, 130, 131
 of curvature measures, 90
Information loss, 72, 73
Intrinsic curvature, 33, 40–42, 45, 52, 63, 67, 69, 74, 80, 91, 95
Invariant, 40, 42, 63, 67, 69, 74, 80, 95, 137
Inverse Gaussian nonlinear model, 18, 22, 27, 111, 144, 146, 152–155, 160

Kullback-Leibler distance, 10, 196

Law of large numbers, 6, 171
Least squares estimator, 24, 25, 117
Likelihood
 distance, 104, 121, 124, 126, 129, 134, 140
 for covariance structure models, 180
 for embedded models, 175
 for exponential family, 7
 for exponential family nonlinear models, 19, 20
 for multinomial nonlinear models, 171
 for varying dispersion model, 143
 in perturbed model, 127

profile, 22, 23
ratio statistic, 9, 22, 86, 88, 94, 143, 159
region, 86, 92, 94, 95, 124
Linear approximation, 33, 92–94, 100
Linear regression, 90, 104, 106, 110, 114, 117–119, 136, 139, 142, 146, 152, 154, 155, 159, 198
Link function, 14, 50
Local influence, 123, 131, 132, 136
Logistic regression, 28, 51, 140

Maximum curvature, 36, 38, 46, 50, 52, 92, 125, 131
Maximum likelihood estimator
asymptotic normality, 58, 199
computation, 24, 208
consistency, 58, 199
for covariance structure models, 180
for curved exponential families, 167
for embedded models, 176
for exponential family, 7, 8
for exponential family nonlinear models, 17, 19, 21, 58, 195
for multinomial nonlinear models, 172
for varying dispersion model, 143
generalized leverage, 137
geometric interpretation, 32, 44, 167
in case deletion models, 102, 107
in case-weights models, 105
in mean-shift outlier models, 105, 107
in perturbed model, 124
information loss, 72, 73
stochastic expansion, 60

Model
case deletion, 102, 105, 109, 111
case-weights, 105, 108
mean-shift outlier, 105, 108, 109, 111, 112
on diagnostics, 105, 106
varying dispersion, 142, 144
Multinomial nonlinear models, 170, 172

Natural parameter, 1, 31, 44, 45, 52, 93, 166, 167, 179
Nonlinearity, 35, 36, 52, 64, 70, 91, 93, 138, 139
Normal
distribution, 60, 153, 159
model, 18, 22, 111, 144, 146, 152–154, 159
nonlinear regression, 14, 32, 35, 47, 49, 63, 69, 74, 80, 85, 89, 99, 108, 110, 119, 139, 146, 152, 154, 155
Normal space, 33, 45, 60
Nuisance parameter, 2, 93, 99, 144, 147

Observed information, 7, 20, 21, 24, 76, 80, 86, 126, 127, 131, 138, 139, 166, 171, 174, 175, 178
Orthogonal parameter, 8, 21, 138, 147
Overdispersion, 141

Parameter subset, 23, 93, 104
Parameter transformation, 40, 43, 70, 80, 85
Parameter-effects curvature, 33, 40–43, 45, 63, 67, 69, 85, 91
Parameterization, 40, 63, 69, 95
Path-independent, 178
Pearson residual, 25
Perturbation, 124, 126, 129, 130, 136

Profile likelihood, 22, 23, 94, 147, 148, 159
Projection matrix, 89, 97, 115, 118, 132, 144, 185, 187

QR decomposition, 33, 45, 46, 50, 172, 176, 187
Quadratic approximation, 85, 86, 94, 96–98
Quasi-
    Fisher information, 178, 182
    likelihood, 80, 84, 178
    likelihood estimator, 129, 178, 181
    likelihood models, 63, 177–179
    likelihood ratio statistic, 179
    observed information, 178, 182
    score equation, 129
    score function, 178, 181, 182

Random error, 3, 7, 56, 129, 166, 171, 176, 177, 184
Random perturbation, 123, 128, 129
Relative information loss, 75
Reparameterization, 34, 137
Residual
    deviance, 119
    Pearson, 25
    studentized, 104, 114, 117, 121, 155
    studentized standardized, 104
    sum of squares, 109, 110, 116, 119
    vector, 32, 44, 60
Riemannian metric, 35

Score function, 7, 20, 166, 171, 175, 177, 180, 181
Score statistic
    adjusted, 149, 150, 159
    for varying dispersion, 144, 150, 159
    in confidence region, 96, 98
    of outliers, 112, 115
Simulation study, 158
Solution locus, 32, 45, 85, 86, 167, 172, 176, 179
Stochastic expansion, 56, 60, 62, 73, 167, 173
Sufficient statistic, 60, 76, 80, 166

Tangent space, 32, 44, 45, 85, 86, 89, 91, 95, 98, 126, 167, 170, 172, 176
Transformation, 28, 40, 85, 86, 88, 90, 93, 94, 147, 153, 170, 177

Underdispersion, 141

Variance function, 3, 19, 177
Varying dispersion, 142, 144, 146, 155, 156, 158

Weighted
    exponential family, 12
    inner product, 31, 33, 44, 185
    least squares, 116